SPINOZA AND THE SCIENCES

BOSTON STUDIES IN THE PHILOSOPHY OF SCIENCE

EDITED BY ROBERT S. COHEN AND MARX W. WARTOFSKY

VOLUME 91

SPINOZA
AND THE SCIENCES

Edited by

MARJORIE GRENE
University of California at Davis

and

DEBRA NAILS
University of the Witwatersrand

D. REIDEL PUBLISHING COMPANY

A MEMBER OF THE KLUWER ACADEMIC PUBLISHERS GROUP

DORDRECHT / BOSTON / LANCASTER / TOKYO

Library of Congress Cataloging-in-Publication Data
Main entry under title:

Spinoza and the sciences.

(Boston studies in the philosophy of science; v. 91)
Bibliography: p.
Includes index.
1.   Spinoza, Benedictus de, 1632—1677.   2.   Science—
Philosophy—History.   3.   Scientists—Netherlands—
Biography.   I.   Grene, Marjorie Glicksman, 1910—
II.   Nails, Debra, 1950—   .   III.   Series.
Q174.B67 vol. 91   001′.01 s   85—28183
[Q143.S725]   [001]
ISBN 90—277—1976—4

Published by D. Reidel Publishing Company,
P.O. Box 17, 3300 AA Dordrecht, Holland.

Sold and distributed in the U.S.A. and Canada
by Kluwer Academic Publishers,
101 Philip Drive, Assinippi Park, Norwell, MA 02061, U.S.A.

In all other countries, sold and distributed
by Kluwer Academic Publishers Group,
P.O. Box 322, 3300 AH Dordrecht, Holland.

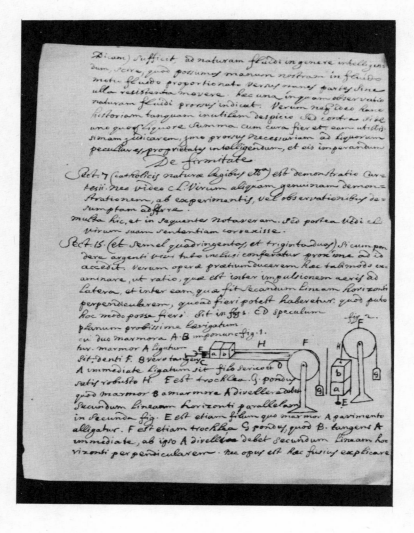

FROM SPINOZA'S LETTER TO OLDENBURG, RIJNSBURG,
APRIL, 1662

(Photo by permission of Berend Kolk)

TABLE OF CONTENTS

IV. SCIENTIFIC-METAPHYSICAL REFLECTIONS

V. SPINOZA AND TWENTIETH CENTURY SCIENCE

VI. BIBLIOGRAPHY

ACKNOWLEDGEMENTS

As this volume approaches readiness, it is simultaneously heartening and staggering to realize how many people (and some of them anonymous) were essential to its reaching its readers. I want to express my appreciation to a few who were not only essential but encouraging: I am grateful to Katie Platt who administered a cumbersome transworld collaboration; to Amélie Rorty who provided excellent advice in the planning of the original Spinoza Sesquitercentenary Symposium; to Reidel's redoubtable Annie Kuipers for unfailing good humor; and to Berend Kolk for helping in innumerable invisible ways. To Joe VanZandt, I am grateful not only for advice about the present volume, but for a decade of stimulating conversations about Spinoza's philosophy of science. Thanks also to the Unmoved Mover, Bob Cohen; and especially to the First Cause, Marjorie Grene.

Most of the work on this volume took place while I was Research Fellow at the Center for Philosophy and History of Science at Boston University, but was completed after I became University Post-doctoral Research Fellow at the University of the Witwatersrand. I gratefully acknowledge the financial support of both institutions.

DEBRA NAILS

Permissions: J. Thomas Cook quotes from George Santayana's unpublished notes by permission of the Houghton Library, Harvard University.

Hans Jonas's article was originally published as 'Parallelism and Complementarity: the Psycho-Physical Problem in the Succession of Niels Bohr' in Richard Kennington (ed.), *The Philosophy of Baruch Spinoza*, Studies in Philosophy and the History of Philosophy 7, Catholic University of America Press, Washington D.C., 1980. It is reprinted here by permission of the author and publisher.

Michel Paty's quotations from unpublished manuscripts in the

Einstein Archives (Spinoza folder, reel 33, no. 7) are reproduced here with permission of the Hebrew University of Jerusalem.

MARJORIE GRENE
DEBRA NAILS

MARJORIE GRENE

INTRODUCTION

Prefatory Explanation

It must be remarked at once that I am 'editor' of this volume only in that I had the honor of presiding at the symposium on Spinoza and the Sciences at which a number of these papers were presented (exceptions are those by Hans Jonas, Richard Popkin, Joe VanZandt and our four European contributors), in that I have given some editorial advice on details of some of the papers, including translations, and finally, in that my name appears on the cover. The choice of speakers, and of additional contributors, is entirely due to Robert Cohen and Debra Nails; and nearly all the burden of readying the manuscript for the press has been borne by the latter.

In the introduction to another anthology on Spinoza I opened my remarks by quoting a statement of Sir Stuart Hampshire about interpretations of Spinoza's chief work:

All these masks have been fitted on him and each of them does to some extent fit. But they remain masks, not the living face. They do not show the moving tensions and unresolved conflicts in Spinoza's *Ethics*. (Hampshire, 1973, p. 297)

The double theme of 'moving tensions' and 'unresolved conflicts' seems even more appropriate to the present volume. What is Spinoza's relation to the sciences? The answers are many, and they criss-cross one another in a number of complicated ways. I shall not attempt here to enumerate all these interconnections; the arguments that follow speak for themselves. But a glance at a few of the 'tensions' and 'conflicts' may serve as introduction to a rich and, I hope, fruitful group of studies of this transcendent and enigmatic thinker. I started to say "transcendent yet enigmatic", but caught myself: the transcendence is the reason for the enigma(s): if the living face behind the mask, the living person behind the printed texts, evades our questions and our answers, that is not because Spinoza was confused or self-contradictory. The texts, indeed, are often confused or self-contradictory for our understanding.

xi

If their writer escapes us, however, it is not by reason of those superficial contradictions; it is, if we glimpse it at all, the grandeur of the
vision behind them that dazzles, and dazes, would-be interpreters. Or
so it seems to me; that is why I have the habit (if twice makes a habit)
of introducing anthologies on Spinoza, but dare not attempt, myself, to
put on paper any would-be explication of his thought. Others, however,
fortunately, are more courageous than I, and the present collection
adds, I believe, important aids toward our reading of Spinozistic texts,
and hence of Spinoza, both in general and in the special context of the
sciences, past and present.

Again, our authors are asking, in one context or another, the basic
question: what is Spinoza's relation to the sciences? In the spectrum
of possible answers to our general question, Maull at the start and
VanZandt and Paty at the close provide the alpha and omega. For
Maull, Spinoza is a stranger in the age of science: we find in him no
kinship with, indeed, hostility to the groping, experimental, cumulative
and critical approach from which modern science springs. To put
Maull's thesis perhaps too crudely, Spinoza is a rationalist, and modern
science is empiricist. For VanZandt and Paty, on the contrary, Spinoza
exhibits a deep and moving kinship with the very archetype of The
Scientist, Albert Einstein. Nor is this just because Einstein, like Spinoza,
was given to metaphysics, as distinct from science. His very science was
metaphysically rooted, as Spinoza's was. And more than that, as
VanZandt argues, some particular and important doctrines of relativistic science in this century have shown analogies with Spinozistic
tenets. So, it seems, in the history of science, rather than being an
outsider, Spinoza assumes, three centuries in advance, something like a
culminating place.

This startling contrast needs to be further complicated; but meantime, and parenthetically, I must comment on one paradox that arises in
connection with Maull's paper. As an author who denies Spinoza's
rationalism she cites E. M. Curley, who in his book, *Spinoza's Metaphysics* (1969), had given a superb exposition precisely of Spinoza's
'rationalism', at least as he then understood it. Yet it is true that in his
'Experience in Spinoza's Theory of Knowledge' (Curley, 1973), Curley
declared: "The view that Spinoza was a rationalist, in the sense we are
concerned with [that is the sense in which knowlege is viewed as purely
a priori] is not only mildly inaccurate, it is wildly inaccurate" (Curley,
1973, p. 26). On the other hand, in his book, referring to Meyer's

statement of Spinoza's denial of the Cartesian view that "this or that exceeds human grasp", Curley declares: "If rationalism consists in having this optimistic view of man's ability to comprehend the world around him, then Spinoza was plainly and unequivocally a rationalist" (Curley, 1969, p. 157). 'Rationalism' is a weasel word; yet it is clear what Curley means in each case. Spinoza is a rationalist in that he sees "the basic structure of science as being ideally that of a deductive system" (Curley, 1973, p. 58). That is what he argued in his book, and one could find elsewhere, in David Lachterman's essay on Spinoza's physics (1978) for example, a powerful metaphysical grounding of that argument — a grounding I believe it needs. Yet in itself, too, the argument is a careful and convincing one. And the idea of science as a deductive system is not so far, either, from one conspicuous strand in scientific thought. Curley's point in the 'Experience' paper is that Spinoza's deductive system, which culminates, in the third kind of knowledge, in the understanding of an individual thing, needs, at that juncture, explicit grounding in experience. And this thesis is expanded and substantiated in David Savan's contribution to this volume. Yes, Spinoza is a rationalist in his emphasis on science as a system of necessary truths in which the consequences follow necessarily from necessary first principles; but yes, he emphatically recognizes a role for experience and experiment in science, and perhaps, one might even want to suggest, less naively than his correspondent Oldenburg or Oldenburg's friend Boyle. That's another question still open for debate.

In addition to the sharp contrast between Maull on the one hand and Paty and VanZandt on the other, thirdly, yet another position must also be mentioned, and that is the view put forward by Hans Jonas in the brief but provocative essay reprinted here. His view, unlike either that of Maull or of VanZandt-Paty, finds an intimate relation between Spinoza and science, but not the one denied by Maull nor yet that asserted by the last two authors. Spinozistic determinism, Jonas insists, far from anticipating the tenets, or tenor, of twentieth century science, may have furnished the fundamental theme implicit in the view even of those smugly empiricist Newtonians. For determinism was in a sense the theme of classical science until quantum mechanics threatened its sway. Jonas sees in Spinozism, therefore, not an anticipation of twentieth century science, but the dogmatic underpinning of classical mechanics, which only the crisis of quantum epistemology can undercut. And only such a philosophical revolution, he argues, can in turn restore our everyday awareness of mind-body interaction to its rightful place. Not of course that it was Spinoza himself who spread this baleful determinism: on the whole, nobody read him and everyone condemned him. But as so often happens with truly rigorous philosophical thinkers, it was he (or so I assume Jonas must be arguing) who put into canonical form the radical implications of the then new science that its more respectable exponents not only did not recognize, but failed to notice that they ought to recognize. And there is something in this argument, too. The Cartesian foundation so influential in early modern thought, including scientific thought, was in itself unstable: the clear and distinct idea and its claim to truth, substance and mode, mind/body, will and intellect, God and the world: none of these could remain in the precarious juxtaposition Descartes assigned them. Two outcomes were fated: the philosophies of Spinoza and of Hume — and for neither is the human individual an ultimate metaphysical reality. Yet for VanZandt and Paty, one supposes, as for Cook and Lloyd also, that very Spinozistic determinism epitomizes the grandeur of a free spirit: liberated, as Einstein too was, by insight into the whole of nature. 'Unresolved conflict' or 'moving tension'? Take your choice, as nature, or your nature, determines you to take it. Is Spinoza an outsider to science, a forerunner of the profoundest expression of twentieth century science or even of science as such, or is he the proponent of the true foundations of classical physics, only now overthrown by quantum mechanical indeterminacy?

Short of these wider issues, our authors are concerned, in the main, with special questions bearing on Spinozistic science. True, Spinoza failed to add — indeed, except for the *Treatise on the Rainbow*, made, so far as we can tell, no effort to add — to the increasing body of concrete knowledge of nature that seems to us so characteristic of his century. At the same time, short of the broader themes of Part V below, an examination of Spinoza's place in what we call the scientific revolution proves rewarding in at least four respects.

First, as I have already suggested in my digression on Maull on Curley, Spinoza was the first and most rigorous proponent of a view of scientific knowledge as a comprehensive and deductive axiomatic system. Even though we would deny the logically or ontologically necessary nature of such axioms, some such ideal of an axiom-based, unified science has surely not been a negligible component of theoretical science as many have conceived, or perhaps idealized, it. I have heard an eminent philosopher of science declare that, of course, in her view, *every* scientist in all his (her) research is hoping to contribute to a unified science. Being — despite Spinoza — an irredeemable pluralist, I don't believe this for a moment. But it is surely a recurrent theme, not only in the philosophy of science, but among (some) scientists themselves. Indeed, the incomparable Sir Isaac Newton himself excelled not so much in experimental technique or detailed observation, as in unifying what had been disparate problems. Henry Guerlac writes:

... what he gave the world in the body of his scientific work was like the steel frame of some great building. The mathematical laws of optics and celestial mechanics are the girders and supporting members; other men will come with the bricks, the mortar and the cut stone to fill in the walls and lay out the partitions. (Guerlac, 1977, p. 143)

Of course, it was Newton's vision, not that of the Dutch recluse, that set classical physics on its course. But the very difference in vision is worth reflecting on. Newton's unifying vision, Guerlac points out, was basically Gassendist and atomistic. And again that is in the main the way the unity of science movement has gone: for explanation, the least is always the most! As Oldenburg adjured Spinoza:

In our Philosophical Society, we indulge, as far as our powers allow, in diligently making experiments and observations, and we spend much time in preparing a History of the Mechanical Arts, feeling certain that the forms and qualities of things can best be explained by the principles of Mechanics, and that all the effects of Nature are produced by motion, figure, texture, and the varying combinations of these . . . . (Wolf, 1928, p. 80; Gebhardt IV, p. 12)

The New Corpuscular Philosophy is the way to go. Yet even in the seventeenth century, that is by no means the whole story. What gave Newton's achievement its unprecedented power was not, or certainly not only, its atomism, which, in advance of the nineteenth century, was still speculative; it was the unification of the phenomena achieved through the principle of universal gravitation. As Adam Smith put it in an early essay on the history of astronomy:

Such is the system of Sir Isaac Newton, a system whose parts are all more strictly connected together, than those of any other philosophical hypothesis. Allow his principle, the universality of gravity, and that it decreases as the squares of the distance increase, and all the appearances, which he joins together by it, necessarily follow. (Smith, 1967, p. 107)

Scope, the sweep of explanation, has always been a scientific value; as traditional interpreters and as some of our contributors have seen him, it was the overriding value on which Spinoza's conception of science was founded.

Second, Spinoza's careful criticism of Cartesian physics forms an important step in the shaping of his own mature science, in what Lachterman (1978) has called "the physics of Spinoza's *Ethics*". This criticism, touched on briefly by Siebrand, is analyzed in depth in Lecrivain's careful study. For the examination of Cartesian science as well as for Spinozistic scholarship it merits close attention. However outlying Spinoza's own life and thought may have been in the advancing physics of his own time, Cartesian science was certainly a major force in the development of classical physics, and Spinoza's criticism of it should not be neglected.

Third, although it was ethical concerns that motivated his work, so that he never focussed as such on the problem of scientific method, Spinoza did hold definite, and complex, views about that question, and did in fact in his own writing put these tenets to work. The essays included in Part II are concerned, in very different directions, with aspects of this problem. Matheron's investigation of the complex relations between Spinoza's arithmetical analogy and the subtleties of Euclidean proofs lends unexpected precision to the distinction between what most of us think of as the second and third kinds of knowledge, or, in terms of the *Tractatus de Intellectus Emendatione*, between the third and fourth. The grandeur of the whole, dazzling though it be, is only confirmed by the delicacy and precision with which its parts have been articulated. But there is another side to the story of Spinozistic methodology: the 'moving tensions' are unending. Savan goes straight to the topic of 'Spinoza: Scientist', with respect to concrete questions of scientific practice. His meticulous examination of Spinoza as working scientist sheds, in my view, floods of new light on this previously obscure topic. In particular, his distinction between what he calls *the principle of detachment*, the (tripartite) *principle of hypothetical explanation*, and the *modelling principle*, allows us to look more concretely and fruitfully than traditional interpretations had permitted at this aspect of Spinoza's thought, in which he stands revealed as a careful, self-conscious practitioner of scientific method in his own time, and even, in agreement with and amplification of Curley's epistemological thesis, as a careful *empirical* worker. Even if the theme of unity, of understanding through and in the whole, was overriding, especially in the *Ethics*, it was balanced by a scrupulous attention to empirical detail in his specialized scientific work.

What scientific work? the reader may ask. As Maull and Savan agree, Spinoza did little in the way of physical experiment and *qua* experimenter contributed nothing to the history of physics. What traditional history of science has generally overlooked, however — and this is the last of my four points — is that Spinoza *was* a practitioner of science, though not of experimental physics. That is what Savan's paper and the essays of Part III together emphasize. In terms of Agassi's overview, Spinoza was a practitioner of what was to be classical political science. In terms of Popkin's paper — whose consonance with Savan's contribution forms one of the major harmonies of this volume — he was a practitioner of scientific hermeneutics. And that *was* science. Science in this spirit needed, on the one hand, especially for the major intent of an ethics *more geometrico demonstrata*, the excursion into basic physical theory of Part Two, but it needed, on the other hand, the kind of accurate empirical methodology that Savan has analyzed and described so carefully.

Philosophers of biology complain that despite Vesalius, Harvey, Linnaeus and a host of others, the advances in biology from 1543 to 1859 are usually passed over silently by accounts in the history and philosophy of science, and science is equated with the physical sciences. But what about 'social science'? It seems that both those who now propose rigorous programs for these disciplines, subsuming them under 'science' as such, and those who distinguish the 'human sciences' from physics by defending their need for an added interpretive component (allegedly) missing from the more 'objective' fields: both these groups of spokesmen agree that the social sciences are late comers, doing their best, like third-world countries, to overtake their more 'advanced' competitors. After all, as we all know, the *Geisteswissenschaften* were split off from the *Naturwissenschaften* in Heidelberg in about 1900. So as independent disciplines they are recent. What does that suggest for our current subject-matter? That before Rickert and Windelband there was only nomothetic science, which was the science of nature, or perhaps that until that Teutonic secession the study of human nature could be held to flow from, and belong to, the study of nature as such? Surely the latter. Not only Spinoza, but Descartes himself, with his metaphor of the tree of knowledge, believed that the new method, freed

of the shackles of syllogistics, would move straight on from the study of
the natural world to the equally precise and reliable study of the
human. In Descartes's case the two top branches were to be morality
and medicine — if only he had lived so long. In the case of Spinoza, it
was two areas of special concern to ourselves and our destinies to
which he applied his scientific skills: politics and the interpretation of
scripture. But Descartes and Spinoza were by no means the only ones.
There was Hobbes; there were Grotius and Pufendorf. There was
Richard Cumberland, with his *Philosophical Disquisition on the Laws
of Nature*, which grounded its refutation of the iniquitous Hobbes on a
method by which, its author declared, the dictates of universal bene-
volence and all that follows from them, "are reduced to a proper
similarity with the propositions of universal mathematics, concerning
the effects of mathematical computation, through which all quantities
are brought together with one another" (Cumberland, 1672, 2b). These
hopes persist, indeed, well into the eighteenth century, with Hume's
attempt to introduce the experimental method into moral subjects. The
fruit of the new method was to be a new science, not only of nature, but
of human nature, not only of those hard solid impenetrable particles
out of which God probably formed matter, but of ourselves — and
a fortiori of those sacred texts on the study of which much of our
conduct had been grounded. It is not the would-be scientific study of
man that is a late comer, but rather the notion that such a study would
not form part of the seamless whole made possible by the new method,
mathematical-times-experimental, that leading minds in the seventeenth
century, including Spinoza, were struggling to articulate and to apply.
Once we rid ourselves, in imagination, of the distinction in kind between 'natural' and 'human' science, we can take seriously, as contributions to the science of his time, Spinoza's research in areas that, though
outside the 'exact sciences' in our view, formed perfectly legitimate
components of a research program characteristic at least of one very
significant group of progressive thinkers in his own day.

In short, while Spinoza contributed to the advance of scientific
knowledge neither laws of refraction, nor gas laws, let alone the law of
gravitation for which, it seems, the world was waiting, his relation to the
sciences, both then and now, is rich and varied. The metaphysical
foundations necessary to support an adequate scientific method, the
vision of a unified science entailed by such foundationist propositions,
the criticism and, partly, correction of Cartesian physical theory, original

use of the mathematical tradition, anticipations of twentieth century doctrines of space and time, the application of a complex investigative method in the emerging field of scientific hermeneutics: all these features are to be discovered when we look at Spinoza in the context of the history of the sciences, from his own time to ours.

April 1984   MARJORIE GRENE

REFERENCES

Cumberland, R.: 1672, *De Legibus Naturae Disquisitio Philosophica*, N. Hooke, London.

Curley, E. M.: 1969, *Spinoza's Metaphysics*, Harvard University Press, Cambridge, MA.

Curley, E. M.: 1973, 'Experience in Spinoza's Theory of Knowledge', in Grene (1973), pp. 25—59.

Grene, Marjorie (ed.): 1973, *Spinoza: A Collection of Critical Essays*, Doubleday, Garden City, New Jersey.

Guerlac, H.: 1977, *Essays and Papers in the History of Modern Science*, Johns Hopkins University Press, Baltimore.

Hampshire, Stuart: 1973, 'Spinoza and the Idea of Freedom', in Grene (1973), pp. 297—317.

Lachterman, David R.: 1978, 'The Physics of Spinoza's *Ethics*', in Robert W. Shahan and J. I. Biro (eds.), *Spinoza: New Perspectives*, University of Oklahoma Press, Norman, pp. 71—112.

Smith, Adam: 1967, 'The History of Astronomy', in *Early Writings*, ed. by J. Ralph Lindgren, A. M. Kelley, New York.

Spinoza, Baruch: 1925, *Spinoza Opera*, ed. by Carl Gebhardt, 4 vols., Carl Winter, Heidelberg. (Cited as 'Gebhardt' in the text.)

Spinoza, Baruch: 1928, *Correspondence of Spinoza*, transl. and ed. by Abraham Wolf, Cass, London; and Russell and Russell, New York.

PART I

SPINOZA AND
SEVENTEENTH CENTURY SCIENCE

NANCY MAULL

SPINOZA IN THE CENTURY OF SCIENCE

1. I would like to suggest a context, both historical and philosohpical, for the papers that will follow on detailed aspects of Spinoza's science. My aspiration, however, is not to situate Spinoza among the natural philosophical giants who opened the way to modern science. I cannot conscript him into the ranks of Descartes and Boyle, Leibniz and Newton. Spinoza does not, alas, fit comfortably in the lineup of scientific 'greats', either theoretically or by virtue of some concrete scientific achievement. He was, of course, a great thinker and a great philosopher. But his philosophy was strikingly disconnected from the sifting and interrogating science that went on around him. His own interest in experimental science is well-documented, but it was carefully bracketed from his larger metaphysical concerns. Philosophically, as opposed to biographically, he was as remote from elementary 'doing' of science and especially from the idea of learning by experience as Plato was.

This, of course, is an interpretation, and one that I can support only by pointing to lacunae in Spinoza's writings about science. It is a view I have reached reluctantly, in part because it resuscitates the old clichés about Spinoza's rationalism and his 'God-intoxication'. Bolder and more imaginative interpreters have recently advanced the thesis that Spinoza never was a rationalist (in any 'interesting sense') and that his views about the value of experiment were very close to Descartes's.[1] This, it seems to me, is very far from the truth. Refurbishing an old cliché may even help us to right the balance. However, I remain open to instruction on this point. I shall offer no complete interpretation of Spinoza, but instead a brief list of questions that I believe any interpreter must answer as well as a set of tentative suggestions about how I would go about answering them.

2. The first question is this: Why, if Spinoza sustained a lively interest in experimental science, was he so estranged from it philosophically?

Spinoza first commented extensively on Part II (and a bit of Part III) of Descartes's *Principles*. Only later, at the urging of Lodewijk Meyer

3

Marjorie Grene and Debra Nails (eds.), Spinoza and the Sciences, 3—13.
© 1986 *by D. Reidel Publishing Company.*

and others, did he turn his attention to Part I, the philosophically
fundamental opening of the *Principles*. (This is not to say that Part II of
the *Principles* is 'experimental', at least in the sense that Part III is. But
it does not engage the reader in thought experiments where principles
are applied to particular physical situations.)

Again, we know of Spinoza's sustained and scientifically sophisticated correspondence with Oldenburg on various experimental and
mathematical matters. In that exchange, Boyle's notions about the
chemistry of nitre are discussed, making clear Spinoza's own experimental adventures with potassium nitrate (KNO_3). Huyghens is also
mentioned in several letters, and from both letters and biographies we
know that Spinoza and Huyghens had many scientific conversations.
Leibniz, we also know, recognized Spinoza's mastery of optics, and sent
him his 'Note on Advanced Optics' in 1671. Spinoza not only ground
lenses for a living, but he also had a thorough understanding of
theoretical optics. He wrote an optical treatise on the rainbow and a
mathematical work on probability and chance.

Like Descartes, Spinoza lived in seclusion. But he too was sur-
rounded by an astonishing variety of scientific effort, both experimental
and theoretical. I have already mentioned Huyghens, who lived nearby.
Spinoza who also knew Hudde and Dewitt and must have been acquainted with their mathematical and statistical investigations. Indeed,
Spinoza lived and worked in an extraordinary time and place. This
Golden age of the Dutch Republic (1585—1695) boasted not only
Rembrandt, but Swammerdam, De Graaf, van Leeuwenhoek, and
Stevin. Even without the institutional framework for science that
developed in England with the Royal Society and Oldenburg's ubiqui-
tous correspondence, the Dutch Republic managed to tolerate and
occasionally to encourage a wide range of individual experimental
interests.[2]

Spinoza's own daily round of activities, what little we know of it,
looks very similar to Descartes's non-philosophical leisure time in the
Dutch years: a little experiment, a great deal more thought, some
mathematics, and a consuming curiosity about the inner workings of the
perceived world. These appear to be precisely the passions and the
habits needed to support an active investigation of nature. But the
similarity is all surface. While we can credit Descartes with analytical
geometry and certain advances in optics, Spinoza made no discoveries,
mathematical or natural. The little work on the rainbow, if I am not

mistaken, is not at all experimental; it merely elaborates mathematical calculations made by Descartes. The essay on chance, similarly, is an attempt to solve some problems set by Huyghens. Despite his philosophical differences with Descartes, Spinoza was ostensibly a supporter of Cartesian 'normal science'. Yet consider the character of Spinoza's own experimental work on nitre. It does not seem to have been undertaken in order to understand the internal composition of nitre, but in order to confront Boyle with a philosophical lesson, namely, that (as we would put it) hypotheses are underdetermined by experiment.

3. The critical exchange with Boyle takes the following form: Oldenburg sends to Spinoza Boyle's treatises *On Salt-Petre* and *On the History of Fluidity and Firmness*. Spinoza replies with a rather daunting series of criticisms which he wishes poor Oldenburg to communicate to Boyle. The core of his criticism of the work on salt-petre is this: (1) Boyle concludes, too hastily, and on the basis of experiments alone, that nitre is a heterogeneous body; and (2) a simpler hypothesis, compatible with nitre's homogeneity, can just as easily explain the Boylean results, as well as some experimental outcomes adduced by Spinoza himself. To this Boyle replies, not directly but through Oldenburg, that he never intended a "philosophic and perfect analysis of Nitre", but wanted to show that the "common Doctrine of Substantial Forms and Qualities ... accepted in the Schools" is weak and indefensible. (This too, is a philosophical thesis, but it is philosophy in the service of experimental science, not in the service of human well-being or of philosophy itself.) Boyle adds, in reply, that Spinoza's own suppositions for the opposite, homogeneity-hypothesis are "gratuitous and unproved" (*Letter XI*, Spinoza (1966) pp. 110—111; Gebhardt IV, pp. 48—49).

In his essay on salt-petre, Boyle had tried to make plausible (or perhaps to demonstrate conclusively) the mechanistic doctrine that heat, color, odor, firmness and the like are the results of bodies in motion. In short, Boyle's explicit concern was the distinction between primary and secondary qualities, and he gave a series of suggestions for how the distinction might be employed in organizing experimental research. (Here and here and here, he points out, it will be possible to explain the secondary phenomena of sensation in terms of the primary qualities of unseen bodies.) Spinoza responded to Boyle's reassertion of a practical and experimental purpose with utter bafflement: if *this* is your idea, Boyle, why are the examples so complicated? Just point out that water changes its sensory properties when it is steam. In any case, you only illustrate your doctrine and prove nothing. You provide no mathematical proof for this corpuscular philosophy. No proof of mechanistic principles will ever be possible, according to Spinoza.

It might be helpful here to distinguish two strategies of justification, which I shall call (for lack of better terms) functional and demonstrative. Boyle knew that the primary/secondary quality distinction was not susceptible to proof in the q. e. d. sense. It was not a theory to be tested, but a suggestion for the generation and testing of a whole family of theories. He felt that it could be justified in practice, by its fruitfulness in setting out heuristic guidelines for the investigation of nature. Fundamentally indifferent toward the details of nature, Spinoza, by contrast, felt no need to embrace any standard of justification that fell short of deductive proof.

Oldenburgh tried (with some exasperation) to win Spinoza to a more generous understanding of Boyle's purpose: "Our Boyle", he reminds Spinoza pointedly, "belongs to the number of those who have not so much confidence in their reason as not to wish that the Phenomena should agree with their reason" (*Letter XI*, Spinoza (1966), pp. 112—113; Gebhardt IV, p. 50). Spinoza persists: his own hypothesis, opposed to Boyle's, agrees so effortlessly with the experiments and with the mechanical philosophy — what can experiments really tell us? Spinoza's message, conveyed unmistakably in his pesky insistence throughout the exchange, is that the experiments (because they admit to different interpretations) decide no unique hypothesis and that a mechanical hypothesis about the sizes, shapes, and motions of unseen bodies may only be justified by rigid mathematical proof from higher principles. The experiments "which I adduced to confirm my explanation", he writes, do so "not absolutely, but, as I expressly said, *to a certain extent*" (Letter XIII, Spinoza (1966), p. 126; Gebhardt IV, p. 66).

In this passage we may hear (falsely, I think) echoes of Descartes's reminder, at the close of the *Principles*, that for hypotheses about the unseen working of nature we can attain only moral certainty. However optimistic Descartes may once have been about the mastery of nature by reason, he was conspicuously less so by the time he finished the *Principles*. Spinoza, by contrast, seems to have been unswerving in his commitment to proof-as-discovery, to the idea that deduction is the only avenue to an increase in knowledge. From the idea of a thing, he writes, all of its properties can be deduced (*Letter IX*, Spinoza (1966), pp. 105—109; Gebhardt IV, pp. 42—46). If we are at all interested in the mixing, sniffing, burning, and mucking about of the learned Boyle, it will be only for the compilation of the histories of things. Writing of Boyle's liquids, he says that we do not need fancy experiments to show that the underlying motions of the unseen components of things are rarely detected by human sense. He continues:

> But I do not therefore look down upon this account as useless; but on the contrary, if of every liquid there were an account given as accurately as possible with the highest truthworthiness, I should consider it of the greatest service for the understanding of the special features which differentiate them: which is to be most earnestly desired by all philosophers as something very necessary. (*Letter VI*, Spinoza (1966), pp. 96—97; Gebhardt IV, p. 34)

In sum, Boyle's experiments do not prove the claim that the secondary qualities of seen things are to be explained in terms of the primary qualities of unseen things, nor do they prove any particular hypothesis about those unseen workings in nature. Experiments are unnecessary even as a spur to preliminary assent about the causes of perceived events, for commonplace observations are enough to convince us that there is much hidden from us by the limitations of our sensory equipment.

But the histories or accounts of the experimental phenomena, says Spinoza, will be useful. Why? Descartes would say that they are necessary in order to ascertain which of all possible law-consistent entities and processes actually obtain in the world. But Spinoza's much tighter, deductively bound system (in which all possibilities are actual) seems to require no active investigation into the phenomena.[3] To repeat my leading question: Why, given Spinoza's apparent interest in experiment, is he so estranged from it philosophically? Can the two be reconciled? And if so, is there strong enough textual evidence that Spinoza himself (on his own and not just at our exegetical urging) sees this as a problem requiring solution?

4. When I turn from Descartes's *Principles* to the *Ethics* or to *On the Improvement of the Understanding*, I am struck by Spinoza's ongoing failure to provide us with more than a few, stock examples of what it is to know something.

In *On the Improvement of the Understanding* there is a celebrated passage (echoed later in the *Ethics*) recounting the "four modes of perception or knowledge" and giving a series of examples to illustrate the modes (Spinoza (1955), pp. 8—9; Gebhardt II, pp. 10—11) "By hearsay [mode one] I know the day of my birth, my parentage, and other matters about which I have never felt any doubt." Secondly, "By mere [vague or vagrant] experience I know that I shall die . . . that oil has the property of feeding fire and water of extinguishing it . . . that a dog is a barking animal, man a rational animal, and in fact nearly all the practical knowledge of life." These first two categories comprise what Descartes called in the first Meditation knowledge "of the senses or by the senses". The second mode seems to correspond to the experimental or observational histories that Spinoza calls useful.

Thirdly, we "deduce" that mind "is united to the body, and that their union is the cause of a given sensation; but we cannot thence absolutely understand the nature of the sensation and the union". Or, Spinoza continues,

after I have become acquainted with the nature of vision, and I know that it has the property of making one and the same thing smaller when far off than when near, I can infer that the sun is larger than it appears and can draw other conclusions of the same kind.

This third mode of knowledge seems to involve causal thinking or causal attribution. It stands, as a category, rather uneasily between habitual inductions (dog as barking animal) and the truths of intuition. It is an uneasy 'middle ground' that involves deduction, or reason (as opposed to intuition.) This third sort of knowing is the locus of Descartes's mere 'moral certainty' and it is the category in which we find the good experimental hypotheses about bodies unperceived or not fully perceived. We find, for example, Descartes's assumptions about light explicitly acknowledged as hypotheses in the opening pages of the *Dioptrics*. To such hypotheses we can attach only moral certainty.

But for Spinoza, the sun's size, once properly understood, is a statement necessarily true. The sun is not the size it first seems when seen. When we understand the inner workings of vision, we also understand that the sun is larger than its perceived image would lead us to believe. The necessary truth is deduced from the right rules but this 'deduction' is further distinguished by Spinoza from the fourth mode of knowledge which also affords necessary truth: "the perception arising when a thing is perceived solely through its essence or through the knowledge of its proximate cause". If "from knowing the essence of the mind I know that it is united to the body", I have knowledge of the highest, intuitive sort. Notice how easily the third mode of knowledge might be confused with the fourth. Descartes purposely elides deduction and intuition — intuition is a very fast deduction. But Spinoza, as I shall explain later, suspects that these two operations of the mind may be very different. He insists on the qualitative psychological and epistemological difference between having the right rule that gives the right answer (mode 3) and knowing the essence of a thing (mode 4).

This taxonomy of knowledge is taken a step further both in *On the Improvement of the Understanding* and the *Ethics*. Spinoza rehearses the four ways of knowing by reference to four ways in which a mathematical proportion may be calculated. The range of knowing (from hearsay to rule of thumb, to deduction or having the right rule, and finally to intuition or 'just plain knowing') is repeated.

And in a later passage of *On the Improvement of the Understanding* Spinoza (1955, p. 26; Gebhardt II, pp. 26—27) tries to give us a better notion of the object of intuition, or better, of the mental event that constitutes "just plain knowing." He explains that "a true idea is distinguished from a false one, not so much by its extrinsic object as by its intrinsic nature", and "If an architect conceives a building properly constructed, though such a building may never have existed, and may never exist, nevertheless, the idea is true ...". Truth is not a relation between ideas and things but depends "solely on the power and nature of the understanding".

This, it seems to me, is a perfectly familiar account of mathematical objects. But it is a queer avowal for a man who insists on the usefulness of compiling histories of phenomena. Is Spinoza interested only in what was then called the objective reality of ideas? What about the multiple connections between mathematical objects and the phenomena of experience?

Contrast Spinoza's attitude with Descartes's worries about the application of mathematics to nature. Descartes tries to solve the problem of applying mathematics to nature with the primary/secondary quality distinction and with his theory of perceptual judgment. By present standards, these attempts may be philosophically misguided. But for early modern science, the effort was developmentally crucial. All this is patently absent in Spinoza, who merely alludes to the problem in the so-called Physical Digression of the *Ethics*. The lengths to which Descartes goes to ensure a fit between our knowledge and the world are lacking in Spinoza because he is concerned only with getting the ideas right. And he is concerned with getting the ideas right because he believes, on philosophical grounds, that nothing else matters. If there are few examples of knowledge in his writing, it is because knowing, or rather getting to know — learning and discovery — is not among his central preoccupations. By contrast, self-knowledge and the mastery of confused perceptions that stem from the emotions, powerfully compel his attention in Part III of the *Ethics*.

Spinoza may, then, be guilty of the charge (so unfairly leveled at Descartes) that one can have scientific knowledge *a priori,* without specific kinds of sense experiences. All this leads us the exchange of letters with Tschirnhaus,[4] who asks all the right questions. Tschirnhaus first remarks

. . . I find it exceedingly difficult to conceive how the existence of bodies having motion and figure can be proved *a priori,* since there is nothing of this kind in Extension when we consider it absolutely. (*Letter LXXX*)

Of course, Tschirnhaus is thinking of Cartesian extension, as Spinoza reminds him in his rejoinder (*Letter LXXXI*). Cartesian extension is passive and quiescent, altogether the wrong notion of body. Then Tschirnhaus asks, just as we want him to, "I should like you to do me the favor of showing me how, according to your thoughts, the variety of things can be deduced a priori from the conception of extension" (*Letter LXXXII*). It has been Tschirnhaus's observation that, at least in mathematics, the definition of a thing yields one property and "that if we desire more properties, then we must relate the thing defined to other things; then, if at all, from the combination of the definitions of these things, new properties result". For example, he continues, I know only that the circumference of a circle is a uniform curve and if I want to know more about it, I must relate it — say, to radii or intersecting lines. This, claims Tschirnhaus, seems to oppose Proposition XVI of Part I of the *Ethics*. (Propostion XVI reads "from the necessity of the divine nature must follow an infinite number of things in infinite ways, that is all things which can fall within the sphere of infinite intellect" and Spinoza reminds us in the proof that "from the given definition of anything the intellect infers several properties" and "It infers more properties in proportion as the definition of the thing expresses more reality, that is, in proportion as the essence of the thing defined involves more reality.

In response to Tschirnhaus's second question, whether the particular bodies of our experience can be deduced from extension alone, Spinoza (who was admittedly weary and ill) replies, no, not if it is Cartesian extension but yes, if it is extension properly conceived. He contiunues, "But perhaps, if life lasts, I will discuss this question with you some other time more clearly". As to Tschirnhaus's allegations about the limits of definition, Spinoza says:

This may be true in the case of the most simple things, or in the case of things of reason (under which I also include figures) but not in the case of real things. For from the mere fact that I define God as a Being to whose essence belongs existence I infer several of His properties, namely, that He exists necessarily, that He is unique, immutable, infinite, etc. And in this way, I might adduce several other examples which I omit at present. (*Letter LXXXIII*)

In Spinoza's last months there is only the answer "Yes" to Tschirnhaus's question about the possibility of *a priori* knowledge of determinate things. Did Spinoza truly believe that scientific knowledge could be gained without experience?

5. My skepticism about Spinoza's philosophical interest in the problems cast up by the development of modern science is, I think, open to two kinds of responses. One important sort of answer would resort to the central issues of truth and adequacy in ideas, and would compare Descartes and Spinoza in this respect. Ultimately, I think, such a response would reconstruct for us the vision of science that Spinoza *might* have conceived, had he our questions in mind.

A second answer would *start* by accepting the premise of Spinoza's radical departure from the epistemological concerns of his natural philosophical contemporaries and address itself to the larger goals of his enterprise. It would note that there is an epistemological agenda associated with the rise of modern science, an agenda to which Descartes, Locke, Newton, Leibniz and even Berkeley and Hume adhere: the primary and secondary quality distinction, the theory of perception and of correct perceptual judgment. About these questions, the learning-about-nature questions, Spinoza has conspicuously little to say. But because he is so passionately concerned with the idea that human well-being is afforded by knowing the whole, he has a good bit to say about causality, possibility, essences and the central metaphysical issues of concern to science. For Spinoza, God may well inhabit the architectural details — the details of determinate natural things — but scrutiny of the details will never lead to understanding. Nor, for that matter, will understanding ever need to be tested against the facts.

Consequently, what we get from Spinoza is a physical theory (or theory-sketch) that is neither drawn *from* nor applicable *to* a concrete embodiment in particular natural phenomena.

Spinoza, then, is a Cartesian normal-scientist; when he is writing about the rainbow and chance, he works from mathematical exemplars set down by Descartes and Huyghens. At the same time, he has a distinctive physical theory that he never develops into a separate program of research. My suggestion, of course, is that he lacks (for a deeper reason) the epistemological apparatus to make the necessary connection between his theory and the world.

It has been left, then, to present-day commentators to suggest how, given Spinozistic physical theory and ignoring the epistemological gap, we might generate a Spinozistic program of investigation. It has become a commonplace in the secondary literature to mention a conceptual link between elements of Spinoza's physical theory and rather more recent scientific notions — comparisons with potential energy, with fields of force, and even geometrodynamics.[5] No one, of course, suggests any direct historical link between Spinoza and late physics, for Spinoza has not been an influential presence in the history of science. But these commentators do applaud Spinozistic metaphysical concepts, which (in the ebullient words of Professor Bennett (1980, p. 397)) are still "brimming with health and vitality". The details of these comparisons, and their tentativeness, need not concern us. The comparisons point rightly to the fact that Spinoza's physical theory remains a possibility for thought. They wrongly suggest that the theory is less remote from experience than Spinoza himself imagined it to be. He intended a theory that carried with it no epistemological directions for implementation, no risk of failure, and no promise of success.

Finally, it is possible, and a bit more generous, to view Spinoza as an object lesson about the philosophical conflicts attending the development of modern science. On the one hand there was a powerful desire — strongest, perhaps, in Spinoza — to replace the Aristotelian system with another system as whole, complete and as attentive to the final goal of human *eudaimonia*. Against this yearning must be set the idealizing, abstracting, and dissecting tendercies that accompany the new ideas about experiment in Galileo, Bacon, and Boyle. The investigation of nature was to be accomplished, these experimentalists believed, by isolating items from disturbing influences. On the ancient and medieval view of science, which Spinoza shared, the principle beneficiary of knowledge was the knower. Bacon, Descartes, Galileo, and Newton all had a different view. They believed that the benefits of scientific knowledge would be distributed equally among scientists and non-scientists. This shows that they had a narrower, less morally-encumbered, idea of knowledge than that advanced by the great philosophers of an earlier age.

University of Chicago

NOTES

[1] "The view that Spinoza was a rationalist, in the sense we are concerned with, is not just mildly inaccurate, it is wildly inaccurate," writes E. M. Curley (1973, p. 26).

[2] These scientific developments are discussed in Struik (1981).

[3] Indeed, Spinoza suggests that the scientific understanding of nature is by no means an amplitive process, but rather a precarious attempt to apply principles already known by the light of reason. In the *Tractatus Theologico-Politicus* he likens the interpretation of nature to the interpretation of scripture, with all the attendant pitfalls.

[4] *Letters* referred to in the following section (*LXXX—LXXXIII*) are found in Spinoza (1966, pp. 361—365; Gebhardt IV, pp. 331—335.

[5] See, for example, VanZandt's essay in this volume.

REFERENCES

Bennett, Jonathan: 1980, 'Spinoza's Vacuum Argument', *Midwest Studies in Philosophy* **5**, 391—399.

Curley, E. M.: 1973, 'Experience in Spinoza's Theory of Knowledge', in Marjorie Grene (ed.), *Spinoza; A Collection of Critical Essays,* Doubleday, Garden City, New York.

Spinoza, Baruch: 1925, *Spinoza Opera,* ed. by Carl Gebhardt, 4 vols., Carl Winter, Heidelberg. (Cited as 'Gebhardt' in the text.)

Spinoza, Baruch: 1951, 1955, *Chief Works of Spinoza,* trans. by R. H. M. Elwes, 2 vols., Dover, New York.

Spinoza, Baruch: 1966, *Correspondence of Spinoza,* ed. by Abraham Wolf, Russell and Russell, New York.

Struik, Dirk: 1981, *The Land of Stevin and Huygens,* D. Reidel Dordrecht.

ANDRÉ LECRIVAIN

SPINOZA AND CARTESIAN MECHANICS

This text is a condensed redevelopment and considerable extension of a study initially published in the first two issues of *Cahiers Spinoza* under the title of 'Spinoza et la physique cartésienne' (Lecrivain, 1977—78). My idea there was to propose a literal and exhaustive commentary on Spinoza's exposition of the second part of the *Principles of Descartes's Philosophy*. In the short space of an article, a few dozen pages, it would of course be vain to pretend to the same ambitions. I prefer to give an exposition of some of the major themes of the mechanistic conception of matter, particularly of those that especially captured Spinoza's attention and on which he reflected critically and demonstratively. The necessarily limited objective that I impose upon myself is, first, to determine the significance of the fundamental concepts, the articulation and systematic organization of which constitutes what is generally called the mechanistic theory of nature; and second, to try to explicate the forms and range of the theoretical and epistemological recasting in which Spinoza was engaged with regard to concepts he had inherited from Cartesianism. The difficulty of this enterprise arises essentially out of the fact that, in such a theoretical edifice, traditional metaphysical categories and scientific principles or concepts issuing from the most recent developments in the new sciences of nature are permanently intertwined. To compensate for the schematic and allusive character of my presentation, and to facilitate the comprehension of these points, which I give so briefly, it may be useful to introduce something not included in the first study: an account of the genesis of the presuppositions and consequences of the Cartesian conception of mechanics. In any case, I invite the dissatisfied reader to refer to the detailed analytic passages of the text published by *Cahiers Spinoza*.

1. SPINOZA'S ATTITUDE TOWARD THE PROBLEMATIC AND THE AIMS OF A PHYSICO-MATHEMATICAL SCIENCE OF NATURE

In this section it is important to discern what conception Spinoza may

15

Marjorie Grene and Debra Nails (eds.), Spinoza and the Sciences, 15—60.
© 1986 *by D. Reidel Publishing Company.*

have had of the science of nature, insofar as he may actually have
acquired it through his recent continuation of the work of Galileo and
Descartes, without excluding the research of the man who was his
neighbor from 1663 — the date of publication of the *Principles of
Descartes's Philosophy* — the Dutch physicist Huygens. This question
also encompasses comprehension of the relation between this type of
epistemological description and the whole philosophical project that
nourished and provided originality to the evolution of Spinoza's
thought.

1.1. *Reasons for Spinoza's Interest in Cartesian Mechanics*

Without going into detail, let us just say that three interests may have
motivated Spinoza's orientation. First, his choice was probably determined by the methodological efficaciousness of the Cartesian conception. It appeared to Spinoza that Cartesian physics and the theory of
mechanics expounded there could be considered as useful intellectual
instruments for guiding and sustaining the elaboration of his own
thought. Probably more profoundly, Spinoza's interest could have been
awakened by the importance of the new field of rationality constituted
as much by the results of Galilean research as by the shift toward
universalization conferred on them by Descartes's forging of a new
mathematical tool — analytical geometry — and his efforts to provide
metaphysical legitimacy. However, access to his new field of knowledge
occurs for Spinoza in the years 1663—1665 *via* a set of mediating
abstractions. Abstract, first, insofar as it is a system of conceptual
elements already elaborated through which the idea of nature displays
a new significance that definitely removes it from the Aristotelian-
Scholastic context, now relegated to the prehistory of science. But
equally abstract in the sense that mathematics and physics were not for
Spinoza — despite his undeniable competence in the field of optics —
sciences that he practiced for themselves, but were more useful as tools
in conditioning his theoretical apprenticeship. As he says in his last
reply to Blyenbergh (*Letter XXVII*), "... Ethics ... must be based
on Metaphysics and Physics" (Spinoza, 1966, p. 199; Gebhardt IV,
p. 160). This is because the fundamental project of the philosopher
remains that of constituting a theory of the affects and human behavior
inspired by the model of rationality furnished by the sciences of nature.
One can add to these diverse reasons a final motivation from the

perspective of historio-criticism insofar as Spinoza's considerations occur at a determinate moment in the evolution of ideas provoked by the development of mathematical physics. In fact, the work of Galileo not only revolutionized the science of motion, but raised the problem of the composition and cohesion of matter and gave a great echo to paradoxes in the notion of the infinite. The works of Torricelli and Pascal, in the field of hydrostatics, contributed to a resurgence and revision of the question of the vacuum. For his part, Huygens, reflecting on the rules of impact proposed by Descartes, was not content merely to show their falsity, but worked out, between 1652 and 1658, the correct formulation of the laws of impact of bodies. But this result was obtained without recourse to any assistance other than that offered by the geometry of Cartesianism and with a sort of indifference to the infinitesimal methods just then beginning to be developed. Spinoza was thus able to get the impression that the combination of geometry and mechanics would conserve not only scientific efficaciousness but epistemological validity.

1.2. *Conditions for the Recasting of the Theoretical Concepts on which Cartesian Physics Was Constructed*

Not only does Spinoza insist on a model of science issuing directly from geometry, but the anthropology corresponding to his ethical and political project is rooted in an outline of physics which takes as essential the grand line of Cartesian mechanics, notably the rejection of teleology and substantial forms. Nevertheless, this integration into ethical discourse of the principal elements of the mechanistic conception implied neither metaphysical submission nor theoretical blindness to what remained obscure and contestable in Cartesian thought. On the contrary, the deliberately constructive objective of Spinoza's deduction can be attained only by setting up a rigorous and vigilant critical authority quick to detect and to cast out all the ideological survivors that the mechanistic hypothesis had been unable to eliminate definitively. From this comes the idea of a reform of the understanding intended to purify it of all the prejudices remaining in spite of the radicalization of the Cartesian enterprise. Spinoza's effort is aimed at two targets. The first directly concerns the conceptual systematic inherited from Cartesianism, the cornerstone mechanistic conception of nature. The second, to which I shall return later, raises the question of modalities and conditions of validity for the abstract procedures used in the sciences of nature.

Indeed, since physico-mathematical science focuses on real beings and on things physically existing, not merely geometrically possible, it is important to elaborate an appropriate concept permitting apprehension and enunciation of the absolute essence of these things. As a corollary, this is enough to move aside the whole empiricist attitude; it reduces to the observation of the temporal and contingent succession of beings subject to change, together constituting what Spinoza called the common order of nature (*Ethics* II, Prop. XXIX, Corollary and Scholium). The correspondence with Oldenburg, which notably includes a severe and sometimes unjust critique of the experimental empiricism of Boyle, aims to expose and to avoid this danger.[1]

Moreover, the essence of Spinoza's thought is associated with recasting some of the categories inherited from the Scholastic tradition already reworked in the critical steps taken by Descartes. For if access to truth occurs *via* knowledge of the laws of nature, the task of philosophy is to universalize, by means of a rigorous and adequate conceptuality, not only rational knowledge of nature but knowledge of the relation of man to nature as well. The homogeneity of the concept of truth and that of being renders impossible any general or formal ontological project and allows one to set aside vain and illusory research into an external criterion for truth. On the other hand, the processes of the understanding can be considered as interior to the very structures of being. As opposed to Cartesianism, which still maintained an ontological distinction among possible, probable and real — corresponding to the stages of the faculties of knowledge: feeling, imagination, and understanding — Spinoza proposes a differential theory of the genres or modes of knowledge that allows the renewal of significance to the concepts of substance and attribute just as easily as to the discernment of these concepts, yet without introducing any limitation or partition in substance, since the attribute is nothing other than the manifestation of the essence of substance.[2] But Spinoza's reflexive project also consists in the explication and validation of the mediations that allow the physical side of things to be elevated to the universal concept of substance. This enterprise is attainable only if one takes into account the internal articulations and stratifications of the new domain of rationality, the Galilean-Cartesian mathematical physics. Indeed, Galilean physics confers on the algebraic operations of Cartesian mathematics a basis in things. It treats at least a few aspects of perceptible objects and does not deny itself recourse to experimental procedures to give an account of the motion of the pendulum, falling bodies, or the resistance of materials to rupture (Galilei, 1952, pp. 131—177). For its part, the Cartesian analysis already attempts to universalize the results of Galilean science in affirming, for example, that the laws of nature are homogeneous and immutable and in demonstrating, by means of analytic geometry, that all the elements of space can be composed and ordered effectively in accordance with the laws that regulate algebraic operations. Combining the one with the other, Galilean physics with Cartesian mathematics, one glimpses why and how extension* can be considered as the ultimate framework of things, decipherable with the aid of geometrical figures and algebraic symbols.

Spinoza clearly marks, within this evolution of thought, the presence of two themes that he concentrates on removing. First, we can know only the relations among extended things or among the ideas of these things. And this is indeed why the understanding, defined as the order and connection of ideas, gives an effective content to the category of attribute. In the second place, it is clear that the universal concept of substance must be recognized as the principle of intelligibility of the laws of nature. Under these conditions, the concept of the infinite becomes expressive of the immanent presence of a rational order actually within things. This implies not only a complete revision of the meaning of the infinite, just as much in relation to the Scholastic tradition as to the Cartesian heritage, but the identification of substance with the infinite productive power of nature — in short, the demonstrable and verifiable affirmation that the infinite is nothing but the actual being of nature. Without being able to develop these two points in detail, let us just say that the concept of the infinite loses henceforth the meaning that previously had been conferred on it by faith or theology — which established the character of incomprehensibility that even Descartes gave it — to attain henceforth an integrally natural and rational significance. Nevertheless, there remains the problem of this new concept of infinity's mode of relation to and compatability with the diverse procedures operating in the sciences of nature and especially in mathematical physics. Indeed, as indicated by *Letter XII* (Gebhardt IV, pp. 52—62), one must acknowledge a careful and rigorous distinction between the different senses of the term 'infinite' according to what is legitimately and adequately known by the understanding or apprehended only by the imagination. If substance and its modes constitute the totality of things really existing, the being of nature is not entirely reducible to the indefinitely divisible magnitudes in accordance with numbers, or to the durations expressed by the relations of time. Thus the question arises of the determination of the status of the operating instruments used by physicists and mathematicians engaged in the practice of the mathematization of nature.

1.3. *Modalities and Validity of the Process of Abstraction in the Sciences of Nature. Determination of the Status of the Abstract Operators: Time, Number and Measure as Auxiliaries of the Imagination*

The very object of physics is bodies in motion, striking and repelling one another, and their natural being would be destroyed if one did not begin by recognizing the infinite unity of productive motion in its inexhaustible diversity. If one did not admit this preliminary condition, the meaning of the physical law would be compromised because the successive positions of the moving object could no longer be linked to one another and, by the same blow, ineluctably, the paradoxes of Zeno would reemerge. Undoubtedly, Descartes himself did not completely escape this, for he was tempted to combine motion with rest, the dynamic with the static. But, to the contrary, without the auxiliary notions of time, measure and number, no physico-mathematical science of nature is possible. Thus it is appropriate to determine their functions rigorously and to recognize the role strictly appropriate to each, an investigation clearly involving the objectivity and rationality of natural law.

Consequently, it is important to explicate the process of abstraction that corresponds to the intervention of these operators, those that permit the clear and exact expression of motion in itself inseparable from the object studied. So one must take care to make these abstract instruments real or natural and to convert them into real properties of things: if one yielded to that fiction, which would mark the triumph of imagination over understanding, the modes could be considered independently of substance, from which they hold all their power of existence, and they would then be found unduly substantified. The principal teaching of *Letter XII* consists in making us comprehend how imagination perverts our knowledge of things by transforming operative signs — relations of correspondence or proportionately determining the successive positions of the moving object — into the actual reality of the body in motion. One would then illegitimately substitute for the infinitely diversified modes of a substance that is infinitely one, only simple modes of thought or even simple auxiliaries of the imagination. If that were so, those indispensable and effective instruments of physico-mathematical knowledge would be converted into epistemo-logical obstacles sufficient to create an impenetrable screen behind which nature would escape all investigation and all intelligibility. At stake in such a problematic is, quite simply, the critical appreciation of nature and of the degree of validity of the appropriate discursiveness of the constitution of this new science, Galilean-Cartesian mathematical physics. The interest of Spinoza's approach is a proposed elucidation and resolution that is both epistemologically pertinent and metaphys-ically unbiased.

The source of these confusions of these errors resides in the mis-understanding of a double process of abstraction improperly converted into one explanatory model of natural reality or, even worse, sub-stituted for it. One begins by detaching the modes from substance and conferring on them a characteristic autonomy; then one identifies in these modes the operative and abstract instruments that assist the imagination in establishing some indicators or in instituting relations between objects. That is the point of departure of the formalist and nominalist illusions of inverting the rational order and judging "things by their names, not names by things" (Spinoza, 1963, p. 109; Gebhardt I, p. 235). On the contrary, one who really knows — that is, by means of the understanding and not only through the imagination — what substance and its modes are, the infinite and the finite, eternity and duration, will not be duped by the apparent contradiction between the continuity of motion of moving bodies and the divisibility of the infinity of space and time, because he will find its resolution in the affirmation of the unity and effective indivisibility of an infinite productive power that has in itself the ultimate ground of law. For Spinoza, mathematical physics can be theoretically valid only if it admits that the real link is between moving bodies and not between numerical indicators that merely symbolically set out some of their determinations or properties. Under this condition, the utilization of abstract operators, of these auxiliaries of the imagination — number, time and measure — again becomes perfectly legitimate without ceasing to be fruitful. From its previous consideration as merely negative, this use recaptures its full positive sense, and these auxiliaries now deserve that name since they function as operative signifiers that imagination proposes and subordinates to the activity of the understanding. *Letter XII* shows clearly, with the example of the infinite, privileged because of its unique complexity, how it is now possible to unravel this web of confusions. In the first chapter of *Thoughts on Metaphysics*, Spinoza had already accorded to time, number and measure, a power and an explanatory status that gave them a sort of priority among the modes of thought. Subject to the native power of the understanding, these signs or symbols, recognized and validated in their operative and expressive functions, are easily integrated into the rational order of the process of the knowledge of nature.

This both operative and symbolic character finds its source in the free and hypothetical use that is made of them by scientists. If one wishes the laws of nature to reach universality, it is important to relate the infinity and diversity of the motion which is produced in the universe to a system of reference, namely, in the case of the Galilean-Cartesian physics, the inertial system of reference. In other words, the choice of an origin of space and time will acquire universal validity only under conditions to be determined arbitrarily by the physicist.

A text of *On the Improvement of the Understanding* confirms the interpretation that I propose with respect to the role of the auxiliaries as indispensable instruments for the knowledge of singular changing things (Gebhardt II, pp. 36—37). Under this last expression, it is necessary to understand perceptible objects, accessible to sense and apprehended by the imagination, that experience presents as just so many indicators to the attention of the understanding. From this point of view, the auxiliaries constitute guides for the senses, and the imagination is allowed to order and to direct the ceaseless and diversified production of empirical phenomena.

The difficulty and the complexity inherent in the correct determination of the function and status of these operators, the auxiliaries of the imagination, result directly from the fact that the modes of extension are known at the same time by the imagination and by the understanding. The confusions and illusions that Spinoza itemizes and elucidates come precisely from an absence of discernment and discrimination between the respective functions of the one and the other. Any reduction or substitution of one of these modes of knowledge for the other generates error. Further, it is the effectiveness of these abstract instruments — time, number and measure — that increases the risks of misinterpretation because the precision and the exactitude with which they allow us to determine the relations among objects inclines one almost inevitably to see nature as a mere assembly of numbers and figures. This was Galileo's conviction, confirmed, moreover, by the Cartesian conception of an entirely geometrizable extension. In this respect, an integral mathematization of nature justified by the convenience and the usefulness of employing these operative and abstract instruments could not but have appeared to Spinoza as an ideological subversion transformed into prejudice:

> For accident is only a mode of thinking, which does no more than denote an aspect. For example, when I say that a triangle is moved, motion is not a mode of the triangle but of the body which is moved; consequently in respect of the triangle, motion is called accident, but in respect of body it is a real being or a mode, since motion cannot be conceived without body but can indeed be conceived without triangle. (Spinoza, 1963, p. 111; Gebhardt I, pp. 236—237)

This passage, however, is sufficient to make it clear that what is in question is less the set of objects and mathematical procedures considered in themselves than the mode of utilization and application of some of the abstract instruments, mathematical in origin or construction, in the domain of physical experience. The uniqueness of this new process of experimental knowledge, combining the properties of really existent beings and the abstract determinations of symbolic operators, did not escape the perspicacity of Spinoza's reflection. The ascription and delimiting of the conditions of validity of such a convergence, the necessity and success of which was each day affirmed by the progress of the nascent physics, constituted one of the essential tasks brought to the attention of the philosopher.

1.4. *Significance and Range of Spinoza's Conception of a Physico-mathematical Science of Nature*

Such a science must be considered the only authentic science of nature; consequently it could be neither ignored nor rejected. Nevertheless, because it concerns really existent beings, it can never be totally detached from the apprehension that the imagination is able to have of it, even if understanding reveals itself as uniquely apt to make intelligible the links established among the various finite modes of extension. Thus, reconstituting piece by piece the articulations of reality, the concepts of mathematical physics give us access to the natural and objective domain *via* the mediation of local and regional laws that for this reason always conserve a determinate degree of abstraction and relativity.

One must not neglect therefore to reintroduce what the process of operative abstraction begins to set aside or place in parentheses. In fact, in nature no element could be durably and legitimately isolated from all others, and each part of nature has its existence, like its essence, in the infinitely infinite productivity of substance.

Cartesian mechanics is certainly able to give an account of the elementary states of matter, but the simplest elements are also the least specific and the least concrete. We shall measure the explanatory capacity of mechanics to reduce progressively these various degrees of abstraction. But, for the moment, nothing permits the affirmation that this alone will be enough to make plainly intelligible such complex objects as living beings, social bodies or political institutions. . . . For in nature there are not only bodies in motion that fall and collide. In spite of certain similaities, we shall have occasion to see what distinguishes the law of inertia from the conception of *conatus*.

2. PRESUPPOSITIONS, EVOLUTION AND CONSEQUENCES OF THE CARTESIAN MECHANISTIC CONCEPTION

2.1. *The Presuppositions of Cartesian Physics*

". . . Although my physics is nothing other than mechanics", Descartes wrote to Debeaune in his letter of April 30, 1639 (Adam and Tannery II, p. 542); by 'mechanics', it is necessary to understand the attempt to explicate the totality of material phenomena by reducing them to themselves, to what characterizes them correctly, that is, their positions and their shape. This results in two effects: first, the reduction of matter to geometrical extension and, second, geometry being the science of understanding, the epistemological subordination of matter to mind.

From these initial principles, one can deduce at least three consequences that strikingly determine the place and function of the theory of impact in Cartesian physics:

2.1.1. *Since the action of matter is referred to its position is space, it can act only where it is.* Immediately we see that any idea of action at a distance, of attraction or repulsion, must be dismissed. Descartes remains a total stranger to the stream of thought that runs from Gilbert to Kepler and from Galileo to Newton.

2.1.2. *Action can have no other form than contact and encounter.* But the mode of this action can only be motion, more precisely, a displacement in space. So the true material phenomenon is the meeting of two or more bodies in motion. This is why all Cartesian physics must be reduced, in the final analysis, to a theory of the impact of bodies and of the communication of motion. This constitutes a systematic set of basic laws from which it becomes possible to analyze, bit by bit, the totality of even the most complex material phenomena whose investigation must permit the restitution from those initial laws of the increasing degree of composition.

2.1.3. *Furthermore, Descartes seems to recognize as properties of matter, able to be clearly and distinctly set forth, only motion, position and shape.* However characteristic they may be, he keeps and distinguishes neither hardness nor elasticity, the propensity that certain bodies have to return to their shapes, reacting against deformations that they have undergone. Also, in the exposition of the laws of impact, Descartes will explicitly formulate the hypothesis of integrally hard bodies while interpreting this hardness in a contradictory way, in terms of elasticity.

2.2. *Mechanistic Principles of the Communication of Motion*

It is in *The World: Or Essay on Light* that one encounters the first explicit statement of the principle of inertia, five years before Galileo's *Dialogues Concerning the Two New Sciences*; because of the role recognized for gravity or weight, inertia had not received a satisfactory exposition. In *The World: Or Essay on Light*, this position is set forth twice in what Descartes calls respectively the first and third laws of nature. The first presents the conservation of speed (Adam and Tannery XI, p. 38); the third, that of direction (XI, pp. 43—44). But between the two, Descartes inserts a second law: "When one body pushes another, it cannot give the other any motion without at the same time losing the same amount of its own motion; neither can it remove motion from the other except by being augmented proportionally ... (XI, p. 41). In any case, it is not yet a question of the principle of conservation of the same quantity of motion (mv), as it will be definitively formulated in the *Principles of Philosophy* II, Article 36, in 1644. In the meantime there will have been an evolution.

To reach the concept of quantity of motion, he had to identify volume and mass, that is, to consider matter as perfectly homogeneous and to exclude the consideration of density. In his development of the principle of inertia, Descartes begins by denying any inertia or natural laziness, as he writes to Mersenne in December of 1638 (Adam and Tannery II, pp. 466—467). The principle of inertia implies, of course, that there is no force in bodies that opposes motion. However, this conclusion will be obtained only four months later. But at the same time, Descartes returns to the proposals that he had sent to Mersenne and appears to accept the idea of natural inertia:

... I think that in the whole of created matter there is a certain quantity of motion which never increases or diminishes. So, when one body moves another, it loses as much of its own motion as it gives away; ... So, if two unequal bodies receive the same amount of movement as each other, this equal quantity of movement does not give the same velocity to the larger one as to the smaller. In this sense, then, one can say that the more matter a body contains the more 'natural inertia' it has. One can say too that a body which is larger is better able than a small one to transfer its motion to other bodies; and a larger body is harder to move than a smaller one. So there is one sort of inertia which depends on the quantity of the matter, and another which depends on the extent of its surfaces.** (Descartes, 1970, pp. 64—65; Adam and Tannery II, p. 543)

So the evolution of Descartes's thought has led from his setting forth the principle of inertia and the negation of all natural laziness, to a first position that what one body loses in speed, another gains, that the largest bodies receive speed less than others, and finally to the position that the distribution of speeds operates in inverse proportion to the respective sizes of the bodies, i.e., of their volumes, hence, matter being taken as perfectly homogeneous, of their masses.

This evolution is achieved at the moment of the publication of the *Principles of Philosophy* (1644), and the principle of conservation of the same quantity of motion passes to the first rank (Article 36). It is defined as the first cause of motion and is founded on the metaphysical principle of the immutability of God and the constancy of his action. The three following laws of nature are then set out:

2.2.1. *The first (Article 37) is the return to the law of inertia.* It is applied now not only to motion but to the figure itself and, one might add, to any state of a body.

2.2.2. *The second affirms the conservation of direction.* That is, it establishes the rectilinear character of uniform motion (Article 39). Combining this with the preceding law constitutes the complete statement of the principle of inertia.

2.2.3. *The third law defines the general principle of communication of motion.* It puts in order the exposition of the rules of impact (Articles 40 and 42).

2.3. *The Stages of the Constitution of the Laws of Impact of Bodies*

The two first laws of impact are elaborated at the end of 1639 and the first months of 1640. The first is progressively refined in three letters, the one of April 30, 1639, to Debeaune, already cited, one of November 13 to Mersenne (Adam and Tannery II, pp. 622—623), and finally, one of December 25 of the same year, also to Mersenne (II, p. 627). Descartes reaches the conclusion, perfectly correct, that if a body in motion strikes a body at rest, the first will drive the second with a speed such that the quantity of motion of the first is divided by the two sizes or masses:

$$v = v' = \frac{mv}{m + m'}.$$

The second rule is vaguely stated in the letter to Mersenne of March 11, 1640. It explains the example of the impact of balls in the game of *paume* (Adam and Tannery III, p. 37).

The two rules, which were later abandoned by Descartes, include some interesting points. The first shows that two hard or soft bodies will not react against one another and will not separate after impact. The second rule indicates that it is the larger body that determines the direction of motion after impact.

This second rule contains other remarkable elements as well. Here Descartes specifies that the ball that rebounds, the smaller one, goes backwards at a reduced speed. On the other hand, he affirms that the force of motion and the point towards which it tends — its determination or direction — are "completely different things", which means that one can separate the direction of motion from its speed and consider it as a scalar or absolute quantity, and not as a vector or relative quantity. From this will follow most of the errors of the laws of impact that later thinkers will criticize and correct. It will result likewise in a restriction of the field of application and validity of the Cartesian mechanistic hypotheses.

2.4. *Nature of and Reasons for Descartes's Errors*

2.4.1. *Quantity of motion and vital force.* The fact that one can have, in Cartesian mechanics, neither negative speeds nor negative motion implies two consequences: first, that motion is never the contrary of motion, and that two opposing motions cannot destroy one another; second, that the speeds of motion in opposite directions are composed by Descartes arithmetically and not algebraically, which will force him in setting forth the laws of impact to violate the principle of conservation.

It is true that the principle of conservation will be not only that of the quantity of motion (mv), or the quantity of impulse, but of the vital force (mv^2) as well, as will be demonstrated by Huygens and Leibniz. In any case, the introduction of this new principle simply legitimates and reinforces the Cartesian idea of conservation of the positive. Of course, insofar as the product mv^2 is always positive, it perfectly expresses the conservation of forces engaged in the phenomena of impact, i.e., the respective directions of motion of colliding bodies. From this perspective, Huygens and Leibniz will not have to refute Descartes but to grant him his full significance — scientifically and experimentally verifiable — in the affirmation of the very principle on which Descartes understood his mechanics to be founded. For Huygens, it will be only a matter of reintegrating direction in the consideration of speed of motion and, consequently, of treating the problem of bodies in impact on the model of a composition of vectors with opposing orientations. For this, it will be enough to resort to a process of translation while remaining strictly within the framework of Cartesian algebraic geometry to determine the respective velocities of moving objects after impact through an operation of adding and subtracting vectors. For Leibniz, it is even simpler and easier, for the requirements of conservation and continuity find their immediate expression in the appropriate mathematical instrument represented by the infinitesimal calculus.

Thus Huygens and Leibniz do not contradict Descartes, but merely manage to elaborate the solution conforming to the principle he stated, a solution that he unhappily missed. So, to express the laws of impact correctly, it suffices to hold to the general formula:[3]

$$v = v' = \frac{mv_0 + m'v_0'}{m + m'}$$

on the condition, however, of interpreting it algebraically and not, as Descartes had done, arithmetically.

2.4.2. *Mechanics and the problems of time.* One must in fact take into consideration the incidences of the problem of time in mechanics. In his *Treatise on Mechanics*,[4] as well as in his letter to Mersenne of September 12, 1638, Descartes not only succeeded in elaborating the notion of 'work', in the modern sense of that word (the product of weight times the distance travelled) but also succeeded in distinguishing this product from the product of weight times speed. Well before Leibniz (who attributed this to himself in the *Brevis Demonstratio* of 1686 while, to make matters worse, accusing Descartes of confusion), Descartes had clearly established the distinction between work and quantity of motion. Unfortunately, this distinction — which is without effect in statics, where it might have been omitted without difficulty — Descartes abandoned where it was indispensable: when he took up the problems of dynamics. Leibniz was warned of his ignorance by Arnauld and did not fail to make this observation in reply.[5]

Actually, Descartes was always inclined to avoid considering speed because it appeared to him both to have implications for time and to send him back to the more general problem of weight. From this perspective, Descartes found himself in a polemical situation with the Galileans and especially those who had gravitated into the orbit of Mersenne. Methodologically, Descartes's attitude appeared to be founded only in the consideration of space and not of time, saving one dimension in the determination of force, as is recalled by the letter to Mersenne of September 12, 1638. In accordance with the demands of method, Descartes logically preferred the simpler hypothesis.

However, we know that the preference accorded to space over time in the determination of the law of falling bodies will actually prevent Descartes from arriving at a solution and a correct formulation of the law, although Beeckman had furnished him with all the elements. These errors must, of course, be put in relation to the themes of the discontinuity of time and the instantaneous propagation of light, themes which served as postulates in the *Dioptrics*. And it is in the *Dioptrics* as well that one sees Descartes composing the direction of vectors with the aid of the geometric procedure of the parallelogram, while treating the arithmetic values in a purely additive manner (Adam and Tannery VI, pp. 94—98).

On the problem of impact, such premises are even more incompatible with the complexity of the phenomenon. Contrary to the perspective of Descartes who wanted to view events as simple and instantaneous, the collision of two bodies is an extremely complex process which lasts a certain time, because it consists of a succession of transformations where, after an initial phase of slowing down, there intervene deformations of the colliding bodies, then a stop, a progressive recovery of the shape of each body, and finally a new distribution of motions, speeds and directions.

2.4.3. *The question of the nature of colliding bodies*. Here again, the indeterminate nature of the problems of impact is an obstacle that Descartes was unable to overcome. He began by excluding hardness as a property of matter, reducing it to geometrical extension. The hypothesis of a perfectly homogeneous material must normally lead to a perfectly soft material. But, in Article 45 Part II of the *Principles of Philosophy*, he makes a series of hypotheses to the contrary that are hardly compatible. For one thing, he proposes to study the impact of two bodies considered in isolation, which seemed to imply that they encounter one another in the void, for the fluid environment had been abstractly set aside. For another, these two bodies are supposed to be perfectly hard and, at the same time, which is contradictory, perfectly elastic. In fact, elasticity is the characteristic property of bodies that are not completely hard — since they undergo deformations — and not totally soft — since they can recover their shape after impact.

In reality, these problems of impact make necessary the intervention of experimental considerations concerning the exact nature of colliding bodies, and are conveyed by the introduction of coefficient of restitution that varies from 0 for inelastic bodies to 1 for elastic bodies (Dugas, 1956, p. 184).

2.4.4. *The hypothesis of a force of rest*. In his letter to Mersenne of December, 1638, Descartes denies all laziness or natural inertia in bodies. As early as April, 1639, this attitude seems to have been put in question, but it is especially in the text of Part II of the *Principles of Philosophy* that the reversal will be obvious. Indeed, it is significant that it is in Articles 43 and 44, which immediately precede the statement of the laws of impact, that Descartes defines rest as an obstacle to motion and sees in it a genuine force of resistance. This conception is really achieved in Article 55 where Descartes characterizes cohesion, or the hardness of bodies, as a positive force of rest. This force of rest, combined with the purely scalar representation of speed, allows him to define a hard body as that on which the animated motions of a certain force rebound. Bit by bit, one passes from a perfectly soft or fluid material — which is how Spinoza interprets the motion of matter in the rings of bodies — to a hard material, paradoxically equipped with the property of elasticity.

2.5. *Conclusions Concerning Cartesian Mechanics*

Despite the accommodation that Descartes is able to give — sometimes at just the wrong moment — to certain teachings of experience, his mechanistic conception of communication of motion can hardly accommodate the experimental results. This leads him to forge a complementary hypothesis in the hope of thereby tying together the abstract laws formulated in the second part of the *Principles of Philosophy* with experimental results. The hypothesis is that of fluid matter, which surrounds and envelops hard bodies, acting on them to modify the real conditions of communication of motion appreciably (cf. Articles 53 and 61).

But this is insufficient, and we have already noted some of the reasons why Descartes found it impossible to overcome the difficulties and to correct the errors that his theory of impact presented. Since all his physics resolves, in the final analysis, to the mechanics of impact, it is understandable that he had no real scientific posterity. The only follower in spirit, if not in the letter, of Cartesianism will be Huygens, but this very point will very soon force Huygens to pay all his attention to this unhappy theory of the impact of bodies, to dispel the errors, and immediately to break with the principles on which Descartes stood, which from the beginning, were ruining the theory constructed on their base. But, by the same stroke, the exact and elegant solution to be given by Huygens to the problem of impact — probably as early as 1656 — drives Cartesian mechanics into the prehistory of real science.

Without even waiting for the work of Huygens, these laws of the impact of bodies were contested during Descartes's lifetime, especially by Brother Gabriel Thibaut, a minor churchman of the province of Lyon, who criticized both the Cartesian theory of extension and the sixth rule, in which a moving body (m) collides with a body (m') of equal mass, and transfers to it a part of its speed while itself rebounding to conserve the other part of its motion and reversing its direction. The objection of Brother Thibaut is well founded:

> That is wrong, as one can see by experimenting at a game of billiards. As someone knocks one ball against another, for example B against C, ball B pushes ball C and then does not move. We experience the same when we play *grandes dames* at dining hall tables: as very often one *dame* pushes another and then does not move from the very place where she has touched the other . . . .[6]

Bringing to a close this discussion of the Cartesian theory of mechanics, it is surely unnecessary to reproach Descartes for having built an abstract physics, contrary to experience, or in any case different from what seems to be demonstrated, and consequently allowing it to be circumvented by the precautions uselessly formulated in Articles 45 and 53; yet we must regret that Descartes lacked theoretical rigor and believed he would be able to combine strictly conceptual exigencies with certain suspect teachings of experience. It will not escape the perspicacity of Leibniz that the first rule, the only sound one, seems also, at the experimental level, to support confirmation of the Cartesian principle.

From this point of view, it seems legitimate to oppose the rigor that Descartes used in proofs in his exposition of mechanics, that is, in his theory of statics, to the theoretical weakness which is shown in the presentation of the laws of impact. Despite the reproaches addressed, for example, to Galileo, Descartes does not seem to have been very sensitive himself to the necessities of a scientific theory. One can find proof of this in the equivocal meaning of the term 'force' in the string of Cartesian texts. It is used first and foremost to designate pressure or weight. The same term is also used to express work, as is shown in the letter to Mersenne, already cited, of September 12, 1638, where Descartes speaks of a force of one, two, or three dimensions. We have already noticed how costly it could be in dynamics, this absence of a strict distinction between work, or the motive force (the product of weight times distance traveled) and that of weight times velocity. But in the *Principles of Philosophy* the term 'force' designates nothing more than the product of weight times speed, the quantity of motion. We must still add to this the force of rest expressing, this time, the resistance that a stationary mass opposes to motion.

One undoubtedly needs — and this, it seems, is the lesson one can learn from the work of Huygens — to detour momentarily from experience, to treat the phenomena of impact with the aid of instruments particularly proposed by algebraic geometry, thus in a purely theoretical manner, and only afterwards to return to experience to try to incorporate into the theory the experimental results, most notably the notion of elasticity, insofar as one attempts to distinguish it from hardness and fluidity. One can thus say, in summary, that it is by his own failure to respect the methodological principles — which he had clearly formulated and justly used against the Galileans, for example — that Descartes fell into error. If he had distinguished in his dynamics, reduced to a theory of impact, that which is in the realm of the concepts of the understanding from that which is the realm of experience, or of what he called the 'testimony of the senses', perhaps Descartes would have been able to avoid this accumulation of errors. But we have seen also what presuppositions, profoundly ingrained in Cartesianism, were weighing against such an effort. The theory of light, explicated in the *Dioptrics*, was already being assimilated to the phenomenon of impact. At the same time, the theory of impact had to appear as the development of this conception, particularly in the dissociation of the vectorial character from the scalar character of speed. In reality, one must recognize that the critical effort demanded by and inaugurated in the *Rules for the Direction of the Mind* collided with the inevitable measure of intellectual confusion that brought about this age of change, of profound upheaval in modes of thought. Descartes's epistemological move is first a theoretical and basic rupture occurring in what legitimately can be regarded as essential: the mathematization of motion, the enunciation of the principle of inertia, the introduction of the infinite. For the rest, Scholastic remnants and scientific novelties still resided together, perhaps not without some overlapping.

These summarized points have no aim other than to render easier access to the reading and transcription that Spinoza gives to the principles and laws of Cartesian mechanics, especially in the second part of the *Principles of Descartes's Philosophy*. Since my objective remains the comprehension of the specificity and originality of Spinoza's attitude toward these major themes in Cartesian mechanics, I will retain in my exposition the structure that is revealed and suggested by the internal order of the Cartesian text. We know, of course, that the second part of the *Principles of Philosophy* is organized around two great moments. The first, corresponding to Articles 1 to 35, constitutes a genuine eidetic of extension and motion. One is assisted in a progressive deployment of the elements and implications of the conception that Descartes had of extended substance and its principal attribute, motion. This exposition remains purely descriptive, geometrical and kinematic. The second moment (Articles 36—64) is the presentation of Cartesian mechanics through the principles, laws and rules that define it. The exposition then takes on an explicative character in its introduction of considerations of dynamic order. Correlatively, one will thus distinguish a first part in the presentation that Spinoza makes, devoted to the essentials of the deduction of the essence of extension and of motion. A second part is devoted to the deduction of the causes of motion and the laws that regulate its communication.

3. SPINOZA'S DEDUCTION OF THE ESSENCE OF EXTENSION AND OF MOTION. ASSIMILATION AND REDEVELOPMENT OF THE CARTESIAN CONCEPTS

3.1. *Deduction of the Essence of Extension*

In the *Principles of Descartes's Philosophy*, Spinoza makes a critical revaluation and redevelopment of the significance of some principal concepts of Cartesianism — and first of extension — which he undoubtedly adopts and validates, but he confers on them, by a radical and deepening move, a new orientation and different range.

3.1.1. *The definition of extension*. In the exposition of the *Principles of Descartes's Philosophy*, the definition of extension precedes that of substance, the opposite of the order of exposition in the *Ethics*.[7] This discrepancy merits some consideration. First, extension is identified as 'quantity' (*quantitas*), which is the term associated with speed that Descartes employs in defining motion. This term is again found in *Letter XII*, then in the Scholium to Proposition XV of Part I of the

Ethics. However, the definition of extension from the *Principles of Descartes's Philosophy* is still far from the one that will be presented as the attribute of substance. It comes closest to the common notion of extension in that it is here characterized by its extendedness in three dimensions and is consequently considered *partes extra partes*. This common notion is actually the preliminary condition for the establishment of a physics. The definition of a real thing remains nominal, however, insofar as it sets out the specific properties of extendedness while abstracting causality and productivity from the thing that it designates. Finally, linked with the second definition, that of substance, it renders possible the definition of motion as local motion.

3.1.2. *Spinoza's ratification of the Cartesian identification of matter and extension.* The first propositions of the second part of the *Principles of Descartes's Philosophy* are not without interest, even if Spinoza seems to take literally the Cartesian conceptions, as soon as what is at stake is the status, strictly material or not, of extension. This theme, which Descartes himself claims touches the very foundation of his physics, had been the object of a correspondence with Henry More.[8] There are three aspects to this polemic. First, the identification of matter and extension brings about a limitation of the concept of extension to what can be the object of the senses and the imagination, this qualification being the condition of the distinction of physics proper from geometry. If such a condition were not admitted, extension could be defined as the property of infinite space and by the same stroke attributed to God. Further, the discussion is concentrated only on the infinite or the indefinite extent of matter and the universe. The notion of infinite/indefinite, functioning in relation to the conception of a created world and the incomprehensibility of the action of God, could exclude consideration of teleology and could avoid the affirmation of the necessity of the universe (since only the *actually* infinite being, God, is necessary). Finally, the polemic concerns the validity of the explanatory Cartesian hypothesis by means of turbulent motion (God's initial introduction and conservation of the vortices) as it is set out in the second part of the *Principles of Philosophy* (Articles 33—35) and developed in the third part (Articles 47 sq.).

The first proposition of the second part of the *Principles of Descartes's Philosophy* truly radicalizes the Cartesian conception of extension. The exclusion of all the sensible qualities of the essence of bodies shows that Spinoza is taking Descartes's part against More, since the latter claimed that matter could not be defined except in relation to sensation, especially tangibility and impenetrability (Adam and Tannery V, pp. 267—268). But Spinoza's objective goes quite beyond the merely Cartesian perspective, insofar as this determination of extension is intended to make perspicuously precise the conditions of a physical science by demonstrating that it requires the transformation of all the sensible and imaginative elements, in other words the empirical results, into purely intellectual elements and concepts. It becomes difficult, then, to maintain the hypotheses or metaphysical implications proposed by Descartes in Meditation VI. We know, in fact, that the distinction of the physical sphere from the pure domain of objects and mathematical operations cannot receive its full significance except under the conditions articulated in the two proofs of the existence of bodies and the incomprehensible union of the soul and the body, postulates so unacceptable to Spinoza that he ridicules them in the preface to Part V of the *Ethics*.

The presentation of the significant content of the category of extension continues in the second proposition, which is only, after all, the positive counterpart of the preceding one. Yet it is doubtless from this moment that one is better able to appreciate the originality of Spinoza, especially in relation to the discussion in which Descartes opposes More.

3.1.3. *From three dimensional extension to the attribute of extension.* Let us recall that More rejects the Cartesian identification of matter and extension and considers the attribution of this property to the whole group of immaterial substances: the soul, the angels and God himself. Further he judges contradictory the negation of atomism with the correlative affirmation of the infinite divisibility of matter, and the elaboration of a theory of the elements and the acceptance by Descartes of the corpuscular conception.[9] A last objection concerns the range and significance of the differentiation between the infinite and the indefinite. For Henry More, the Cartesian distinction is acceptable only if it bridges, on the one hand, the infinity of God or divine space and, on the other, the finitude of matter or purely material extendedness, an inadmissible interpretation for Descartes, who clearly ends up explicitly affirming that by indefinite extension he now understands the unlimitedness of material extension in a positive way, and not only negatively. This is a discrete but indisputable way to integrate into Cartesianism the results of the Copernican theoretical revolution and those of Galilean science.

If one wants to understand the original position of Spinoza toward the two competitors, it is necessary to represent it in the following manner.

With Descartes and against More, Spinoza maintains the distinction between thought and extension and so rejects the idea of a spiritual extension of divine space. But convinced, like More himself, of the unsatisfactory character of the Cartesian concept of extension, he prepared to rework it thoroughly. So, in the *Ethics*, as opposed to what he had seemed to support in the *Principles of Descartes's Philosophy* and the *Thoughts on Metaphysics*, extension would no longer be excluded from the nature of God but, to the contrary, integrated into the notion of *natura naturans*. In any case, differing from More, this advance of the category of extension implies no distinction between divine extendedness and material extendedness, properly speaking. Such a division has no basis in things, but is only an effect of the process of knowing. Apprehended by the imagination, extension is no longer manifest as infinitely divisible materiality in finite modes that could be attributed to God only by a sort of anthropomorphic perversion. Alternatively, apprehended by the understanding, extension continues, homogeneous and indivisible, able to be constituted as one of the attributes of infinitely infinite substance. So Spinoza interprets in epistemological terms what More had proposed in ontological and theological terms. On this point, the originality of Spinoza would consist in the elaboration of a concept of extension in rigorous accord with the rejection of the theme of creation. Further, in his polemic with Descartes, More had ended with a move to the idea of a real finitude of the world. From the *Principles of Descartes's Philosophy* to the *Ethics*, Spinoza followed a more or less opposite path to arrive at the affirmation of the infinity of extension and necessity of the universe.

3.1.4. *Epistemological implications of the identification of extension and matter.* In the *Principles of Descartes's Philosophy*, this moment corresponds to the deduction and the demonstration of Propositions III—VI. Adherence to Cartesian mechanics not only brings about the interchangeability of the concepts of body and space but in another way marks the exclusion of both the void and atomism. In any case, the rejection of the void aims less at showing its physical impossibility than at establishing the absurd and contradictory character of the term, that is, its conceptual impossibility. On precisely this point, the pronouncements of the *Ethics* confirm the text of the *Principles of Descartes's Philosophy*.[10] The negation of atomism is more delicate and raises a number of problems, as given in the epistolary controversy between More and Descartes that touches directly on the question of the infinite and its indivisibility. In fact, the renewal of the term 'indivisible' in the seventeenth century, particularly striking in the works of Cavalieri and Galileo, must be connected to the atomistic stream represented by Beeckman, Gassendi and More. The Cartesian refutation of atomism aimed at that time to preserve the mechanistic conception from the difficulties and paradoxes issuing from the formidable concept of the infinite. To surmount these, Galileo had been led to anticipate certain holistic themes (Galilei, 1952, pp. 131—177). But the Cartesian argument invoked metaphysical considerations such as the incontestable validity of clarity and distinctness or the incomprehensibility of the infinite, inaccessible to human understanding, affirmations that Spinoza could not admit and that, in any case, the preface of the *Principles of Descartes's Philosophy* had already declared contrary to his own thought (Gebhardt I, p. 132). If Spinoza accords with Descartes in rejecting all forms of atomism, and in indivisibility at the moral and physical level, still, none of the reasons presented by Descartes can be preserved after one accepts the perspective of infinity and substantial indivisibility. *Letter XII* first, but even more the *Ethics*, is engaged in a discrimination and stratification of the concept of the infinite, forcing a distinction between the infinity of nature and that of varying magnitude. In this respect, the sixth proposition of the second part of the *Principles of Descartes's Philosophy* achieves the integration in the concept of extension of the various properties that characterize it in itself. We note that Spinoza's presentation distinguishes, and not without reason, the admission of the definiteness of the extendedness of matter from the affirmation of its unity and of its identity, that is, its homogeneity, a thesis commonly shared by thinkers of the seventeenth and eighteenth centuries.

Until now, the essence of extension, geometrically defined by its three-dimensionality, had not been questioned. Now one must consider the modifications or variations that are produced there, and which can be only changes of figure or displacements of place coming from the motion of the parts that compose it. The purely geometrical conception of extension now forces considerations of a kinetic order which will in turn lead to the formidable problem of the rigorous conceptualization of essence and the modalities of motion.

3.2. *Deduction of Essence and of the Modalities of Motion*

3.2.1. *The definition of motion as local motion* (motus localis). In the text of the *Principles of Descartes's Philosophy*, Definition VIII is just the literal transcription of the definition of motion proposed by Descartes at the beginning of Article 25 of the second part of the *Principles of Philosophy*. But here again as in the case of the definitions of extension and substance, although motion is itself a real thing to him, we have before us only a nominal definition that does not provide knowledge of its real cause, extension itself having been defined beforehand as a common property and not as that infinite productive activity engendering motion and rest at the same time (*Letter LXIV*, Gebhardt IV, p. 278). This is undoubtedly the reason why Spinoza judged it appropriate to clarify this definition with a few remarks, some of which will lead to a significant revision of the Cartesian conception.

3.2.1.1. The first remark recalls the interpretation that Descartes proposed for the individualization of material bodies: a body is defined by the different parts that are found transported together (*Id omne quod simul tranfertur*).[11] We shall see that Spinoza will not stop at this elementary and abstract kinetic principle of individualization, but will substitute for it an original conception having double significance: external, by the application of the flat parts of bodies, one on top of the other, and in a general way by the pressure of surrounding things — which also seems to have been Huygens's point of view; internal, by the theory of the singular essence of *conatus*, the degree of complexity and cohesion characteristic of each individual. One finds the question posed again of the confrontation between the law of inertia and the principle of *conatus*. And one will see in greater detail that this relative and external kinetic interpretation of physical individuality by Cartesian mechanics differs appreciably from the internal dynamic of *conatus*, expressive of the modality according to which the infinite productivity of substance is deployed in each finite mode.

3.2.1.2. The second remark explicitly underlines not only the purely nominal and kinematic character of the definition of motion as local motion, but equally the positive nature and legitimacy of practical abstraction. So the separation of the concepts of motion and force is situated in the strict continuation of the Galilean project and proves that force can no longer be considered in the Aristotelian way, as the cause of motion, but only as the cause of changes that affect motion: acceleration, deceleration, modification of direction.

3.2.1.3. The third and fourth remarks determine the conditions of the assignment of motion, recalling Articles 28 and 29 of Descartes. The first states that motion must be understood as the transport of a body from the vicinity of those that touch it towards other bodies, and not only as the passage from one place to another. This affirmation substitutes the condition of contiguity for the theory of natural place expressive of Aristotelian kinematics. The other observation adds to contiguity a second condition, that of the immobility of neighboring bodies, which leads to acceptance of relativity of the assignment of motion.

3.2.1.4. Finally the fifth and last remark allows one to pass to Definition IX, which characterizes motion in terms of circles or rings of bodies. In unique fashion, the concept of motion receives added complexity through the introduction of the themes of its composition and its turbulent modality, implying the hypothesis of an infinity of motion at the level of the parts, without which, for one thing, the assignment of the proper motion of each body would be compromised.

Before going even further with the exposition and the deduction of the essence of motion and its modalities, it should be said that, in 1665, Spinoza, still situated within the perspective of the Cartesian conception, defines each mode of extension, and in a general way each individual body, by a formula expressing the determinate and optimal proportion of motion and rest compatible with the play of the parts that consitute it, but within certain limits subject to variation: see *Letter XXXII* (Gebhardt IV, pp. 169—176). This certainly puts him beyond the properly Cartesian conception of external physical individualization by a geometrical type invariant to a principle of internal individualization and dynamic nature. But does such an innovation not irreversibly transgress the domain of extendedness and the validity of the fundamental concepts of Cartesian mechanics?

3.2.2. *Space, time and motion: problems of the infinite and the continuum.* Considering only extension, each body is now defined in its material identity by a constant of geometrical order corresponding to the volumetric invariant that Descartes had rashly assimilated to the mass of the body. It is necessary now to introduce a second determination, that is, the adherence of this body to only one motion, the explication of which will lead initially to the theory of the rings of bodies which will in turn constitute the abstract model of the Cartesian metaphor of turbulence.

But such an explicative chain cannot persist without difficulty within the Cartesian conception as presented in Articles 33 to 35 of the second part of the *Principles of Philosophy*. Actually, the considerations of infinitesimal order, explained in Article 34, are immediately counterbalanced by the affirmation, in Article 35, of the incomprehensibility of the divisibility of the infinite. Inevitably, two difficulties arise:

3.2.2.1. The mode of correspondence between the affirmation of the indivisibility of instants of time and that of the divisibility of the infinity of extension.

3.2.2.2. The possible compatibility between the indefinitely continuous divisibility, in other words, the divisibility "at least potentially" to which Definition VII explicitly refers in the second part of the *Principles of Descartes's Philosophy* (Gebhardt I, p. 181) and an infinite divisibility really realized, or enacted, that would constrain one to leave the traditional framework of the mathematics of Euclidean origin, a move which Descartes never considered making, at least not officially and publicly.

These particularly delicate questions permit us to glimpse the reasons that motivated the long Scholium to Proposition VI (Gebhardt I, pp. 191—196), which Spinoza used as a transitional text between the exposition of the essence of extension and that of motion. The solution that Spinoza will advocate, of which we have here the first trace, will consist in the articulation of the infinite divisibility of modes above the indivisible infinity of substance and attributes (*Letter LXIV*, Gebhardt IV, p. 278). True, it is not certain that Spinoza ever achieved an entirely satisfactory conception of these points, at least if one accepts the last letters to Tschirnhaus (*Letters LXXXI* and *LXXXIII*; Gebhardt IV, pp. 332 and 334—335). Nevertheless, one can try to indicate the conditions for a clear and distinct comprehension of the divisibility of the infinite. It suffices to consider it only at the level of modes and carefully to avoid any confusion between a real distinction and a modal distinction. Certainly Descartes expressly differentiated them, but was constrained to admit the real divisibility of matter in giving an account of its circulation in turbulent constrictions; he could arrive only at numerical discontinuity — an infinite represented in the imagination — in excluding the continuous divisibility of modes which is legitimate only because it is rationally controlled by the understanding. He should have been forced to break with the idea of the instant conceived as an indivisible entity of time.[12]

3.2.3. *Instantaneous propagation and the hypothesis of contiguity.* It is clear that the modalities of motion come directly from the preliminary conception of extension, from the affirmation of a full world, without a void and without atoms. So Proposition VII of the *Principles of Descartes's Philosophy* (Gebhardt I, p. 196) transfers to motion what was set out from a static perspective in Proposition IV: the principle of the substantial and geometrical invariance of extension and its con- stituent parts, particular bodies. In so passing from the geometrical plane to a kinetic perspective, the validity of the principle gains at the same time in universality, in complexity and in concreteness. On the other hand, it continues to be made in abstraction from temporal considerations, the Cartesian conception of motion supposing instan- taneous propagation and simultaneity.

Descartes adds to these two conditions the hypothesis of contiguity. This new determination is nothing but the transcription on a spatial plane of the preceding two. But it serves also as the substitute for a genuine continuity. So Spinoza was very careful in the demonstration of Propostion VIII to separate contiguity from circularity (Gebhardt I, p. 197). He proves in fact that there is no motion that does not require time, however small the time might be. The impenetrability of matter and the contiguity of bodies seem to have implied a new conception of the instant. It is no longer considered as a temporal atom, a very brief time, but as a vanishing quantity, a genuine infinitesimal.

One is now in a position, it seems, to evaluate more precisely the Cartesian thesis of the 'subordinate substantiality' of particular bodies that can be indifferently considered as modes or as parts of extension. But Spinoza cannot accept this equivocation of usage and significance of the term 'substance'. With regard to extension, particular bodies are only modes, defined by their figure and motion, and able to exist and subsist only by the regulated play of the motion which affects all the others. But these different bodies, which infinitely and eternally divide extension could not contravene its substantial unity and pulverize it into a multitude of particular irreducible substances. Only bodies considered in relation to other bodies appear physically determined — the kinetic point of view — and no longer only geometrically, so that they can be characterized as invested with an appearance of substantiality. Only this prevents their always being reduced to geometrical configurations since they issue from combinations of motions. So this duality of aspect is not founded in things but results solely from the distinction between the perspective of the imagination and that of the understanding. But temporarily retaining the notion that this duality conditions an entirely mechanical interpretation of the phenomena of nature *via* a series of impulses 'in a chain', one is introduced directly to the second modality of motion: its circularity.

3.2.4. *The necessarily circular character of motion.* In reality, this new characteristic of motion results from the combination of the two conditions of instantaneous propagation and contiguity. If in fact there is no void in the universe, it is easier to conceive the plenum as at rest than in motion, as Descartes did. But if motion is produced, it is necessary that it be instantaneous, that it concern all the bodies in motion simultaneously, and that these bodies be in contact with one another. The affirmation of the circularity of motion is a logically necessary consequence of the properties that have already been recognized in it. But by the same stroke, the conception of motion becomes uniquely complicated. Actually, the abstract image of the circle of matter gives an account of the motions of solids without insurmountable difficulties because all the parts concerned move at the same speed. It is quite different when one passes from a perfect circle to an imperfect circle — from the study of the motion of solids to that of the motion of fluids.

The well-known example is that of circling matter in a channel the diameter of which varies constantly and continually. The analysis of such a phenomenon can be made only by introducing the principle of inverse proportionality of cross-sectional area and speed. This requirement presents a sufficient degree of generality, for it can be applied later to the whole set of problems of communication of motion and impact. But it is necessary to add to this initial principle a second condition. It is not only that the cross-sectional area and the speed of the circling particles are inversely proportional, but that the variation of the one from the other is *continuous*. The circular conception of the

motion of fluids in a channel of continuously variable diameter consequently demands the recognition of the decline of the speed by
infinite degrees, as Galileo had admitted, but as Descartes had refused
to admit.

As Descartes had attempted to maintain a precarious equilibrium
between discontinuity and continuity, Spinoza's conclusions, in radicaliz-
ing the Cartesian conception and placing it within a rigorous demonstrative process, intended to affirm the incontestable primacy of continuity. The provisional terminus of this path will take the form of the
redevelopment of the concept of matter; the divisibility of extension,
heretofore considered valid, will be transferred to bodies in motion.
This path leads us from the abstract to the concrete, from the geometric
to the kinematic. But it remains for us to make one more step, that is,
to go from the kinematic to the dynamic taking into consideration the
causes of motion and studying the effects of the intervention of the
causes.

4. THE CAUSES OF MOTION AND THE LAWS OF ITS COMMUNICATION: PRINCIPLES AND RULES OF THE MECHANISTIC CONCEPTION

Following the internal order of the second part of the *Principles of
Philosophy*, we arrive now at properly mechanistic concepts, the systematic articulation of which constitutes what one might call Cartesian
dynamics, at least if one accepts the traditional terminology that signifies by the word 'dynamics' the study of motion and the rules of its
communication from the perspective of the causes that produce it.

4.1. *The Causes and the Principles that Determine the Motion of Matter*

This essential theme, fundamental to Cartesian physics, can be further
subdivided into three moments that correspond respectively to the
attribution of the origin of motion to God, the affirmation of the
conservation of the quantity of motion, and finally the establishment of
the principle of inertia.

4.1.1. *God as the cause of motion.* As already indicated, Descartes
represented extension — substantial but geometrized — as an inert mass
at rest, where only the efficient and eminent intervention of God is

capable of introducing motion. It is not the same for Spinoza. Extension, as an attribute manifesting the absolute essence of substance, is given an infinite productive power which is the totally immanent cause, creating in and by itself those effects that are the infinite immediate modes of motion and rest eternally dividing it. In other words, moving from the theme of creation to the idea of production, Spinoza is in a position to propose a perfectly immanent and rational interpretation of mechanics, removing from it the transcendental metaphysical presuppositions to which Descartes had subordinated it. This change of perspective is radical and fundamental, finding illustration and confirmation in the formulation of the first principle of conservation, that of the quantity of motion.

4.1.2. *Principle of conservation of the same quantity of motion and rest.*
The change that Spinoza implements is apparent in the terminology he uses. Proposition XIII of the second part of the *Principles of Descartes's Philosophy* (Gebhardt I, p. 200) substitutes in the presentation of the principle of conservation the term '*to impart*' (*impertire*) for '*to create*' (*creare*) which Descartes had used. Spinoza also speaks of the same quantity of motion *and* rest, situating the two modes of extension on the same plane without any priority of one in relation to the other. Spinoza's reading reestablishes the equilibrium and the equality that the Cartesian conception had compromised. Yet he removes nothing from the ascribable, determinable and, so to speak, quantifiable character of motion; quite the reverse. In any case, the expression '*certam et determinatem quantitatem*' does not mean that motion is reduced to an essemble of numerical results that measure it abstractly. The size in question cannot be specified except by means of the relations of proportionality, as will be shown later. Under this condition one can be assured of the advance and the autonomy of a rational physics appropriate to the knowledge of extension and its modes, motion and rest. But the significant shift made by Spinoza will later — from 1665 — find its sanction in the definitive substitution for the Cartesian version of the affirmation of the conservation of the same *proportion* of motion and rest. These are a pair, as one can now see through the effort to make the conception of the individual bodies more complex and concrete.

4.1.3. *Principle of inertia and theory of* conatus. The setting out of the law of inertia, properly established by Descartes in 1633, in *The World: Or Essay on Light*, is one of the elements of mechanics, the theoretical precision of which can be considered only as an advance. In any case, Spinoza presents it as an axiom although he attempts to present a demonstration of it. By comparison, one will observe that in the *Ethics*, the principle of inertia is presented in a corollary, though previously signaled as self-evident (*per se notum*).[13] The demonstration is not lacking in interest in that it makes reference only to the power of God and his eternity, affirmations that are not uniquely Cartesian since they will be found later at the basis of Part I of the *Ethics*. This leads us to say that the validity of the law of inertia is subject to no particular metaphysical hypothesis, but offers a character of universality that permits its legitimate consideration as an axiom or a common notion through which the absolute power of God is expressed. Nevertheless, because it is the manifestation of this power on the plane of extension, motion and matter, it must be demonstrated and deduced.

Having said this, we note that there exist certain similarites between the presentation of the principle of inertia and that of *conatus*. One finds in them the same subject (*unaquaeque res*), the same determination (*quantum in se est*) and the same verb (*perseverare*).[14]

But it would be wrong to conclude immediately their pure and simple identity and I want now, on the contrary, to indicate some of the differences that appear to me essential.

4.1.3.1. The identity of determination (*quantum in se est*) specifies the characteristic proper to all existing modes, to wit, the degrees of power and being that qualify them individually. From this perspective, there is continuity and coherence between the law of inertia and the theory of *conatus*.

4.1.3.2. However, the text of the *Principles of Descartes's Philosophy* explicitly sets out the conditions of applicability of the principle of inertia by providing a limited definition of the sphere of extendedness, even though the universal validity of *conatus* does not seem to be qualified by any reservation.

For one thing, the law of inertia is applied to a simple and undivided thing. It clearly seems that these bodies cannot be defined except by their degree of motion and rest, and not differentiated from one another except by the relative speed of their motion. So the essence of these things is reduced to their speed, and a change of direction cannot be considered as compatible with their simplicity and their undividedness except on the condition of leaving this speed unchanged. But any change of direction is accompanied by a phenomenon of acceleration or deceleration. This allows one to specify the principle of inertia in terms of uniform rectilinear motion.

Further, the principle of inertia implies that the body in question is considered in isolation, abstracted from the environment, which underlies the strictly theoretical, rather than experimental, character of the formation. This is why the text of the *Principles of Descartes's Philosophy* speaks of persevering in the same state (*semper in eodem statu perseverat*) while the *Ethics* evokes the effort to persevere in one's being (*in suo esse perseverare conatur*). From one expression to the other, one passes from a purely theoretical and abstract condition to a real and individualized principle that regulates the behavior of any modal being.

4.1.3.3. The text of the *Principles of Descartes's Philosophy* presents a physics that is still very abstract and which does not go as far as the *Ethics* (Part II, Proposition XIII, Lemma 3, Axiom 2) in considering the theory of the simples bodies (*corpora simplicissima*). As long as we remain within the conditions prescribed in the *Principles of Descartes's Philosophy*, it will not be possible to speak of '*conatus*' with full rigor. As indicated explicitly in Proposition VI of Part III of the *Ethics*, this term designates the *effort* to persevere in one's *being*. But, the notion of effort supposes a resistance, a power of opposition or negation. On the contrary, the principle of inertia makes reference to a *condition* that reduces to the conservation of a speed and a direction, and implies that there is no obstacle or opposition of any sort (resistance, friction, etc.).

Actually, the law of inertia, as long as it expresses the essence of the material body — the degree of power of the body considered in isolation — attests in fact that the *conatus* of such a body reduces to the strict conservation of its quantity of motion and rest, in other words to the conservation of its speed and direction. The loss of its speed signifies, if not its pure and simple disappearance, at least the loss of its individuality. But in entering into more and more numerous combinations, these simple and undivided bodies will continue to move with the speed and the direction that corresponds to their quantity of motion and rest, subject to whatever encounters and impacts occur. Further, *via* composition and integration, these bodies will generate and be affected by innumerable variations of matter and motion, and increasingly greater complexity of individualized and diversified bodies. It is in this sense that one can evoke a continuity of the principle of inertia with the concept of *conatus* insofar as considerations other than mechanical ones never intervene.

On the other hand, the law of inertia determines the minimum variation of *conatus* by reducing the being to the condition, the complex individual to the simple and undivided body; it eliminates what founds and correctly determines *conatus*, namely the effort that the infinite productivity of substance imposes on each of the modes, through which it expresses the constraint to struggle — not only to preserve its being, but to increase and expand it in opposition to other modes that surround it, press against it, collide with it or threaten it, in proportion to their respective degrees of power. Considered from this perspective, the principle of inertia is, so to speak, the reverse and the extreme limit of *conatus*, as if, with these simple and undivided material bodies, one were presented with a simple state that is nothing but an ultimate limit of being. It indicates no more than the exhausted forms of *conatus*, its theoretical and abstract support, the point where being is barely distinguished from nothingness. Here, the '*quantum in se est*' attains its minimum, below which there is mere nothing. In sum, the simplest bodies are all exteriority, which is why *conatus* vanishes in becoming the simple and abstract law of inertia.

4.1.4. *The inifinitely complex hierarchy of particular bodies.* In truth, even the simplest bodies never exist in isolation. It is necessary therefore to complete the explication of the principle of inertia by reintroducing different degrees of complication and relative concretization. It is in the interval opened by Propositions XIII and XIV of the second part of the *Principles of Descartes's Philosophy* (Gebhardt I, pp. 200—202) that the hierarchy of the infinity of particular bodies appears. Each is defined by the combination of two principles, one characterizing the total individual determined by the conservation of the same global quantity of motion and rest, and the other characterizing the simple inertial motion of isolated and undivided bodies. The physical individuality of a particular body is thus effectively fixed by a quantity determined by motion and rest.

Because simple bodies regroup into more and more complex collections, the hierarchy becomes such that the individuals include a vast quantity of individual parts, a quantity that rapidly surpasses any assignable number. So the divisibility of infinite extension as well as the decrease of the infinite degrees of speed send us finally back to the divisibility of the continuum, and we are obliged to pass from the parts to the whole. This is why Spinoza recognized only a distinction of reason, and not a real distinction, among the various parts of an even slightly complex body. From then on, however, everything is physically and mechanically expressed in a determined or determinable quantity of motion and rest; this expression is not always numerically ascribable, but only exceptionally so. It appears to require a type of size other than mere extensive size, which one can call size or intensive quantity, including the notion of 'degree'. It is effectively by this condition that one can be assured of the continuity — by transition — from the principle of inertia to *conatus*, which designates in fact the actually existing manifestation of an individual essence, or again, a degree of power to actualize in an optimal manner. But here we reach the limits of the scientific application and verification of the theoretical conceptions of Spinoza, in that the field of mathematical knowledge was still characterized in that epoch, by two deficiencies: it lacked both the infinitesimal calculus, uniquely capable of permitting the determination of an infinite bounded by two limits, upper and lower; and the concept of real number, rendering possible not only the ability to manipulate but rational intelligibility of the spatial continuum.

4.2. *General Laws of the Communication of Motion*

These laws are explained and demonstrated in Propositions XVIII to XXIII of the second part of the *Principles of Descartes's Philosophy* (Gebhardt I, pp. 207—211). They are accompanied by an attempt to clarify the principal concepts of dynamics, notably the concepts of motion, i.e., of force — identified as a quantity of motion or a quantity of rest — and finally of speed. This effort at terminological refinement must be considered as the phase preparatory to the development of a theoretical and scientific problematic and as an operation of delimiting the principal context into which the study of the impact of bodies and the presentation of the rules that determine it will be fitted.

4.2.1. *Spinoza's exposition shows clearly that the domain of the laws of the communication of motion already has norms and is limited by two theoretical principles.* The first is the principle of conservation of the same quantity of motion and rest already set out in Proposition XIII. It

appears in fact as the fundamental generalized and abstract requirement, to which all derived laws are subordinated, and which regulates
the state of global motion of the mechanistic system constituted by the
ensemble of moving bodies that encounter one another.

   The principle of inertia intervenes when one no longer considers only
the mechanical state of the system considered as a whole, but considers
the internal transformations that are produced within the system after
the impact of two or more bodies, transformations that affect their
respective states of motion and rest. These modifications bring about a
new distribution of motion, speeds and directions among the various
bodies concerned. But this distribution can be performed only in relation to the constraints demanded, in part, by the law of inertia and, in
part, by the rule of proportionality, which together constitute and
determine the law of motion of each body.

4.2.2. *One can add to these two essential principles of the mechanistic
conception of motion a hypothesis that appears as an application of the
law of inertia, such that the change of state of a body exhibits minimal
variation.* Descartes himself had already formulated such a hypothesis
in a letter to Clerselier of February 17, 1645, in order to legitimate the
established distinction between the speed of motion and its direction
(Adam and Tannery IV, pp. 183—188). Yet he had not considered it
useful to include it in the French translation of the text of the *Principles
of Philosophy* in 1647. This hypothesis is obvious from the principle of
the minimum which Fermat used as a basis for producing a critique of
the Cartesian demonstration of the laws of refraction in the *Dioptrics*.
Spinoza's presentation and use of it shows an original orientation, the
implications of which have only barely been seen, in that this principle
of economy and of least action (or minimal change) can perhaps be
understood as the beginning of a principle of internal regulation — of
an almost statistical nature — which will make possible a dynamic
definition of individuality, notably in the *Ethics*.

4.2.3. *Suffice it here to indicate two consequences of the methodological
and terminological revision given by Spinoza's presentation.*
   4.2.3.1. The first consists in that the separation of direction from
motion, properly speaking, is deduced directly from the demands of
invariance of the quantity of motion and rest in a determined mechanistic system. This separation, the origin of which Spinoza had justly

noted as the second book of the *Dioptrics*, can be admitted only because the change of direction of a body in motion which rebounds against a stationary body nevertheless leaves unchanged the quantity of motion and rest characteristic of the ensemble constituted by the two bodies. Yet this implies that at no time can this state of motion of a body be transformed into a state of rest, even were it for the smallest period of time. This observation aims to set aside the Scholastic thesis of the *media quies*, an intermediate pause between the two phases, constrained and normal, of all motion produced in the sub-lunar world. Unhappily, this legitimate theoretical concern is met in return for a misunderstanding of the complexity of the phenomenon of impact. Leibniz, who will not have the same scruples and will not pronounce the same interdictions, will reconstruct this complexity much better and will succeed in explicating the succession of phases that compose the totality of the phenomenon.

4.2.3.2. The second consequence corresponds to the determination of effects of the variation of speed in relation to the quantity of motion. This can be defined as directly proportional to speed, the respective masses of the bodies being assumed to be equal. The interest in such a procedure is that it allows one to construct the concept of quantity of motion without bringing in, as Descartes did, a transcendent and transitive action of God on matter. Quantity of motion, interpreted in terms of force, then becomes a function of speed. However, when speed slows down until nullified, one obtains the concept of quantity of rest or force of resistance to motion. These concepts determine the limits of variation of the effect of speed so that quantity of motion and quantity of rest appear as strictly correlative notions, conceived only in relation to one another, according to a scheme of inverse proportionality. The difference is obvious with the Cartesian thesis of a force of rest presented as the internal principle of cohesion of hard bodies, in relation to which motion can be conceived only under the condition of an intervening external causality. Finally, the mode of construction of these concepts supports the affirmation of the relativity of motion and rest on the one hand, and that of speed and slowness on the other. The ultimate outcome of all this work of terminological and conceptual clarification is a differentiation not only of the notions of motion and speed, but of two concepts of speed (Gebhardt I, pp. 209—211). In fact, speed does not so much express motion itself as the quantity of motion and rest in a body; in this way, it is the result and the expression of internal transformations that are the exchanges or communications of quantity of motion and quantity of rest produced in it. As to the distinction of the two concepts of speed, it takes the following form: the first is dynamic in relation to motion and rest, corresponding to the instantaneous relative speed; the other defines it as the space covered in a unity of time and expresses a kinematic point of view which differentiates it from motion. This is the divergence in perspective between Galileo and Descartes.

4.3. *Theory of Impact*

We know that the laws of impact of bodies that must constitute the heart of Cartesian mechanics are unhappily erroneous with the exception of the first. One might thus be surprised to observe that Spinoza continues to admit these rules although certain of them were contested even during Descartes's lifetime, and although Huygens, who had been Spinoza's neighbor from 1663, had for his part resolved the problem by 1658. Recall also that in *Letter XXXII* to Oldenburg of November 20, 1665, Spinoza does not deny the validity of six rules out of the seven, only the sixth appearing to him unacceptable (Gebhardt I, pp. 174—175). Thus I shall not linger on this part of the exposition and shall simply put into evidence the new elements that are comprised in Spinoza's presentation of these rules of the impact of bodies.

4.3.1. *The problematic of impact.* These rules can be considered *qua* rules only on the condition that they satisfy the different principles that have already been enunciated. In other words, these particular laws of the communication of motion have significance only if they constitute the strict application of these principles. They must therefore be rigorously deduced, and one must avoid allowing empirical considerations to interfere with theoretical demands and concepts. The fact that one remains in a context strongly marked by abstraction is what highlights the study of the general modalities of the meeting of two bodies, considered as isolated and independent of the forces possibly exerted on them by the bodies surrounding them. Here the problematic of impact is restricted to a two-part investigation.

It is a question, first, of explicating the modalities according to which modifications of motion are brought about, the speed and direction of colliding bodies. But the different possible modifications will be inscribed within a limited frame of the study of three possibilities:

(1) modification of direction only,
(2) modification of motion only,
(3) modification of direction and motion at the same time.

But one must also determine the cause, the force, that produces these transformations, especially in distinguishing between force of motion and force of determination.

4.3.2. *First group of rules*. This first group can be considered as homogeneous in that all the cases studied compose only a modification of the direction, which can affect both of the bodies or only one of them. In the text of the *Principles of Descartes's Philosophy*, the possibility under consideration corresponds to Propositions XXIV to XXVI (Gebhardt I, pp. 211—213). Observe that Spinoza's explication proceeds from a first case of perfect equality to a second case, of inequality and asymmetry, leading to a restored equality by means of a mechanism of compensated inequality. It is, in a way, a sort of process and, at the same time, a complexification, generalization and concretization, even if the reestablishment of the initial equilibrium is obtained only by the modification of direction alone. It is well known that the third case considered by Spinoza has no equivalent in the Cartesian text and that this thoroughly pertinent addition anticipates Leibniz's observation in the *Animadversions* concerning the case where the impact "has taken place between two unequal bodies coming from opposite directions at unequal velocities . . ." (Gerhardt I, p. 380).

4.3.3. *The particular case of the third rule*. The third rule can be detached from those that precede it and from the three that follow insofar as, for one thing, it conveys a case of inequality obtained from the first rule, by an excess of speed of one of the bodies over the other while the two remain equal in size; for another, it imposes a modification at the same time of both the motion and the direction in such a manner that only the direction of the slower body is modified and the motion of each is rendered equal to that of the other; thus the equilibrium can be reestablished. These, then, are the conditions for a return to equilibrium from an initial inequality that Spinoza's demonstration will attempt to make precise and to explicate.

Spinoza retains the possibility of introducing a new conceptual specification, that of the 'force of speed' permitting the assimilation of the state of a body to the degree of speed that characterizes it. This force of speed is the quantity to which direction appears as the associated and corresponding variable. This implies that the direction followed by a body is a function of its degree of speed and that, in the case of the collision of two bodies, it is the body that moves fastest that determines the direction of motion under the hypothesis of equal masses.

Finally Spinoza considers the hypothesis of an oblique impact in order to distinguish between the different meanings of the term 'force' and to incorporate reasoning based on the composition using the parallelogram of speeds, inspired by what Descartes had attempted in his *Dioptrics*. The interest of this example is to adjust to the various domains of the new physics, all of which it encompasses simultaneously — the fall of bodies (studied with the aid of an inclined plane), the laws of the motion of the pendulum (associated with the preceding, as demonstrated in Galileo's *Dialogues Concerning the Two New Sciences* and in the work of Huygens), the laws of reflection (as they are set out in the *Dioptrics*), and finally the laws of communication of motion, properly speaking.

4.3.4. *Third group of rules: rules 4, 5 and 6.* These rules correspond to modifications occurring in the direction and/or the motion itself, according to three possibilities that Spinoza explicitly indicates:

(1) One body carries the other with it and both are animated by the same speed.

(2) One body, the stationary one, remains at rest and the other rebounds.

(3) The body that moves rebounds, but transfers a part of its motion to the body formerly at rest.

Rule 4 recalls the first case, rule 5 the second, and rule 6 — the validity of which Spinoza will come to suspect in 1665 — the third.

4.3.5. *Rule 7.* This can be set aside in the sense that it rests on a new hypothesis, that of the encounter of two bodies in motion, moving in the same direction, but at unequal speeds. As for the preceding rules, the solution proposed by Spinoza consists essentially in reducing inequality and reestablishing equilibrium by admitting as small a variation as possible. The two moments of the rule lead then, in the first case, to a modification only of the speed of motion of the bodies in order to reach an equalization of their respective speeds without modification of direction, and in the second case, to a modification of the direction of the continuing body without modification of its speed.

Despite the number and the diversity of the cases considered, the phenomenon of impact remains blemished by abstraction because it has been studied previously only in the easily exemplified but too simplified case of the collision of two isolated bodies. It is appropriate then to place it in the natural frame of bodies surrounding it, and thereby to integrate all the previous results into the perspective of a generalized communication of motion and of equilibrium among the different forces exerted within the totality of the matter.

4.4. *Towards a Generalized Theory of Equilibrium and of Change: the Role of the Surrounding Bodies*

With this theme, we reach the ultimate state of the more or less redeveloped presentation of the elements fundamental to a mechanistic physics. Conscious of the proven difficulty of reconciling the abstraction proper to the theoretical hypothesis — that of two bodies meeting in isolation, supposed to be perfectly hard while improperly identified as elastic — with the experimental observations linked to the development of the statics of fluids and the dynamics of solids, Descartes had left the necessity to insert the phenomenon of impact into a more concrete and infinitely more vast context: the rest or motion of a solid body plunged into fluid, the particles of which can be considered either at rest or in motion. Here notions of solidity and fluidity intervene that have not yet been clarified. For his part, Spinoza, in his presentation of Cartesian mechanics, takes great care not to voice the slightest hypothesis concerning the nature of colliding bodies.

4.4.1. *Recalling the Cartesian problematic.* This consists in considering three possibilities in succession:

(1) The case of rest, immobility or equilibrium of a solid body within a fluid which is itself immobile.

(2) The placing in motion, by the action of an external force, of a solid plunged into a fluid considered as immobile.

(3) If the fluid is itself in motion, it takes with it the solid body that is immersed there, which in its turn can be considered as at rest or immobile within the fluid.

It is not difficult to observe that, by the mediation of a dual complexification, one is finally brought back to the initial case, to a position of equilibrium of the solid in relation to the fluid which surrounds it.

4.4.2. *Innovations contributed by Spinoza.* These are of a diverse nature. Certain of them, having a methodological and demonstrative character, consist in recasting the Cartesian order of exposition, for example by inverting the order of consideration of the quantity of motion, on the one hand, and of the speed or, more exactly, the force of determination, on the other. More generally, Spinoza's exposition ignores the hypothesis of turbulent motion on which the Cartesian text of the *Principles of Philosophy* constantly leans, but seeks to demonstrate that the equilibrium of a body depends on the equality of action of the corpuscles that act on it. It is an implicit anticipation of the principle of virtual speeds, which gives a sort of theoretical basis to the last propositions and demonstrations of the second part of the *Principles of Descartes's Philosophy*.

4.4.3. *Solidity and fluidity.* These last texts tell us something about the topic. However, in his preface Louis Meyer emphasizes the necessity to complete them with "several propositions regarding the nature and properties of fluids" (Spinoza, 1963, p. 7; Gebhardt I, p. 131). So we can thus allow ourselves to be assisted here by other texts, particularly the corespondence and the *Ethics*.

In fact, in *Letter VI* to Oldenburg, Spinoza links the notions of fluidity and hardness to the use of the senses and distinguishes them formally from motion, rest, and their laws, that is, from those "notions which are pure and which explain Nature as it is in itself" (Spinoza, 1966, p. 93; Gebhardt IV, p. 28). A little further on, Spinoza says precisely that the term 'fluid' is only an 'extrinsic denomination' by which one tries to transcribe the experience that one can have of imperceptible bodies moving with a speed inversely proportional to their mass. Since the proportion considered by Spinoza is susceptible to variation *ad infinitum* by insensible and continuous transition, there results a relativity of the notion of fluidity and solidity. It is only by comparison that one body can be said to be more solid or more fluid than others. Since all the consituent parts of a body are animated by a determinate degree of speed, however weak, far from having to explain the production of motion from an initial state of rest, it is preferable to admit that matter is constituted of particles in motion, differentiated by the force and degree of speed which are proper to them. The most universal hypothesis is thus that of fluidity, and the hardness or solidity is only the expression of the decline *ad infinitum* of this degree of speed or again of the increased participation of the body in slowness. It is now better understood why the example of the imperfect circle necessarily imposes the hypothesis of fluidity.

So *Letter VI* contributes the proof that, by 1661, Spinoza conceived the comparative relativity of motion and rest, and was already representing it in terms of proportion, before that expression had been established in *Letter XXXII* as the universal principle of conservation.

Moreover, another passage of *Letter VI* explains the cohesion of a body by the pressure that is exerted on it by surrounding bodies (Gebhardt IV, p. 31). This acts contestably as the first trace of the explanatory principle on which will later be articulated the definition of the individual as the union of bodies in the *Ethics* (Part II, Prop. XIII, Definition). This principle is substituted for the Cartesian interpretation of the cohesion of bodies through the reciprocal resting of their parts, which is the affirmation of a force of rest.

5. CONCLUSIONS

I will summarize here very briefly a few of the more interesting aspects of the reading and of the reinterpretation that Spinoza proposes for Cartesian mechanics.

5.1. *System, Rigor and Conceptul Revision*

The effort of systematization and rigor demonstrated by Spinoza seems intended to meet the demands that Descartes could not or did not know how to satisfy. It is accompanied by a conceptual revision of some major notions of Cartesianism: matter, motion and rest, force, speed and direction. Finally, concerning the nature and cohesion of material bodies, a redevelopment is mounted that advances particularly by abandoning the turbulence model, even though that is one of the dominant aspects of Cartesian physics.

5.2. *Statics and Dynamics*

In the second place, Spinoza's thought appears oriented towards the project of conceiving a dynamics that — without renouncing positive and legitimate elements (Descartes's analytic geometry), and deliberately perpetuating mechanics — would no longer be subject to the limiting constraints that weighed it down: the principles and laws of statics from which it was, so to speak, extrapolated. The path was narrow and difficult between fidelity to the results of the new physics and refusal to resort to metaphysical hypotheses of supra-geometrical or extra-mechanical order, as Leibniz would choose to do. But the elucidation of the effective articulation of the dynamic beyond the static passes from the articulation of the attribute of extension, to the immediate infinite modes (motion and rest), and finally to the mediate infinite mode. It is only under this condition that the presentation of the principle of inertia and the admission of mechanics could be reconciled with the internal dynamics of *conatus*, the living and concrete dialectic that characterizes it. Here mechanistic physics and rigorous meta-physics reciprocally imply and interpenetrate one another.

5.3. *Physics, Ethics and Politics*

If one can admit that the writing of the *Ethics* attests to the primacy and permanence of ethico-political preoccupations in Spinoza's thought, one can consider that reflection on the principles and laws that define the mechanistic conception of matter is correlatively unwound in the project of elaborating a political science intended, for its part, to give an account of the phenomena of equilibrium and disequilibrium, of stability and evolution, determining the different social and political systems considered in their historical development. From there, it is not impossible to think that this effort, accorded to the explication of a scheme of the evolution of societies by not only internal but external transformations, was accommodated with greater and greater difficulty by a mechanics as abstract and rigid as that of Descartes, and left apparent the exigency and urgency to recast it from its foundations.

Paris

NOTES

* The author's word '*étendue*' has been translated by the English word 'extension', and his '*extension*' by 'extendedness'. — DN

** Whereas Kenny translates Descartes's '*vitesse*' with the English word 'velocity', the author's '*vitesse*' is translated by the English 'speed' except in Leibnizian contexts, where it is translated by 'velocity'. — DN

[1] Cf. Spinoza's *Letters VI* and *XIII*, Gebhardt IV, pp. 15—36, and 63—69.

[2] Spinoza, *Letter IX*, Gebhardt IV, pp. 44—46.

[3] In this formula, m and m' designate the respective sizes of the bodies, v_0 and v_0' their speed before impact, and v and v' the speeds after impact of each of the bodies.

[4] See also his *Letter to Huygens* October 5, 1637 (Adam and Tannery I, pp. 431—447).

[5] See Leibniz's *Letter to Arnauld* November 28—December 8, 1686 (Gerhardt II).

[6] See Descartes's *Letter to Mersenne* April 1, 1647 (Adam and Tannery V, p. 70; cited by Dugas (1956), p. 184, Note 2).

[7] Spinoza, *Principles of Descartes's Philosophy* II, Definition 1 (Gebhardt I, p. 181). Cf. *Ethics* I, Definition 5, and Part II, Proposition II.

[8] Descartes, *Letters to More* February 5, 1649, and April 15, 1649 (Adam and Tannery V, pp. 267—278 and 340—348).

[9] Descartes, *Principles of Philosophy* III, Art. 47 sq. (Adam and Tannery VIII, pp. 101 sq.; IX—2, pp. 125 sq.).

[10] Spinoza, *Principles of Descartes's Philosophy* II, Proposition XIII (Gebhardt I, p. 188). Cf. also *Letter XIII* cited above and *Ethics* I, Proposition XV, Scholium.

[11] Descartes, *Principles of Philosophy* II, Article 25 (Adam and Tannery VIII, pp. 53—54; IX—2, p. 76). Spinoza, *Principles of Descartes's Philosophy* II (Gebhardt I, p. 182).

[12] To compensate for the too abstract and elusive brevity of these remarks, one may refer to the detailed commentary that I have given on the Scholium to Proposition VI in Lecrivain (1977—78, pp. 131—145).

[13] Cf. Spinoza, *Principles of Descartes's Philosophy* II, Proposition XIV, (Gebhardt I, p. 201); and *Ethics* II, Proposition XIII, Corollary to Lemma 3.

[14] Cf. Spinoza, *Principles of Descartes's Philosophy* II, Proposition XIV (Gebhardt I, p. 201); and *Ethics* III, Proposition VI.

REFERENCES

Descartes, René: 1879—1913, *Oeuvres de Descartes*, Edition Adam and Tannery, 13 vols. Vrin, Paris. (Cited as 'Adam and Tannery' in the text.)

Descartes, René: 1970, *Descartes; Philosophical Letters*, transl. by Anthony Kenny, Clarendon Press, Oxford.

Dugas, René: 1956, *La Mécanique au 17e siecle*, Éditions du Griffon, Neuchâtel, Switzerland, and Éditions Dunod, Paris.

Galilei, Galileo: 1952, *Diaglogues Concerning the Two New Sciences*, transl. by Henry Crew and Alfonso de Salvio, University of Chicago Press, Chicago.

Lecrivain, André: 1977—1978, 'Spinoza et la physique cartésienne', *Cahiers Spinoza* I, 235—265; II, 93—206. Éditions Réplique, Paris.

Leibniz, Gottfried Wilhelm: 1875—1890, *Die philosophischen Schriften von Gottfried Wilhelm Leibniz*, ed. by C. I. Gerhardt, Georg Olms Hildesheim, Berlin. (Cited as 'Gerhardt' in the text.)

Spinoza, Baruch: 1925, *Spinoza Opera*, ed. by Carl Gebhardt, 4 vols., Carl Winter, Heidleberg. (Cited as 'Gebhardt' in the text.)

Spinoza, Baruch: 1951, 1955, *Chief Works of Spinoza*, transl. by R. H. M. Elwes, 2 vols., Dover, New York.

Spinoza, Baruch: 1963, *Earlier Philosophical Writings: the Cartesian Principles and Thoughts on Metaphysics*, transl. by F. A. Hayes, Bobbs-Merrill, Indianapolis.

Spinoza, Baruch: 1966, *The Correspondence of Spinoza*, transl. by A. Wolf, Frank Cass and Co., London.

Translated by
DEBRA NAILS
and
PASCAL GALLEZ

SPINOZA AND THE RISE OF MODERN SCIENCE
IN THE NETHERLANDS*

The most important development of the scientific revolution in the seventeenth century was the discovery and application of experimental method. In the Netherlands the benefit of this method resulted in far-reaching consequences in the physical sciences such as Christiaan Huygens's construction of the pendulum clock and Antoni van Leeuwenhoek's visualization of the circulation of blood with a microscope. These innovations could be realized thanks to the adoption of the idea of mechanism in physics. On the basis of this essentially technical assumption, the empirical world could be isolated and imitated in artificial demonstrations. It is clear that this could only be done with the help of practical knowledge. Therefore scholars and artisans worked together, as was the case, for example, in Isaac Beeckman's Collegium Mechanicum which he founded in 1626 in Rotterdam. This kind of cooperation was strongly influenced by the logic of Petrus Ramus, which stood on the principle of visual presentation (graphicalness) and operated with the method of natural deduction. In this way, according to Ramus, we can read our conclusions easily from what we see clearly and distinctly. In his conception, scholars and artisans should not differ so much in attitude and viewpoint. We know that Ramist logic was taught at Leiden University, so Beeckman's project of cooperation is not surprising.

Bento de Spinoza, as he was called in colloquial Portugese, belonged to a family of merchants. We know also that he took a positive attitude toward manual labour, because he practised the trade of glass-polishing. He preferred polishing optical glasses by hand, but Huygens tells us in his correspondence that Spinoza also made use of a lathe.[1] Together with the Amsterdam burgomaster and mathematician Hudde, with whom Spinoza corresponded on this subject, he and Beeckman and of course Christiaan Huygens were forerunners in exercising the method of polishing optical glasses for the sake of the improvement of the microscope. That Spinoza was deeply concerned with physics in general, appears from his logical reconstruction of Descartes's *Principia Philosophiae* for the benefit of a pupil (probably Joh. Casearius). It was the

Marjorie Grene and Debra Nails (eds.), Spinoza and the Sciences, 61—91.
© 1986 *by D. Reidel Publishing Company.*

first and only book he published under his own name (Spinoza, 1663): *Renati Des Cartes Principiorum Philosophiae Pars I, & II, More Geometrico demonstratae.*

In the same book we can find Spinoza's reformulation of Descartes's metaphysical thinking, entitled *Cogitata Metaphysica.* With this publication he wanted to make clear that being acquainted with Cartesian philosophy was a necessary condition for understanding his own, as yet unpublished, writings (cf. *Letter XIII,* Gebhardt IV, pp. 63—69). He expected that the reception of the book would make people curious to know what else he had written: if it really would induce an attentive audience, then he saw no danger in bringing his own philosophy before the public. Apparently Spinoza was not satisfied with the reactions he received, because he published his *Tractatus Theologico-Politicus* in 1670 anonymously and with great caution. Further, we cannot fail to note that this particular treatise does not contain anything in the field of physics. Would he rather have published a text that was meant as a logical (and critical) sequel to Cartesian philosophy *if* the reactions had been more positive? The fact is that there are only two other sources concerning Spinoza's involvement in physics: his correspondence with Henry Oldenburg, the Baconian Secretary of the Royal Society, about the experimental work of Robert Boyle, and of course his *magnum opus,* the *Ethica* (1677). Yet, when compared with the explicit discussion of physics with Boyle (*via* Oldenburg), the metaphysical expositions in Part II of the *Ethica* (*De Natura et Origine Mentis*) are implicitly involved with the Cartesian conception of physics. I will discuss some of these references below, paying attention to Spinoza's treatise on the rainbow (Spinoza, 1687) as well. Concerning the historical and textual setting of the problem, the project I have in mind is to show that Spinoza was indeed taking part in the so-called rise of modern science. His role will appear even more interesting, when we see that it was in the Netherlands that Cartesianism and Newtonianism found their first and most exclusive adherents. For the verification of the hypothesis in favor of Spinoza's modest role and function in revising the sciences, it is necessary to define clearly what was understood by physics in seventeenth-century Dutch universities (and outside these institutions). The same should be done with respect to philosophy, and the identification of the professional knowledge philosophers worked with when doing philosophy. (Were they professionally medical men or mathematicians etc.?) In the second place, it would be interesting to know about

Spinoza's reputed Cartesianism: at what points does it differ from Descartes's, and what are the central theses on which both philosophers agree? Finally, I want to show how the reception of Spinoza in the seventeenth century was primarily concentrated upon his philosophy of religion.

1. PHYSICS IN THE NEW UNIVERITIES

During the Dutch revolt against the Spanish king Philip II in 1575, the Protestant leader William of Orange founded a new university at Leiden. It was meant to be a stronghold of Calvinist intellectual education. Before 1648, the end of the Eighty Year War, there were also new universities founded at Franeker, Groningen, Utrecht and Harderwijk. Their position differed widely from foreign universities as, for example, those in Germany and France. They did not at first promote scientific research, but, in accordance with their political and religious hallmark, focused on a pedagogy of intellectuals. Thus the educational programs supplied primarily preparatory courses in logic, physics and ethics, inclusive of practising the classical languages (Dibon, 1954, pp. 107—119). The aim of producing an intellectual vanguard for Protestant and Aristotelian-Cartesian advanced studies seemed to be safeguarded by this broad type of educational system. The emphasis on practical tools and instruments for scientific work was indeed unmistakably due to the humanistic tradition in the Netherlands starting with Agricola (1444—1485), the famous inventor of an ideal scholarly program, De Formando Studio, which strove for a perfectly balanced coherence of physics and ethics in academic education. In the well cultivated academic disputes that were held at the universities, we can perceive the influence of Ramus (1515—1572), whose practical application of logic found its way into dialectics and rhetoric as the indispensable vehicles of thought.

As to natural philosophy in particular, the study of physics proper, theoretical disputation dominated until the thirties of the seventeenth century, when Harvey's discovery of the circulation of the blood, *De motu cordis et sanguinis in animalibus, anatomica exercitatio* (1628), found its first support in the Netherlands[2] and Galileo's ideas on the new sciences appeared in print (Galilei, 1638). Moreover, Descartes, living in Leiden at that time, disclosed his philosophical method in his *Discours de la méthode*. In contrast to the case for theoretical disputes, experimental method was gaining ground. All of this also implied a radical change in the conceptual outlook of physics. If its initial meaning was exclusively oriented toward speculative qualification of matter and form, motion and force, and the description of natural history, our knowledge of nature was henceforth to be obtained in a strictly artificial, and that meaning *mechanistic,* way. Physics, earlier conceived as philosophy of nature (including a cosmological and religious evaluation), was no longer considered opposed to human intervention in nature by art and experiment. The outcome of natural processes and artificial imitations of those processes counted as equivalent (Hooykaas, 1979, p. 128; Dijksterhuis, 1975, p. 457). The revolutionary force of this new paradigm should not be underestimated: for the Dutch universities the rise of the new science meant the beginning of turmoil and of political and religious opposition.

Soon Leiden University was the forerunner in Cartesian physics with such leading scholars as Adriaan Heereboord (who was also an admirer of Gassendi), Johannes de Raey and Frans van Schooten (whose books on mathematics were in Spinoza's library). Other universities such as those at Utrecht, and without hesitation Groningen, followed this example, albeit as a *philosophia novantiqua* in the context of Aristotelianism (Thijssen-Schoute, 1954; Dibon, 1954; Ruestow, 1973; McGahagan, 1976). But theological objections against the pretended limitation of the new science to matters of physics, as if it were neutral with respect to extra-physical or metaphysical assumptions, led to the official decree of the States of Holland (1656) to abstain from Cartesian disputes for the sake of the liberty of philosophizing! Nevertheless Cartesianism could not be stopped. The conviction that reason alone cannot sufficiently achieve the practical assumptions of physics paved the way for the adoption of Descartes's extension and motion as the principal phenomena of physics to which our minds have access. Mathematics and physics, deduction and demonstration, were seen as key and keyboard of the scientific materialism that was to dominate the experimental discoveries of Herman Boerhaave (who practiced medicine on a Newtonian basis), Willem 's Gravesande (who rejected Cartesian hypotheses, favoring an experimental method as disclosing the causes of physical phenomena with mathematical evidence), and finally Petrus van Musschenbroek (who saw logical reasoning and mathematical deduction as an accumulation of experimental knowledge, and as such the source of the growth of knowledge). To be quite clear about this, these three natural scientists did in fact leave Cartesianism, which they exchanged wholeheartedly for Newtonian philosophy. Cartesian physics in the Netherlands was a short and violent storm that formed a prelude to Newtonianism. *In globo* Cartesianism was adopted, and its method was welcomed as a handy instrument, but its followers sooner or later carried their thought a step further, avowing Newton's *Hypotheses non fingo* (stressing the provisional character of empirical knowledge by induction).

1.1. *Cartesianism. Spinoza's Background*

The mainly positive reception of Cartesian philosophy in the Netherlands was partly due to the fact that Descartes lived for twenty years (1629—1649) in several Dutch cities. As we know, Spinoza was impressed by the *Principia Philosophiae,* although he remained at a critical distance from it; more or less the same attitude can be observed in some of his friends, for example Lodewijk Meyer, Jarig Jelles and Pieter Balling.[3]

For instance, Meyer's assistance in correcting Spinoza's manuscript of the *Renati Des Cartes Principiorum Philosophiae* was very valuable. He not only edited the fairly perfect text, but was responsible for the introductory recommendation.[4] In 1664, one year after the *editio princeps,* the Latin text of the *Renati Des Cartes Principiorum Philosophiae* was translated into Dutch by the learned merchant Balling. He produced a clear and more or less explanatory text, which was to be the only Dutch text of the *Renati Des Cartes Principiorum Philosophiae* until the recently published new translation by Fokke Akkerman (1982) and Hubertus Hubbeling. Although it may be the case that Balling's work widened Spinoza's public, we do not know whether it really created a factual audience among Dutch reading citizens. I suppose that Spinoza's account of Cartesianism was no doubt hailed and/or countered exclusively in the Latin speaking intellectual world. One could say that Spinoza's apparent disappointment with the replies to his reconstruction of Cartesianism was partly justified by the fact that he did not wish to become the subject of a troublesome dispute (as Descartes still was). He himself must have been well informed about the polemical writings on Cartesian physics. Living in the stoic-libertine atmosphere of his teacher Franciscus Van den Enden, whose Latin school he visited after being expelled from the synagogue in

1656, he became acquainted with literature hitherto unknown to him. Besides such current academic literature as Heereboord's *Melemata philosophica* (1654) and Burgersdijck's textbooks (obligatory reading even at Harvard and Oxford), he hit upon van Schooten's mathematical treatises as well as the medical works of at least ten different authors (Von Dunin Borkowski, 1933—36, II, pp. 188—189; *Catalogus,* 1965). The impact of Van den Enden's education must have stimulated his interest in an empiricist direction, which was the more significant because his teacher was an ardent Cartesian. Sometimes, one wonders if the so-called Cartesians really understood their master's inclinations towards empirics with respect to their value and function. As far as our evidence reveals, Cartesianism provided some sort of open minded attitude towards new ideas and the will to revise older positions. Whether the hard core of Cartesian philosophy was effective did not matter so much. Therefore, the virtual lack of rigor in what we might call Van den Enden's Cartesian empiricism need not bother us as a presumed failure or confusion of principles. Happily, Spinoza's understanding of Descartes's philosophy and principles of physics was more reliable. His unmistakable clarity in reformulating Descartes's *Principia Philosophiae* (referring also to the *Meditationes*), was a painful experience to modish Cartesians who had never reached the foundation of their beliefs. Netherlands Cartesianism, and in this I fully agree with Thomas McGahagan's thesis (1976, p. 106), was in the greater part a betrayal and trivialization of Descartes's thought.[5]

As to the new science, the Cartesians insisted on an evolutionary picture starting with Bacon *via* Gassendi to Descartes as its culmination. If they had been more cautious, the incompatibility of Gassendi and Descartes would have been understood clearly. But they were confused, as if Gassendi's atomism were not totally different from Cartesian (theistic) 'idealistic' mechanism. For Descartes, God is the efficient cause of all possible, that is mechanistic, structures of nature. Whoever has a clear and distinct idea of the laws of nature and can operate these laws, is able to construct any process to be justified experimentally. Gassendi, on the other hand, took physics as only a nominal description of hypothetical forces. He was closer to the effects of the new science than was Descartes. The latter refused to give in to his rival colleague and defended himself against Gassendi's denial of any metaphysical assumptions for physics. Descartes's defence was notably weak, and Gassendi implied that his rival's method of doubt as well as his concept of God were products of imagination.[6] In short, Dutch Cartesians more or less missed the point concerning differences between these rival philosophers.

Turning back to Spinoza, it goes without saying that the Cartesian issue had been for a long time *en vogue,* and that he himself did in fact reassert critically some central Cartesian opinions that had been discussed previously. His contribution was original insofar as it was an eye-opener to a lot of Cartesians: they were shocked!

1.2. *Spinoza's 'Renati Des Cartes Principiorum Philosophiae' and Cartesianism*

Spinoza opens his introduction to the *Renati Des Cartes Principiorum Philosophiae* with a clearcut statement on Descartes's philosophy that is remarkable from a methodological point of view. He avoids unnecessary formalities and hesitation as he writes:

Before we approach the propositions and their demonstrations it seems wise to set before ourselves briefly why Descartes doubled everything, how he laid the solid foundations of the sciences, and, finally, by what means he liberated himself from all doubts. All this we would have reduced to mathematical order had we not decided that the prolixity required for setting it out might prevent these matters' being rightly understood, since they ought to be seen at a glance as in a painting.

Descartes, then, in order to proceed with all caution in investigation, attemped:

To set aside all prejudices.

To find the foundation upon which everything else might be erected.

To reveal the cause of error.

To understand all things clearly and distinctly.

In order to secure his first, second, and third aims, he undertook to reduce everything to doubt, not like a skeptic who apprehends no other end than doubt itself, but in order to free his mind from all prejudice. Thus he hoped to discover the firm and unshakable foundations of science, which, if there were any, could not escape him as he followed this method. For the true principles of knowledge should be so clear and certain as to need no proof, should be laced beyond all hazard of doubt, and should be such that nothing could be proved without them. Then, after he had discovered the principles, it was not difficult for him to distinguish true from false and to expose the cause of error, and, accordingly, to prevent himself from taking something false and doubtful as true and certain.

To secure his fourth and final aim, that is, to understand all things clearly and distinctly, his principal rule was to enumerate all the simple ideas of which all other ideas are composed, and to examine each of them individually. For when he could perceive the simple ideas clearly and distinctly, without doubt he could also understand with the same clarity and distinctness all other ideas made up ot these simple ones. (Spinoza, 1963, pp. 3—14; Gebhardt I, pp. 141—142)

This kind of procedure is typical for a lesson-book, but not for Spinoza. He seems to be imitating the style of lecturing typical at the universities. It is useful to set Spinoza's account of Cartesianism against the background of the Leiden philosopher Heereboord, who identified himself with an Aristotelian—Cartesian perspective. This synthetic type of philosophy also incorporated a more or less didactic style of lecturing as a preliminary to a more rigid discipline of arguing with axioms and propositions etc. It was accomplished through such scholastic forms as *disputationes, quaestiones physicae* etc. in the fashion of the new science; and Heereboord, according to his synthetic position, never rid himself of the disputational form. Constrained by a scholastic dependency, the same is true for Spinoza. Yet, leaving aside that he refers to Heereboord's *Melemata philosophica* in his *Renati Des Cartes Principiorum Philosophiae,* Spinoza must have picked up more advice in the former's books, for his more mature work bears clear evidence that he was sympathetic to Heereboord's way of systematizing philosophy (Gueroult, 1968, pp. 245—247). But it is clear that the didactic style of the *Renati Des Cartes Principiorum Philosophiae* has disappeared in the *Ethica.* We might evaluate this change in presentation as the outward indication of Spinoza's development away from Descartes. For a better understanding of Spinoza's commitment to Cartesianism, it is therefore necessary to map out the points on which his views differed from Descartes's, as well as to clarify some central positions they had in common. (Despite the most crucial differences, as we shall soon see, there are indeed strong reasons for Spinoza to propagate Descartes's philosophy, albeit in contrast to what was labelled as 'Cartesianism'.)

Particularly in the *Renati Des Cartes Principiorum Philosophiae,* a few differences can be signaled between Descartes and Spinoza, which, for brevity's sake, will not here be contrasted with passages in the *Ethica*:

(1) The *Renati Des Cartes Principiorum Philosophiae* is expounded geometrically, to wit synthetically, whereas Descartes favored the analytic method.

(2) *Cogito ergo sum* is Spinozistically changed to *ego sum cogitans* (Gueroult, 1970, pp. 76—77).[7]

(3) As the first part of the *Renati Des Cartes Principiorum Philosophiae* contains Descartes's physics (from the latter's *Principia Philosophiae,* Part I, and the second part concerns metaphysics, Spinoza gives a high priority to physics in contrast to Descartes's apparent wish to deliver a metaphysical theory.[8]

(4) Spinoza rejects the Cartesian model derived from atomism.

(5) In the *Renati Des Cartes Principiorum Philosophiae*, Spinoza modifies the Cartesian theory of motion: both motion and rest are henceforth conceived as actual modes of extension (Van der Hoeven, 1973, pp. 6—7).

(6) Spinoza rejects Descartes's concept of time (in accord with point 5), and conceives it as an instrument for comparison among objects in motion (or rest).

(7) Spinoza rejects Descartes's concept of God conceived as an external cause.

If we look at the geometrical conception of the *Renati Des Cartes Principiorum Philosophiae* (point 1), it is remarkable — even for pedagogical reasons — that it is presented in synthetic form. As we know, Descartes favored an analytic interpretation of *more geometrico*. According to Curley (1978, pp. 134—136), who also consulted Gueroult's structural interpretation on this point, the synthetic form of the *Renati Des Cartes Principiorum Philosophiae* makes clear that Spinoza misunderstood the contrast Descartes wished to make between analysis and synthesis, for unlike the *Meditationes* the *Principia Philosophiae* had already been written in the synthetic variant of geometrical reasoning. I could agree with Curley if there were more evidence than Meyer's misinterpretation of analysis and synthesis in his introduction to Spinoza's *Renati Des Cartes Principiorum Philosophiae*. I think that Meyer is justified in writing by way of introduction that the synthetic exposition is especially appropriate for teaching, whereas analysis is an instrument for making discoveries. In a superficial way, this distinction between discovery and orderly transferral of the meaning of discovery to an audience is not wrong. So I think Spinoza cannot be blamed for Meyer's commendatory remarks preceding the *Renati Des Cartes Principiorum Philosophiae*. It should be borne in mind that he entitled his recommendation 'Een hartelike groet aan de oprechte lezer van Lodewijk Meyer' (Kind regards to the sincere reader by Lodewijk Meyer). It is quite clear that Descartes himself affords us a rather vague distinction between the methods. In his *Secundae responsiones,* he considers both terms as related to *a priori* and *a posteriori* reasoning:

Analysis veram viam ostendit per quam res methodice & tanquam a priori inventa et, adeo ut, si lector illam sequi velit atque ad omnia satis attendere, rem non minus perfecte intelliget suamque reddet, quam si ipsemet illam invenisset.

Synthesis e contra per viam oppositam & tanquam a posteriori qauesitam (esti saepe ipsa probatio a priori quam in illa) clare quidem id quod conclusum est demonstrat, (...) modum quo res fuit inventa non docet. (Adam-Tannery VII, pp. 155—156)[9] (Analysis shows the true way by which a thing is invented methodologically, that is *a priori,* so if the reader follows this method he can reach everything sufficiently, and he will understand no less perfectly what is supplied by this method than if he had discovered the very thing himself. Synthesis, on the contrary, is the opposite way of inquiry, that is, *a posteriori*, by which clearly is demonstrated what can be concluded (although the proof itself may be often *a priori* by nature (. . .)) but it does not teach us how a thing has been discovered. — *my translation*)

That is all Descartes says about this subject, and it would be too easy to conclude that he merely intended to articulate the reciprocal type of connection between analysis and synthesis. By the thirties of the seventeenth century, philosophers admired geometrical reasoning. Geometry could afford an adequate picture of the anatomy of human nature, extended as well as cognitive. Most of them physician, for example, Locke, Bacon and De Mandeville, they practised medical principles in their philosophies for the benefit of good health. The analysis of the human body enabled men in medicine to fight serious diseases that had taken victims from practically every family. This 'material' background of the analytic instrument, shows us the so-called context of discovery from the scientists' perspective. On the other side, we find the context of presentation (to the public etc.), transferring true knowledge to an audience that is not in a position to verify the alleged truth by experiment. So how do they know that what is presented to them is really true and certain? To begin with, what can be shown to the audience (including fellow-philosophers) is the end and effect of an innovation in a context that suffices for it use. I mean this: the context must answer the benefits of the product of discovery. Furthermore, at a critical distance, one can perceive the origin of this product of discovery when it has been taken out of its context. So the synthetic perspective we started with is better for testing than for discovering.

Now in what way is Spinoza's *Renati Des Cartes Principiorum Philosophiae* synthetic, as we contended above? If we follow Descartes, then Spinoza should not be showing how a thing has been discovered (*modum quo res fuit inventa*). The propositions of the *Renati Des Cartes Principiorum Philosophiae* are all formulated according to, and as deductions from, Descartes's axioms and definitions. For the sake of didactics, the origins of the propositions described is not discussed. I take it for granted that Spinoza proceeds synthetically. But if we compare the composition of Part I with that of Part II of the *Renati Des Cartes Principiorum Philosophiae,* the synthetic label does not fit so well for the latter. For some reason, for example, the style of analytically construing a theory from ground level, one might offer the defense that Spinoza is at least ambiguous in the use of geometry as a synthetic instrument. Therefore, in my opinion, one should not over-estimate the importance of Spinoza's synthetic exposition of Descartes's philosophy. I find myself in agreement with Meyer's preliminary remarks, saying that synthesis presents better the definitions, axioms postulates and theorems; so if one disputes one of the consequences, it is easy to check that everything found in the consequences is already contained in what went before.

Far more interesting than the synthetic exposition is Spinoza's inter-pretation of *cogito ergo sum,* which he changed to *ego sum cogitans* (point 2). To be sure, in talking about Spinoza's interpretation, I do not imply that he consciously took the opportunity to improve the philos-ophy he was teaching to his pupil in his own way. Before treating this point in detail, it is better to take a look at the text. The most useful passage I know is *Renati Des Cartes Principiorum Philosophiae* I, Proposition IV:

Ego sum non potest esse primum cognitum, nisi quatenus cogitamus (Gebhardt I, p. 152). (*Ego sum* cannot be the first thing known except insofar as we are actually thinking. — Spinoza, 1963, p. 25)

And the explanation in the covering Scholium reads:

Unusquisque certissime percipit, quod affirmat, negat, dubitat, intelligit, imaginatur & c. sive uno verbo, Cogito, *sive* sum Cogitans *unicum* (*per Prop. 1.*), *& certissimum est fundamentum totius Philosophiae.* (Gebhardt I, p. 153) (Everyone perceives with certainty that he affirms, denies, doubts, understands, imagines, and so forth; or, that he exists doubting, understanding, affirming, and so forth; or, in a word, that he exists *thinking*; and he cannot doubt these things. Therefore, this statement, *cogito* or *sum cogitans,* is the unique (by Proposition I) and entirely certain foundation of all philosophy. — Spinoza, 1963, p. 25)

In drawing our attention to these contrasting elements — human actions such as understanding or imagining and the underlying indubit-able certainty of the fact that one is thinking — Spinoza tacitly marks a difference between, so to speak, the first order act of thinking and such second order activities as affirming and negating induced by human thinking. I know that he does not explicitly say so, but he surely cannot be held to take the opposite view, namely that thinking provides all other kinds of mental products which are equal *qua* supply of knowl-edge. We need not assume such a move. Spinoza's approach is clearly more existential than methodological. Despite the original meaning of *doubt* conceived as a method, he seems to substantialize the act of thinking into the groundfloor of being. This may sound a bit heavy, but for now it seems more desirable to make a more or less *black and white* distinction. As it is my concern to pick out the subtle but striking point of difference, some magnification of scale is allowable, I think. Since we do not overestimate the impact of all this, but remain close to the text of the *Renati Des Cartes Principiorum Philosophiae,* our procedure does no harm. Yet it is a fact that Spinoza's existential interpretation has many consequences for the whole of the theory he discusses. Let us suppose for example that *cogitare* really were the groundfloor of being. Then Descartes's famous Real Distincton (Williams, 1979, pp. 104ff) between extension and thinking is finished because, as really separate substances, extension and thinking can no longer be considered as separately realizable by God's omnipotence. But this separate realizability is the presumed condition upon which Descartes's theory is built.

Perhaps Spinoza, however much he might have wished otherwise, could not abstain from doing metaphysics, even when he actually meant to be doing physics. But this is not so surprising, when we bear in mind that in a Spinozistic approach the axiom of parallellism teaches us that there is no metaphysics without physics! Therefore, our distinction (point 3) that Spinoza chose as the first part of his *Renati Des Cartes Principiorum Philosophiae* the second part of Descartes's *Principia Philosophiae,* makes sense; at least this fact underlines a high priority for physics. And if we wish to make the most of it, we might interpret this existential way of doing metaphysics as an original and lucky strike to overcome Cartesian imperfections. The building of castles in the air is ruled out because there is always a physical reference for any metaphysical theory whatsoever. Embedded in the existentially reshaped *ego sum cogitans,* Spinoza's approach, we can safely declare, was purely metatheoretical — and this could *also* refer immediately to the material of physics as its ultimate object of inquiry. Indebted to the same principle, he considered the atomistic model to be one of the main mistakes in the *Principia Philosophiae* (point 4). Although in Descartes's opinion too, atomism should be rejected, he nevertheless operates with a scheme that is derived form that theory. This brings him into the rather awkward position that he must deny the possibility of the independent existence of particulars, because they are all parts of a greater substance, but yet must concede that matter shows an interreactive pattern between distinct volumes of itself. To secure his anti-atomism he equates matter and space, leaving no 'room' for a vacuum. So all and everything has the same density; there are no particular atoms crisscrossing or overlapping by impulse. The causal principle that is basic to this model is God's precreative push, which makes matter operative. This is very much against Spinoza, who perceived cause as an 'immanent' principle. In his *Ethica* it is expecially Propositon XIII of Part II (Gebhardt II, p. 96) that testifes to the infinite circularity of motion and rest as the exclusive property of one body in relation to another. If one body is in motion, it can only be stopped by another. So motion and rest are the product of one and the same internal principle, i.e. causal reaction (point 5). By the way, this very Proposition XIII is at the same time an interesting ratification of human perception as inseparably connected to bodily awareness. And this is again an incitement to remain as close to reality as possible; the human body is the door to eternity, if there is any picture we want to have in mind.

Consistently with what has been said about motion, Spinoza rejected the Cartesian concept of time (point 6). As early as April, 1663, when he wrote his famous *Letter XII* on the nature of infinity to Meyer, he emphasized his view with respect to the relativity of time. (Descartes had presented time as perfect and ideally real in that it depicts a *degree* of actualizable duration.)

The concept of time is an aid to the imagination for purposes of comparison. As the imagination produces thought-constructions, their duration can be measured in relation to the interval of light and darkness by day and night. These periods provide a schedule that 'objectifies' duration. Now Spinoza stressed the fact that this type of measuring can count only as approximation. Number, measure and time are nothing in themselves. To know what phenomena such as the eternal particular (Kierkegaard) or actual infinity (Spinoza) really mean, one should first learn the difference between duration and time. For one cannot split up an hour into its parts (however small the parts may be) in order to know how long it takes to let it pass. It takes time to discover the timeless actuality of being.

1.3. *The Rainbow. In Admiration of a Fusion of Mathematics and Physics*

Ever since the beginning of human history the colors of the rainbow have fascinated mankind. The mystery of its appearance made the

rainbow even an ominous prediction of things to come; *Genesis* teaches
that it could also be reevaluated after a tremendous natural disaster,
after which the rainbow serves as God's promise never to impose a
deluge again upon mankind. From an optical point of view, in the
seventeenth century those colors that appeared in the sky after a
shower were an object of speculation. Thus, as science developed more
and more, the rainbow became an object of geometrical investigation as
well as a specimen for doing calculations with respect to the early
theory of probability as developed by Pascal, Fermat and Leibniz.
Spinoza knew and possessed Huygens's *Tractatus de ratiociniis* (1657),
which was to be the first and original exposition of the calculus of
probobility before Bernoulli's *Ars conjectandi* (1713).[10] (The latter,
together with his brother, the Groningen professor of mathematics
Johann Bernoulli, who also published a refutation of Spinozism entitled
Spinozismi depulsionis echo (1702), had many pupils all over Europe.
His school was especially influential in Germany, Switzerland, Italy and
the Netherlands.) That he was indebted to Huygens can easily be
deduced from his correspondence. As early as 1666 we find him
writing about prospective chances in gambling to his friend Johannes
van der Meer; his choice of words clearly depends on Huygens's
vocabulary. As Van der Meer was a financial specialist owning a life
insurance firm, his practical interest in the calculus is understandable.[11]
We have no sign that Spinoza continued to have contacts with his
correspondent, who lived nearby in Leiden, but happily there were
others with whom he could have discussions in the context of theoretical optics, with Johannes Hudde for example, a mathematician
who was, among other things, a burgomaster of Amsterdam. Still more
important were his meetings with Huygens, although the sources reveal
only their mutual interest in grinding lenses with the aim of technical
improvement.

One reason for this absence of communication on the calculus was
probably the fact that Huygens was disappointed with Spinoza's *Renati
Des Cartes Principiorum Philosophiae*. As he was himself an ardent
Cartesian, who aimed at a realization of his master's program, he surely
must have objected to Spinoza's synthetic description for pedagogical
purposes, which differed from Descartes's conception of synthetic
reasoning. The catchword was, for Huygens as well as for Newton and
Leibniz: synthesis! The proof for what had been discovered analytically
was in its synthetic (mathematical) construction. Huygens was also

deeply involved in an atomistic conception of physics, which Spinoza, as we know, did not share. But still another reason might help to clarify the fact that Huygens consulted Spinoza exclusively for technical aspects of lens-grinding. Living in Paris from 1666 on, Huygens was completely preoccupied with conducting the Académie des Sciences. His interests were mainly devoted to developing the pendulum and to his experiments with the double refraction of light in calcite. Most important was that he concentrated on the same problems in mechanics and optics as Newton did! *Via* Leibniz, whose tutor he was for a short time, and *via* Oldenburg, their mutual proceedings received a wider public. It goes without saying that Spinoza, setting aside his philosophical achievements, deserves no place among physicists of this stature.

Meanwhile, Spinoza carried on studying probability and the geometrical composition of the rainbow. But he could not decide to publish the text of his investigations. According to Colerus's biography, some prominent persons read his treatise *De Iride* and advised him not to publish it. Intimates of the late philosopher reported that he put it to the flames. Jarig Jelles is not sure about that; in his *Voorreeden* (preface) to Spinoza's posthumous works (*Nagelate Schriften*) he says that he believes there possibly may exist "a small paper on the rainbow" (*een klein Geschrift van de Regenboog*) "if he did not burn it") *zo hy 't niet verbrant heeft*) (Spinoza, 1677, p. ii). Indeed, his cautious formulation was justified. In 1687 the treatise was published by Levijn van Dijck, the municipal publisher in The Hague; the title reads *Stelkonstige Reeckening van den Regenboog, Dienende tot naedere samenknoping der Natuurkunde met de Wiskonsten* (Algebraic calculation of the rainbow; for the purpose of further connecting physics with mathematics). And, as it was still dangerous to publish or write about Spinoza, in the same year the publisher printed a refutation of Spinozism by J.F. Helvetius (1687, pp. 1—59). For the same reason Spinoza's treatise on the rainbow was published anonymously!

Considering now with what caution Spinoza formulated his thought, he may indeed have been taking to heart the supposedly critical advice of certain prominent persons. At the head of the introduction is a motto derived from Cicero's *Tusculanarum Quaestionum* that says: *In summo apud illos honore Geometria fuit, itaque nihil Mathematicis illustrius. At nos metiendi rationandique utilitate hujus artis terminavimus modum.* (They, the Greeks, held geometry in high esteem, so there were no persons more illustrious than mathematicians. But we, the Romans, have restricted the importance of this art in order to use it only for measuring and counting.)

Thus, giving full stress to the *practical* application of geometry, he even informs the reader preceding the introduction that the hallmark of this treatise is "to inform the amateur" (*om de ongeleerde te hulp te komen*). Ironically, he holds out the prospect of the brilliant careers of Hudde and Huygens "indeed the apple of one's eye for all who love these arts") *voorwaer den Ooghappel van alle die geene, die deze Konsten beminnen*), and Johan de Witt, the late Grand Pensionary of Holland (during the Dutch Republic the most important official pleni-potentiary), who contributed to the edition of Descartes's *Géométrie* and wrote his famous treatise on the analysis of probability in the context of life expectancies and life insurance, which was considered to be a masterpiece far ahead of its time (Rowen, 1978, Chapters 19 and 20). Spinoza does not conceal the fact that he stands in the shadow of these scholars as far as the object of his treatise is concerned; but the reader might nevertheless be stimulated to follow their illustrious example. Whatever one may think of this approach, Spinoza was well aware of his minor position. Now the question forces itself upon us, whether he nevertheless did succeed in arranging a further junction of physics and mathematics in his treatise. Materially, the answer must be negative. Spinoza based himself on Descartes's opinion that light can only be propagated in rectilinear motion. Descartes could not be blamed for that, since Grimaldi's discovery of light diffraction was only made in 1665. Yet Spinoza, apparently uninformed about the latest developments in the theory of light, sticks to Descartes's theory. That theory was almost twenty years old at that time. From a theoretical point of view, more over, our question with respect to Spinoza's success or failure is still more interesting. To answer it, we should first be in-formed about how Spinoza defined 'light'. At least in the *Renati Des Cartes Principiorum Philosophiae* he produces an intriguing equation, saying that the *decreta Dei lumine naturali revelata* is *leges naturae*. More generally he had in mind that *ratio mentis lux est*. As the difference between *lux* and *lumen* (*lux* is associated with the concept of clarity; *lumen* can refer to artificial light as well) is only trivial, we can safely take for granted that Spinoza regarded 'light' (traditionally) as this world's *being actually apparent* to us. If there were no light (and notwithstanding that this point is selfevident I will say) there would not be any appearance of the world. But, and here our second condition for answering an evaluative question with respect to the treatise comes in, what does Spinoza say about the origin of what we can see, or better, *that* we can see? Do we actually see the product of *lumen's* reflections on our retinas, *or* is what we perceive only a product of imagination? (Unfortunately we cannot go into the background of this problem here, so we shall leave out, e.g., the Newtonian revolution and its influence in the Netherlands.) In my opinion, the thesis defended by Filippo Mignini (1981, p. 57) is very important. In his *Ars imaginandi,* on appearance and representation in Spinoza, he says that, epistemologically speaking, there is a perfect and coincidental relation between the image we have and its external cause.

According to Spinoza there is a connection between imagination conceived as the internal sense (*senso interno*) of perception, and its external function (*esoterno*), that is, to procure the impression of physical existence of an actually observed object. This demonstrates sufficiently that one was well aware in that era, of the fact that imagination has a poetical function; generally, its internal sense will produce pictures by force of its exclusive sensitivity alone. But from an *external* point of view, indeed, just blackness will not appeal to my external sense or leave an impression behind, so it simply cannot be reproduced (*riproduzione*)by the internal sense either. (*my summary translation from the Italian*)

Indeed, Spinoza teaches a perfect '*consapevolezza*' (awareness) as the coincident product of seeing and (physical) appearance. Mignini's position implies a reappraisal of imagination in Spinoza's theory of knowledge. (Cf. also Wetlesen, 1979, p. 285.) Recall the highly interesting Proposition XIII in *Ethics* II, where it is said that "*Objectum ideae, humanam Mentem constituentis, est Corpus, sive certus Extensionis modus actu existens, & nihil aliud*" (Gebhardt II, p. 96). (The object of the idea constituting the human mind is the body, in other words a certain mode of extension which actually exists, and nothing else: Spinoza, 1955, p. 92.)

The corporeal nature of any idea whatsoever emphasizes once again Spinoza's 'realistic' approach. Imagination as the first kind of knowledge also operates in a positive way on the second level of *ratio,* and even on the highest level of intuition! The role and function of imagination depends on the coincidental truth value carried by rationality and/or intuition (cf. prophetic imagination as discussed in the *Tractatus Theologico-Politicus*). Therefore only imagination operating independently should be refuted as an inadequate source of true knowledge.

Now let us return to our point of departure and ask ourselves again whether Spinoza really did succeed in arranging a further junction between physics and mathematics. If I am right, there are two viewpoints coming into play. First, we have to face the objection that Spinoza was not up to date in physics, as we concluded earlier. But formally that is no problem for a fusion of physics with mathematics: the rainbow and mathematical calculation in his conception are just a new application of Euclid's classical theory. Newton, for example, made a sharp distinction between the analytic method of discovery (which he did not practise in his *Principia Philosophiae,* 1687) and his synthetic method of presentation. That is to say, he broke with Descartes's modern analysis as defended in the latter's *Géométrie,* calling Cartesian geometry "the analysis of the bunglers in mathematics" (Westfall, 1981, pp. 379—380). It is impossible to ignore the fact that Spinoza's treatise proceeded wholeheartedly in a Cartesian way. There is not a bit of doubt in his exposition. Beyond this, referring to the requirements we supplied above for our question, it cannot be denied that from a metaphysical point of view Spinoza consistently supported (although in considerable shorthand) a new theory of perception and imagination. But that was not enough for the junction desired.

2. INTERLUDE [12]

In a way, there is reason to believe that the crisis of sixteenth century Europe including the outburst of new science following afterwards, was as devastating as only an earthquake can be. To contemporary intellectual circles, the outcome of that revolt was beyond imagination. It had turned the world upside down. From then on the thought has forced itself upon us that the laws of physics are decrees of fate (Whitehead, 1967, p. 11).

For the benefit of natural science, long avowed theoretical assumptions were bereft of their value and judged to be 'out of order'. And although the vanguard occupied themselves with restoring quiet, especially the theologians were suspicious and discontented with what they considered to be mere 'make believe stories'. At last, new scientists could not continue saying that nothing had changed, but that only some revolutionary developments in physics had taken place. That this would not effect 'extra-physica' sciences was indeed unbelievable. The mechanical world picture engendered more and more sympathy, so that its characteristic features were the pattern for practically every theory. An attactive example is Spinoza's *Ethica*, Part III, on the origin and nature of the *affectiones*, which is clearly attempt at a mechanistic psychology. Another characteristic that entered (philosophical) thinking and began to serve as a pattern was the idea of materialism. Previously, no notice had been taken of the principle of materialism as a source of controversy, since in the context of physics matter as such was (divinely) above competition. But with the rise of the new science the technical power of man increased, so that nature could be imitated in scientific experiments. The theological implications of this development varied widely; but there were important overtones articulating the possibility of exclusive materialism in religion as well. And if we look at the early reception of Spinoza's philosophy in the Netherlands, there are many refutations that consider his *Deus sive natura* as religious materialism, i.e. atheism. At any rate, religious belief seemed to be deprived of its ultimate concern, i.e. God. As we know, more than three hundred years of philosophy of religion cannot yet supply an answer that satisfies everyone. It is not so much that there is a lack of logical tools or studious scrutiny — what we need is available — but there is still a big problem with those incredibly puzzling concepts ruling since the Dutch Golden Age: necessity, nature and notion.

Given the scientific revolution, which indeed turned the world upside down, preferring the lessons derived from bare nature over the evidence of reasoning, it would be ironical (semantically) if we did not take into account the impact of this 'turn' with respect to the meaning of words.

Language as a basic tool of communication bears the fingerprints of its users. Therefore, textual evidence for historical events forms part of the creative process responsible for their existence. Someone must have found its appearance a notable datum, otherwise the historical event would have been forgotten. The *reception* of an event — be it a revolution or the publishing of a new philosophical text — depends in great part on the contemporary's subjective observation. Primarily, this kind of subjectivity is a positive attitude because it is due to a temporary lack of information. The newness of any unexpected event can only be abolished by dialectical interrogation. So when the news comes through, and the earlier stage of subjectivity has faded away, a better foundation for more adequate evaluation is laid. Personal choice may be a dominant and even highly colored feature in the reception of, for example, philosophy, but it is (alas) to some degree inescapeable. All *receivers,* i.e. all of us, are just players in a game; the difficulty is that we are very often expected to behave as if we knew the rules of the game we are playing although in most cases we are in the process of switching from one game to another. There are indeed an innumerable number of games being played. And there are events taking place that bring us into doubt, as we find ourselves unable to say what the game might be to which we are introduced, not to mention the rules we should obey. Still more complicating is the situation in which we have to decide about games that were played in times past. How must we deal with the scientific revolution, for example, or the impact of Spinoza's philosophy on Dutch society! It seems that we cannot escape playing two games at once: to play a role as if we were seventeenth century citizens, but also and immediately to screen the object concerned in our operative perspective, i.e. the twentieth century vision. The reader can be assured that we will not go into the highly important assumptions of the theory of reception here, nor will we bring onto the scene Mertonian or Weberian characters. Notwithstanding our high estimation of cultural sociology, its investigations do not explain why people reacted as they did to new developments. Statistics and comparison are just other types of (culturally patterned) interpretation. It appears to me that the most penetrating and obstinate keyword is a one-word judgement, i.e. influence. In my opinion its content in most cases is almost nil. It would be a challenge for philosophy to study what really happens in a literary sense, when we speak of reception and equate this with, for example, influence. With respect to this article, however, I have kept silent about this area; you may have missed Kuhn or Lakatos?! Indeed, I did not take the perspective of philosophy of science because that would imply that I am convinced of Spinoza's role in that field. What I have tried to do, and I hope the effort is not far off the mark, is to show that Spinoza was actively committed to the development of the new science. In the next and final section, I will investigate part of the reception of his work, doing this on the legitimate assumption that physics was viewed as the mirror of truth.

3. PHYSICS AND PHILOSOPHY OF RELIGION.
SPINOZA'S RECEPTION

There is a curious opposition in Christianity between nature and life. In

the Christian system of belief there is, strangely enough, no unam-
biguous definition of nature available; the concept is many-sided and
concentrates on genuine purity (positive and negative) in general, but it
is definitely *not* that nature refers uniquely to physical reality! God
created man analogous to his own image, *Genesis* says, and the early
Christians of the first centuries instituted baptism as a *rite de passage* to
certify the admission of new believers into the church, but also to save
them from nature. Even from a theological point of view, this is an
extraordinary matter. For it seems that to be born in God's image is to
be primarily a displaced person. So baptism is conceived as taking part
in God's grace, which is the same as sharing real life in accordance with
his revelation to one. Life and nature therefore stand in a queer relation
to one another. It is necessary to be reminded of this when we discuss
the early reception of Spinoza's philosophy in the Low Countries.[13] It is
no exaggeration to describe the religious atmosphere in those countries
as dominated by state Calvinism, while allowing numerous denomina-
tions a more or less free position in so far as they acted in accordance
with the pragmatic tolerance executed by the government.

It is acknowledged that one of those religious groups in particular was
closely associated with Spinoza's theological thought: the Rijnsburger
Collegianten. They preached an anti-institutional and anti-confessional
kind of Mennonism, holding meetings of a disputative character. In
1672, one of those Collegiants, the Rotterdam merchant Johannes
Bredenburg, delivered a confidential paper to his fellow members in
which he defended what he took to be Spinoza's thesis that there are
two sources for true knowledge, namely nature and revelation. He
argued for a rational justification of religion, which is a rather modern
practise, but he drove himself into a troublesome position by dealing
inadequately with concepts such as nature and necessity, another indication that Spinoza not only stood in bad repute in the eyes of his
opponents (e.g. Bredenburg's Spinozism was attacked vehemently and
without understanding), but that even his followers could not supply
a solid interpretation of his most accessible book, the *Tractatus The-*
ologico-Politicus. I give Bredenburg's courageous example for its illus-
trative force (Siebrand, 1984). The confusion about Spinoza's concept
of nature *and* his theory that reason operates by force of natural light
(*lumen naturale*) always remained a serious obstacle in debating his
philosophy. Any superficial equation of nature with reason and necessity brings the reader invariably to the cheap charge of belief in fate and
atheism.

Since *nature* was an ambiguous concept at that time (especially in ordinary conversation), the rise of the new science was a welcome occasion to think that the first and foremost criterion for the acquisition of knowledge could be found in the equation of appearance and reality. Spinoza's Bible criticism, which was far more original than that produced by the godfather of historical criticism, Richard Simon,[14] had a special eye for what I would like to call the reception of religious ideas. To give an exemplary quotation from his *Tractatus Theologico-Politicus*:

... philosophers who endeavour to understand things by clear conceptions of them, rather than by miracles, have always found the task extremely easy — at least, such of them as place true happiness solely in virtue and peace of mind and who aim at obeying nature, rather than being obeyed by her. Such persons rest assured that God directs nature according to the requirements of universal laws, not according to the requirements of the particular laws of human nature. It is plain, then, from Scripture itself, that miracles can give no knowledge of God, nor clearly teach us the providence of God. As to the frequent statements in Scripture, that God wrought miracles to make himself plain to man ... it does not, therefore, follow that miracles really taught this truth, but only that the Jews held opinions which laid them easily open to conviction by miracles. (Spinoza, 1951, p. 88; Gebhardt III, pp. 88—89)

Thus, he concludes, it is better to understand phenomena *per Causas naturales* than to speculate consistently with one's own preferences. In literary theory of reception, we would say: the Jews interpreted certain experiences as supernatural and miraculous because of their horizon of expectations. Therefore they were easily to be convinced of the 'evidence' counting for miracles, and so they can receive any textual account in this field positively. Spinoza put himself on the firm base of one exclusive substance, that being God, and including all nature outside of which nothing is possible nor even thinkable. So if one has failed to find this world attractive or interesting, and has sought compensation in an 'unseen world' of one's own experience or imagination, one has surpassed the boundaries of God's nature, which is, according to Spinoza, impossible.

Now, before we pass on to two concrete examples of Spinoza's reception in the context of physics and philosophy of religion, it is appropriate to say that the new science as such was not detrimental to religion. On the contrary, as it filled the English *virtuosi* with still more reverence because of the wonder of God's nature, so the Dutch physicists studied nature to the glory of God. Yet there was no direct

relation between the new science and a particular theological transformation of its principal propositions. Many views got about. The
rejection of Aristotelianism and the rise of modern science together
with the obvious uncertainty of the theologians with respect to the
implications of all this, resulted in a return to biblical exegesis. But, as
we observed above, the Christian concept of nature could hardly profit
from renewed exegetical efforts. And whatever the theologians could
supply, historical Bible criticism or the traditional *philosophia mosaica,*
there was never a fair debate on Spinoza's *Deus sive natura.*

3.1. *Bernoulli's Refutation*

The Groningen professor of mathematics, Johannes Bernoulli (1667—
1748), who was Swiss born, enjoyed great authority in his field. As a
matter of fact, he was the leading mathematician in Europe at that time.
His reputation was compromised because of his involvement in the
celebrated controversy between Newton and Leibniz over the invention
of the calculus. He chose the side of Leibniz, accusing Newton of
having derived his infinitesimal method from Leibniz's. His role in the
whole affair was rather provocative, to say the least, because he did not
reveal his identity for a long time, fighting Newton *via* confidential
instructions to his pen-friend Leibniz. The latter understood very well
how to handle the information to which he had access. Only when
Bernoulli became aware of the fact that Newton had got wind of his
provocations did he write to Newton asking him earnestly to believe
that it was not his custom to issue anonymously what he neither wished
nor dared to acknowledge as his own. Newton accepted this, but he
knew that Bernoulli's denial was a lie. He wrote to Bernoulli (September 26, 1719) with great subtlety, that he was too old now for taking
pleasure in mathematical studies (Westfall, 1981, pp. 787—788; Hall,
1980, *passim*). This background story to characterize Bernoulli is more
interesting because he was a great scholar; in general, the scientific war
of pamphlets and flying sheets in those years shows many comparable
examples. For reading Bernoulli's refutation of Spinozism (as far as I
know forgotten until now), his quarrelsome nature may supply some
understanding for his vigorous argumentation. In 1702, he published
his sixty pages against Spinozism entitled *Spinozismi depulsionis echo.*
It would be interesting to know why a mathematician of his standing
found it necessary to write against Spinozism at the beginning of the

eighteenth century. Well, if we look at his text, he appears to be quite clear-headed about his motivation. He was involved in a theological debate, and had accused a theologian (Paul Hulsius) of Spinozism on the grounds that, if one concedes that the human body can sin, the distinction between body and mind is obliterated (on the assumption of mind-body parallelism). He says he understands his opponent is upset, since *neque vult esse Spinosista, sed Spinosum esse nomine et omine.* Apparently, the predicate of Spinozism was still very compromising. And, as he himself was accused now of Spinozistic thought, he writes an apology in defence of his discredited orthodoxy. He argues that geometry is not applied sufficiently in Cartesianism (implying that this particular stream of thought is responsible for more problems in philosophy of religion?): for example, the methodological doubt in Cartesian philosophy is a mere fiction since it never allows the hypothesis that God in reality does not exist.

More interesting, however, are his deliberations on the tenability of corporeal resurrection in the Christian-Spinozist theory of his opponent. What about the identity of *nutritio* and *numerus,* he asks, by which he means the necessary correspondence of the whole and its parts, since the all-including substance, God, cannot diminish or increase because of a fluctuating number of bodies to be resurrected. We could only support such a view, he says, if we could somehow take for granted that "our bodies are conceived as the fortuitous assembly of atoms, as Democritus and Epicurus thought" (*nostrum corpus ex fortuito concursu atomorum compactum censes cum Democrito et Epicuro*), which he finds a more or less ridiculous assumption (Bernoulli, 1702, p. 53). For then God is supposed to have given preformative power to atom-like fibers ('*stamina*' according to his opponent), operating materialistically according to the way they were programmed. Bernoulli thinks it to be more profitable to ask what is the nature of the human body, and to discuss the question with professional scientists. He believes he has demonstrated his charge of Spinozism very well; and along with his conviction that resurrection implies both body and mind, he testifies to his belief in scripture. Thus, the evidence of the sciences is considered to be remote from religious belief, although the latter counts as decisive.

Bernoulli's refutation, written in passionate style, began as an inspection of the arguments delivered by a theologian. The text is overloaded with rhetoric, and is not very precise. But notwithstanding its persuasive

redundancy, it is interesting that the author reprimands an assumed
Spinozism from an ideal-typical point of view. He says, if you were
really a Spinozist, then you would know that certain Christian principles
of belief cannot be sustained. So there is a remarkable lack of consistency in your position. Secondly, Bernoulli separates theology from
the other sciences. (Spinoza's separation of philosophy and theology
does not seem to have influenced Bernoulli's standpoint.)

I have chosen Bernoulli as an example because he is a mathematician
with a great reputation. That his arguments contrast sharply with his
professional scrutiny in the context of science is at first sight a bit
disappointing. But if we compare his refutation with those of others,
there is not so much difference. There are many refutations available (I
have traced a great many of them), and practically all of them —
bearing the marks of their own eras — fail to make a connection
between physics and philosophy of religion that really makes sense.[15]

3.2. *Van Velthuysen's Criticism*

Lambert van Velthuysen heard of the *Tractatus Theologico-Politicus*
because of its atheistic reputation (*Letter XLII,* Gebhardt IV, pp. 207—
218). It is important to know that he was a Cartesian, and that he
set out to defend the Christian message by philosophical means. He
was a forerunner in the renewal of political theory, in promoting the
Hobbesian ideas borrowed primarily from *De Cive*. Not to dwell too
long on this, his reaction to the *Tractatus Theologico-Politicus* was very
tart for a man of his standing. In a treatise bearing the sub-title
Oppositus Tractatui Theologico-Politico & Operi Posthumo B. D. S.
(1680, pp. 1492ff), Velthuysen declared that Spinoza creates a controversy between philosophy and theology which is unsolvable. For as
theology demands obedience to God's eternal laws and the rules and
virtues derived from them, philosophy separated from theology can
never succeed in having the promised perception of God with the help
of its sole instrument, the function of reason. Natural light, the light of
reason (*lumen naturale*) cannot supply salvation. Velthuysen thinks this
to be the supreme controversy (*maximi momenti*) in Spinoza's philosophy. In his view, philosophy should extend its scope of application to
the field of morals in a very specific way since the *potentia Dei* as such
does not teach us how to suit one's life to one's concept of God (p.
1500). Of course, Velthuysen knew very well that it would be naive to

claim that a concept of God can be directly translated into a norm
conducive to the moral pattern of one's life. We should remember that
he was a Cartesian who tried to arrive at a synthesis with his declared
Hobbesian sympathies. But that was only one of the reasons why he
misunderstood Spinoza. As a Cartesian, he wanted to save the rationality of God, whereas as a Hobbesian he supported political power as the
ultimate source of human wellbeing. The legal order of the state and the
(eternal) order of nature cannot easily be united. Therefore Spinoza's
solution, that the order of nature is identical with the order of thought,
is in Velthuysen's conception the same as saying that rationality could
produce salvation. We are here confronted with the central question of
the seventeenth century. Formulated in brief, it says: given the structure
of physics, how can we make sure that religious morality is its very
antitype? In a wider sense, the scope of the question includes different
problems with respect to the scientific heritage from sixteenth-century
scholasticism. The rise of the new science was made possible thanks to
the knowledge available in the field of logical distinctions (*logica
modernorum*) and the more or less positive approach to the material
world as an object of reason. But scholastic rationalism was not
equipped for making contact with matter directly. So the new atmosphere of the seventeenth century implied that at least two problems
were to be solved, namely a redefining of *rationality,* and the reinforce-
ment of *philosophy* as a discipline (the borderline of the latter lay under
severe attack from the side of science).

   Philosophy had to test conclusions with scientific materialism
(Whitehead, 1967, p. 17). Could philosophy make use of the framework of the new science? If this framework were used, would it
function merely as a projection or could it be genuinely integral?
Although this was not particularly what Velthuysen questioned while
reading Spinoza, he was insistent about related problems in their
correspondence. For example, his scholastic dispute about Spinoza's
concept of necessity: Velthuysen found it unreasonable to accept the
status quo of a necessary natural order; this was, according to him, the
same as giving in to fatalism.[16] Velthuysen's reception of Spinoza's
philosophy may count as an illuminating example of the response of a
scholar whose sincere will was to understand the *Tractatus Theologico-
Politicus* as a book that deserved its place in the atmosphere of intellectual progress, but whose reason could not allow him to go as far as
his will would have wished. Despite the fact that he shared the code of

the new ideas coming from philosophy, he nevertheless had a different horizon of expectations, which was supplied by scholastic material. It is not surprising that Spinoza devoted sympathy to only one of his critics, namely Lambert van Velthuysen.[17] It goes without saying that Velthuysen was on the other side of the bridge, more versed in philosophy of religion than in physics. It is a pity that Spinoza never wrote a substantial contribution to physics. If he had done so, his critics would have been obliged to meet his arguments head-on. But alas, since there is no way to repair the gap between his philosophy of religion and his investigations in physics,[18] any interpretation of Spinoza must necessarily proceed *via* his philosophy of religion. But, in accord with my profession, I do not feel sorry about that.

State University of Groningen

NOTES

* This study is subsidized by the Foundation for Research in the Field of Theology and the Science of Religions in the Netherlands, which is a department of the Netherlands Organisation for the Advancement of Pure Research (ZWO). I am grateful to Prof. H. G. Hubbeling for his kind comments on this article.

[1] Cf. *Letters XXXII* and *XXXIII* in Gebhardt IV, pp. 169—176 and 176—179. See also Huygens (1888—1950, VI, p. 158). It is noteworthy that Spinoza possessed Huygens's *Horlogium Oscillatorium* as well as the *Tractatus de Ratiociniis*.

[2] As Harvey was a Ramist, there are understandably good reasons available for the struggle between Descartes and Harvey on the causes of blood circulation. Cf. also Hooykaas (1977, pp. 88—92).

[3] Lodewijk Meyer (1629—1681) worked in medicine in Amsterdam. He founded the art society Nil Volentibus Arduum (Nothing is impossible for a willing person) in 1669 and was a regent of the Amsterdam theatre. Besides some literary work, the writing of scene plays, he published his well-known study of the interpretation of the Bible (Meyer, 1666). As the title of the work makes clear, Meyer took philosophy to be an interpretative instrument for reading the Bible correctly. This was in contrast to Spinoza, who defended a separation of philosophy and theology. Jarig Jelles (d. 1683) was a Mennonite grocer in Amsterdam. He wrote the introduction to the *Opera Posthuma* (Spinoza, 1677), and earlier paid for the publishing of Spinoza's *Renati Des Cartes Principiorum Philosophiae*. After Jelles died in 1684, his confession of faith entitled *Belijdenisse des algemeenen en christelijken geloofs vervattet in een brief aan N.N.* (Confession of universal and Christian belief contained in a letter to N.N.) was published. He had corresponded about its content with Spinoza, as we know from the letters. Pieter Balling translated Spinoza's *Renati Des Cartes Principiorum Philosophiae* into Dutch. He is considered to be the author of *Het licht op de kandelaar* (The light on the candlestick), a spiritualist text with ethical implications.

[4] Cf. *Letters XII* and *XIII* (Gebhardt IV, pp. 52—62 and 63—69) as well as the introduction to the *Renati Des Cartes Principiorum Philosophiae* (Gebhardt I, pp. 127—
133). See Von Dunin Borkowski (1933—36, III,pp. 95—146).

[5] McGahagan's contention that Spinoza owed no intellectual debt to the university
Cartesians although he sometimes reveals his awareness of their language and
arguments, seems to me beside the point. Intellectually, Spinoza may have adopted
more from his friends, but as an autodidact he certainly took advantage of the topics
discussed by the so-called Cartesians. That most of them defended a misinterpretation of
Descartes was clear to him, as we can deduce from his *Renati Des Cartes Principiorum
Philosophiae*. But this was no hindrance to Spinoza's interest in their disputes. His very
motive for publishing the *Renati Des Cartes Principiorum Philosophiae,* that is, as a test
case in how people would react, shows that he was far from isolated and uncommitted
about the Aristotelian—Cartesian fracture that separated the academic circles. It must
be noted that the application of Cartesianism as an adequate label for Heereboord has
already been questioned by Dibon (1954).

[6] Pierre Gassendi (1592—1655) paid a visit to the Netherlands in 1628—29. He
travelled for eight months, spending most of them in the South, and he stayed only two
weeks in the northern part of the country. We know from Isaac Beeckman's journal that
both scholars discussed Epicurean philosophy. But the very often stated presumption
that it was Beeckman who stimulated Gassendi to extrapolate his Epicureanism in the
field of ethics to the (atomistic) investigation of physics cannot be verified. The fact is
that Gassendi, although he was acquainted with many scholars, especially at Leiden and
Amsterdam, felt more at ease in Louvain where the humanistic tradition created an
atmosphere of scientific progress. As to the criticism of Aristotelism, Louvain was far
ahead of Leiden at that time. As to Gassendi's influence in the Netherlands, he seems to
have persuaded Renerius to do real philosophy, i.e. physics instead of *vulgaris
philosophia* (Aristotelianism).

[7] Hubbeling (1964, pp. 92—93) discovered that Spinoza corrected Descartes's third
proof for God's existence in the *Renati Des Cartes Principiorum Philosophiae* with the
help of the assumption that the *gradus perfectionis* runs parallel to the *gradus
existentiae*. Spinoza favors *a priori* proofs, but he invokes *a posteriori* elements derived
from existence which clarify his apparent chief interest never to allow the existence of
an idea independent from its — according to him necessary because parallel —
existential reality. In my opinion, Spinoza's existential interest can already be deduced
from his reformulation of Descartes in the *Renati Des Cartes Principiorum Philosophiae,* which is closer to reality than the original *methodological* conception.

[8] Hubbeling (1978, p. 66) takes the view that Spinoza gave a one-sided account of
Descartes in that he dropped the physical arguments and concentrated almost exclusively on metaphysics. As Descartes was chiefly interested in metaphysical arguments, I
think Spinoza cannot be blamed for his approach: on the contrary, it indicates that he
understood Descartes perfectly well, and as he had in mind to 'test' his public in order
to find out what the reception of the *Renati Des Cartes Principiorum Philosophiae*
might be, he made a good choice in dropping most of the physical sections. Probably
we could better extend Hubbeling's line in this way: Spinoza's lack of empirical interest
in the context of the *Renati Des Cartes Principiorum Philosophiae* is due to his
judgement that Descartes did not succeed in his plan of constructing a mathematical
conception of physics. Therefore Spinoza chose the most obvious method to fulfill this

aim, i.e. the geometrical one, which Descartes unfortunately did not use. As such, Spinoza's approach presages his own plan tried out in his treatise on the rainbow, namely to connect mathematics and physics.

[9] The French translation so often cited is less correct: for 'analysis', . . . *methodiquement inventée, et fait voir comment les effets dépendent des causes . . . ;* and for 'synthesis', . . *et comme en examinant les causes par leurs effets (bien que la preuve qu'elle contient soit souvent aussi des effets par les causes)* . . . ! Any distinctive feature seems to be mitigated, so they do not appear to be different at all.

[10] Cf. *Catalogus* (1965, p. 30), where it is mentioned that Frans van Schooten's *Exercitationum mathematicarum* (1657), was bound together with Huygens's book.

[11] Cf. Petry (Akkerman, 1982, pp. 501—504) who gives important biographical information on Van der Meer that was unknown until now.

[12] Since mankind never failed in having a general and solid desire for real truth, philosophy has always been either an idolatrous object of compassion, or at least a source of consolation. Nevertheless, the struggle for truth seems to endure.

[13] The Hebrew text of the Old Testament always treats nature as related to history. The verb *qnh* (to create, to make) cannot be found in *Genesis* 2—3, so its meaning with respect to nature is uncertain, as it seems to denote primarily he or she who made something. As to *br'* (to create, to make) the object of creation is intended. In *Genesis* 1 it is clearly derived from oriental sources pertaining to the introduction of light and darkness, water and land etc. produced out of an already pre-existent substance. (Cf. Jenni and Westerman, 1971; and the handbook for theology and science of religion *Die Religion,* 1957).

[14] John Woodbridge (Lake Forest, Illinois) supplied new evidence for Simon's dependency on Spinoza's *Tractatus Theologico-Politicus,* although Simon *never* betrayed the fact that he revised his *Histoire critique du Vieux Testament* with the help of Spinoza's work. Woodbridge announces the publication of the *Brerewood Adnotations,* a manuscript of Simon's that has been lost since the 1730s. I am grateful for having been able to read his article 'The Reception of Spinoza's *Tractatus Theologio-Politucs* by Richard Simon (1638—1712)' *pro manuscripto.*

[15] It would be no problem to give, say, one hundred or more titles that are really interesting from the general perspective of Spinoza's reception. Because I concentrate exclusively in this article on the narrow track between physics and philosophy (of religion), this example fits better than the much disputed criticism by Bernard Nieuwentyt, for instance.

[16] Cf. *Letters XLII, XLIII* and *LXIX* (Gebhardt IV, pp. 207—218, 219—226, and 300); Velthuysen had numerous meetings with Spinoza!

[17] Cf. *Letters LXIX* cited above.

[18] *Vide supra.* Spinoza wanted to publish his own ideas on physics only if the *Renati Des Cartes Principiorum Philosophiae* were given a positive reception.

REFERENCES

Akkerman, F. *et al.* (eds.): 1982, *Spinoza. Korte Geschriften,* Wereldbibliotheek, Amsterdam.

Bernoulli, J.: 1702, *Spinosismi depulsionis echo id est depulsionis imputationes depulsae, irrisionis irrisae, punctiones compunctae, torsiones retortae, argutiae redargutae, ludibria delusa, contemtus contemti etc.,* Joh. Lens, Groningen.

Catalogus van de Bibliotheek der Vereniging Het Spinozahuis te Rijnsburg: 1965, E. J. Brill, Leiden.

Curley, E.: 1978, 'Spinoza as an Expositor of Descartes', in S. Hessing (ed.), *Speculum Spinozanum 1677—1977,* Routledge and Kegan Paul, London, pp. 133—142.

Descartes, R.: 1879—1913, *Oeuvres de Descartes,* Edition Adam-Tannery, 13 vols., Vrin, Paris. (Cited as 'Adam-Tannery' in the text.)

Dijksterhuis, E. J.: 1977, *De mechanisering van het wereldbeeld,* Meulenhoff, Amsterdam.

Dibon, P.: 1954, *L'Enseignement Philosophique dans les Universitiés Nèerlandais á l'Époque Pré-Cartésienn* (1575—1650), n.p., Leiden.

Galilei, Galileo: 1638, *Discorsi e dimostrazioni matematiche intorno a due nuove scienze,* Elsevier, Leiden.

Gueroult, M.: 1968, *Spinoza. Dieu* (*Ethique 1*), Éditions Montaigne, Paris.

Gueroult, M.: 1970, *Études sur Descartes, Spinoza, Malebranche et Leibniz,* Georg Olms, Hildesheim and New York.

Hall, A. Rupert: 1980, *Philosophers at War. The Quarrel Between Newton and Leibniz,* Cambridge Unviersity Press, Cambridge.

Helvetius, J. F.: 1687, *Adams oud Graft, Opgevult met jonge Coccei-Cartesiaenschen en Descartis-Spinosistischen Doodsbeenders etc.,* Levyn Van Dyck, The Hague.

Hooykaas, R.: 1977, *Religion and the Rise of Modern Science,* Scottish Academic Press, Edinburgh and London.

Hooykaas, R.: 1979, *Geschiedenis der natuurwetenschappen,* Bohn, Scheltema and Holkema, Utrecht.

Hubbeling, H. G.: 1964, *Spinoza's Methodology,* Van Gorcum, Assen.

Hubbeling, H. G.: 1978, *Spinoza,* Verlag Karl Alber, Freiburg and Munich.

Huygens, C.: (1888—1952), *Oeuvres complétes,* 22 vols., Société Holandaise des Sciences, The Hague.

Jenni, E. and Cl. Westerman: 1971, *Theologisches Handwörterbuch zum Alten Testament,* 2 vols., Chr. Kaiser Verlag, Munich.

McGahagan, Th. A.: 1976, *Cartesianism in the Netherlands 1639—1676; The New Science and the Calvinist Counter-Reformation,* UMI, Ann Arbor.

Meyer, L.: 1666, *Philosophia S. Scripturae interpres,* Jan Rieuwertsz, Amsterdam.

Mignini, F.: 1981, *Ars imaginandi; Apparenza e rappresentazione in Spinoza,* Editioni Scientifiche Italiane, Napoli.

Die Religion in Geschichte und Gegenwart: 1957, 7 vols., J. C. B. Mohr (Paul Siebeck), Tübingen.

Rowen, H. H.: 1978, *John de Witt, Grand Pensionary of Holland, 1625—1672,* Princeton University Press, Princeton.

Ruestow, E. G.: 1973, *Physics at 17th- and 18th-Century Leiden,* Martinus Nijhoff, The Hague.

Siebrand, H. J.: 1984, 'On the Early Reception of Spinoza's *Tractatus Theologico-Politicus* in the Context of Cartesianism', in C. de Deugd (ed.), *Spinoza's Political and Theological Thought,* North-Holland, Amsterdam, pp. 214—225.

Spinoza, Baruch: 1663, *Renati Des Cartes Principiorum Philosophiae Pars I & II, More Geometrico Demonstratae,* Jan Rieuwertsz, Amsterdam.

Spinoza, Baruch: 1677, *Opera Posthuma,* Jan Rieuwertsz, Amsterdam.

Spinoza, Baruch: 1687, *Stelkonstige reeckening van den regenboog, Dienende tot naedere samenknoping der Natuurkunde met de Wiskonsten,* Levyn Van Dyck, The Hague.

Spinoza, Baruch: 1925, *Spinoza Opera,* ed. by Carl Gebhardt, 4 vols., Carl Winter, Heidelberg. (Cited as 'Gebhardt' in the text.)

Spinoza, Baruch: 1951, 1955, *Chief Works of Spinoza,* transl. by R.H.M. Elwes, 2 vols., Dover, New York.

Spinoza, Baruch: 1963, *Earlier Philosophical Writings; The Cartesian Principles and Thoughts on Metaphysics,* transl. by Frank A. Hayes, Bobbs-Merrill, Indianapolis and New York.

Thijssen-Schoute, C. Louise: 1954, *Nederlands Cartesianisme,* in *Verhandelingen der Koninklijke Nederlandse Akademie van Wetenschappen,* Afdeling Letterkunde. Nieuwe Reeks Vol. 60, North-Holland, Amsterdam.

Van der Hoeven, P.: 1973, *De cartesiaanse fysica in het denken van Spinoza,* Mededelingen XXX vanwege Het Spinozahuis, E. J. Brill, Leiden.

Velthuysen, L. V.: 1680, *Tractatus de cultu naturali et origine moralitatis. Oppositus Tractatui Theologico-Politico & Operi Posthuma B. D. S.,* R. Leers, Rotterdam.

Von Dunin Borkowski, St.: 1933—1936, *Spinoza,* 4 vols., Aschendorffschen Verlagsbuchhandlung, Münster.i.W.

Westfall, Ric. S.: 1981, *Never at Rest, A Biography of Isaac Newton,* Cambridge University Press, Cambridge.

Wetlesen, J.: 1979, *The Sage and the way. Spinoza's Ethics of freedom,* Van Gorcum, Assen.

Whitehead, A. N.: 1967, *Science and the Modern World,* The Free Press (MacMillan), New York.

Williams, B.: 1979, *Descartes: The Project of Pure Enquiry,* Harvester, Harmondsworth.

PART II

SPINOZA: SCIENTIST

DAVID SAVAN

SPINOZA: SCIENTIST AND THEORIST
OF SCIENTIFIC METHOD

Two questions concern me in this paper. First, what is the place and importance of Spinoza's work as a practising scientist? Second, what did Spinoza think were the right rules to follow in carrying out specific scientific investigations?[1] Coupling these two questions will, I believe, throw some new light on Spinoza's work and thought. Our second question will occupy the larger part of this paper, since the answer depends upon Spinoza's theory of the necessary inadequacy of the human mind and our almost total ignorance of how things are "linked together in the universal system of nature (*totiusque naturae ordinem et cohaerentiam*)" (*Tractatus Theologico-Politicus*, Gebhardt III, p. 191). In order to focus the discussion of the second question, I will introduce it by pointing to an apparent contradiction in what Spinoza wrote concerning the study of the emotions. In a well known passage in the *Ethics* he appears to take an a priorist stand, while in an echoing passage in the *Tractatus Politicus* he appears to take an empiricist position. The resolution of this apparent contradiction is central to an understanding of Spinoza's theory of scientific method.

First, then, what is the importance of Spinoza's work as a practising scientist? Let me consider the evidence, item by item. In his correspondence with Boyle, via Oldenburg, Spinoza described some chemical and physical experiments he had undertaken. The experiments appear to have been *ad hoc*, conducted in order to examine critically the work that Boyle had done. There is no evidence that Spinoza was interested in carrying on a continuing program of chemical and physical experimental research. The experiment on pressure described in *Letter XLI* (Gebhardt IV, pp. 202—206) to Jelles likewise appears to have been carried out in order to reply to a question that Jelles had asked "first by word of mouth and then in writing".

Spinoza was an expert lens grinder and a student of optics. He was on the periphery of Huygens's circle, and in his letters Huygens showed considerable respect for Spinoza's knowledge and practical skill in

95

Marjorie Grene and Debra Nails (eds.), Spinoza and the Sciences, 95—123.
© 1986 *by D. Reidel Publishing Company.*

optics. Leibniz too consulted Spinoza with a problem in optics. Nevertheless, Spinoza does not appear to have carried on any original research in this field. He does indeed make some criticisms of Descartes,[2]
but his treatise on the rainbow, while it makes some additions to
Descartes's *Dioptrics*, is not an important advance.

Spinoza wrote an adumbration of Descartes's *Principles*, Parts I, II,
and III, and in Part II of the *Ethics* he gave a brief statement of the first
principles of physics. He did indeed criticise Descartes's sixth law of
motion, but he made no significant contribution to the development of
theoretical physics.

Unquestionably, Spinoza attached great importance to mathematics,
and to geometry in particular. Geometry is a model of clarity, demonstration and of explanation which shows how to dispense with final
causes. However, Spinoza made no contribution to geometry or to the
philosophy of mathematics. Nevertheless, three points are worthy of
mention. One, his view that chance is a measure of ignorance did not
prevent him from sharing the interest of Huygens and others in the
mathematical calculation of chances. *Letter XXXVIII* to van der Meer
(Gebhardt IV, pp. 190—193) and a further short paper on the calculation of chances attributed to Spinoza are devoted to solving some
problems in the calculation of chances.

Two, as Frege remarked in his *Grundlagen der Arithmetik* (III.49),
Spinoza's definition of number is a faltering anticipation of Frege's. In
Letter L to Jelles Spinoza wrote,

. . . we do not conceive things under numbers until they have been subsumed under a
common class. For example, he who holds in his hand a penny and a dollar will not
think of the number two, unless he can call the penny and the dollar by one and the
same name, such as pieces of money or coins . . . (Spinoza, 1928, p. 269; Gebhardt IV,
p. 239).

Frege points out that this definition, however, excludes zero from
number. He could have added that in *Letter XII* (Gebhardt IV, pp.
52—62) Spinoza also denies cardinality to infinite multitudes.

Three, although he denied that number applies to the infinite, Spinoza
held that there are different infinites, that two infinites need not be
equal (see *Letter XII*), and that the infinite attributes are equal to one
another because for any item in one attribute there is exactly one
correspondent in each of the others. He maintained further that one

infinite may be the cause, reason, or source of other infinites, and that these others may then be regarded as together constituting the first.

As far as the natural sciences and mathematics are concerned, I conclude that although Spinoza was thoroughly competent and acquainted with some of the best work of his time, he contributed little of importance to research and theory. However, Spinoza also wrote on politics and on the Bible. The *Tractatus Politicus* was not intended to be an empirical investigation of political behavior and the functioning of political institutions. Spinoza took the basic principles of human passion, action, and politics as established in the *Ethics*. He was fully convinced that historical experience had already revealed all the possible basic forms of political organization. His intention in writing this treatise, unfortunately never completed, was a practical one. As the subtitle of the work indicates, its purpose was to show how political institutions "must be organized if they are not to degenerate into Tyranny, and if the Peace and Freedom of the citizens is to remain intact". The frequent references to historical events are illustrations and examples, not data for hypotheses.

Quite a different situation had faced Spinoza when he prepared his *Tractatus Theologico-Politicus*, first published in 1670. The general purpose of the work is to demonstrate that the stability of the state and its religion is not threatened by intellectual freedom but is enhanced by free philosophical and scientific investigations. Since in the past attacks on freedom of thought had frequently claimed support from the Bible, and since past interpretations of the Bible were often inspired by fear, love of mystery, and superstitious ignorance, Spinoza

determined to examine the Bible afresh in a careful, impartial, and unfettered spirit, making no assumptions concerning it, and attributing to it no doctrines, which I do not find clearly therein set down. With these precautions I constructed a method of Scriptural interpretation, and thus equipped proceeded to inquire . . . (Spinoza, 1883, p. 8; Gebhardt III, p. 9).

In his formulation of his hermeneutic principles, in Chapter 7, he wrote that he proposed to approach the study of the Bible as he would the study of nature, and his methodological principles were essentially the same, in both cases. Spinoza showed that the methods of the natural sciences could be fruitfully extended to the scientific study not only of the Bible but of historical texts generally. Spinoza is the founder of scientific hermeneutics.[3]

The major procedural principles of scientific hermeneutics are set
forth in Chapter 7 of the *Tractatus Theologico-Politicus*. I have already
referred to Spinoza's first principle.

The method of interpreting Scripture does not widely differ from the method of
interpreting nature . . . . For as the interpretation of nature consists in the examination
of the history of nature, and therefrom concluding to definitions of natural phenomena
on certain fixed data (*ex certis datis*) so Scriptural interpretation proceeds by the
examination of Scripture (and) inferring the intentions of its authors as a legitimate
conclusion from certain fixed data and principles. (p. 99; Gebhardt III, p. 98)

Second,

Scripture does not give us definitions of things any more than nature does: therefore,
such definitions must be sought in the latter case from the diverse workings of nature;
in the former case, from the various narratives about the given subject which occur in
the Bible (pp. 100—101; Gebhardt III, p. 99).

Spinoza is not speaking of verbal definition, of course. This is clear
enough from the analogy he drew with the inference to the intentions of
the authors of a text. In the *Ethics* (Part I, Proposition XIX, Demonstration, and XXXIII, Scholium 1) he states that by the definition of a
thing he understands its nature or essence. The definition of a physical
thing is the analysis of its structure, its identifying ratio of motion and
rest. In the case of Scripture, Spinoza gives the example of Moses's
statement that God is a fire. If we compare this with other passages in
which Moses says that God has no likeness to any visible thing and yet
other passages in which the word 'fire' is applied to jealousy we can
conclude that the two propositions, "God is a fire" and "God is jealous"
are identical in meaning. Spinoza clearly recognizes that it is by the
study of the diverse workings of nature — of nitre, for example — that
we arrive at an understanding of its distinctive structure.

A third principle of scientific hermeneutics is that there must be a
good understanding of the nature and properties of the language in
which the text was written and in which the authors customarily spoke.
As far as possible, written expressions must be compared with one
another as well as with the spoken language. In keeping with this
principle, and to make the critical examination of Scripture easier,
Spinoza wrote a Hebrew grammar.

Fourth, each part of the text must be analyzed and arranged according to topic. Note must be taken of ambiguities, obscurities, and contradictory passages.

Fifth, the meaning of a passage must be carefully distinguished from its truth or falsehood. In discussing the history of the Hebrew text, the problems connected with ascertaining the meanings of its terms and phrases, its syntax, and its script, Spinoza proposes that "words gain their meaning solely from their usage" (p. 167; Gebhardt III, p. 160). Hence no single person is able to corrupt or change the meanings, the syntactical structures, or the sacred or secular implications of a linguistic usage unless those changes are broadly accepted by a majority of the language users. Thus Spinoza distinguished between the intentions of a particular language user, the meanings of his sentences, and their truth or falsehood.

Sixth and seventh, the examination of a text must consider

... the life, the conduct, and the studies of the author of each book, who he was, what was the occasion, and the epoch of his writing, whom did he write for, and in what language. Further it should inquire into the fate of each book: how it was first received, into whose hands it fell, how many different versions there were of it, by whose advice was it received into the Bible, and, lastly, how all the books now universally accepted as sacred were united into a single whole. (p. 103; Gebhardt III, p. 101)

Spinoza is well aware that in the transmission of an ancient text many changes may have been introduced, inadvertently or perhaps intentionally.

This review should have made it evident that most of those who have written on Spinoza as a scientist have arrived at an unbalanced and even incorrect interpretation because they have tended to focus on his chemical and physical experiments. But Spinoza's experimentation in these fields was occasional and *ad hoc*. In theoretical physics Spinoza was, with an important exception which I will consider later, by and large a follower of Descartes and Huygens. It was in the extension of the scientific outlook and scientific methods to the study of the historical texts that Spinoza was innovative and influential. He emphasized the importance of the careful collection of empirical data. Variations and changes in the data must be noted, compared, and cross checked. In several cases Spinoza states that our data are quite inadequate and no particular explanation is warranted. In other cases he acknowledges that his explanatory hypothesis is tentative and doubtful. In his most successful and important scientific work, then, Spinoza is an empiricist.

In turning now to our second question we must remember that Spinoza has agreed that the method of interpreting nature is similar to the empirical method of his hermeneutics. What then is his philosophically based general theory of method?

Unfortunately, our direct data are sparse. In comparison with Bacon, Hobbes, and Descartes, all of whom he had read, Spinoza wrote relatively little specifically about the fundamental principles governing the methods of the new sciences. The reason is not hard to find. In all of his books and in most of his letters Spinoza's dominant purpose is religious or ethical or both. I use 'religion' to signify concern with salvation, eternal perfection, and blessedness. The *Ethics*, he wrote (in the Preface to Part II), is intended "to lead us as it were by the hand to the knowledge of the human mind and its utmost blessedness" (quotations from the *Ethics* are from the White translation, Spinoza, 1930a). The origin of the *Tractatus de Intellectus Emendatione*, Spinoza wrote, was in a spiritual crisis in which he was "as a sick man struggling with a deadly disease, when he sees that death will surely be upon him . . .", and was forced to seek some remedy, however uncertain, because all his hope lies in finding that remedy (Spinoza, 1930b, p. 3; Gebhardt II, p. 7). His aim was

to direct all sciences to one end and aim, so that we may attain to the supreme human perfection which we have named; and, therefore, whatsoever in the sciences does not serve to promote our object will have to be rejected as useless . . . (p. 6; Gebhardt II, p. 9).

Of course, the more we understand nature the more we understand the human mind, human society, and God. Nevertheless, the best is the enemy of the good, and Spinoza's paramount interest was in attaining blessedness, salvation, and liberation for himself and for all human beings. To this end science is only a means. His remarks on scientific method are incidental to his major concern.

In order to focus the discussion, let me begin by considering two parallel passages on the proper method for studying human actions and passions. In his preface to the third part of the *Ethics* Spinoza contrasts the common approach to the study of the affects with his own. It has been usual to treat human behavior as if it were a world apart, distinct from the rest of nature. Consequently, instead of trying to understand and explain the causes of human behavior it has been customary to "bewail, laugh at, mock, or . . . detest" human weakness. Spinoza however, maintains that nature allows no exceptions.

(The) laws and rules, according to which all things are and are changed from form to form, are everywhere and always the same; so that there must also be one and the same method (*ratio*) of understanding the nature of all things whatsoever, that is to say, by the universal laws and rules of nature . . . . I shall, therefore, pursue the same method in considering the nature and strength of the affects and the power of the mind over them, which I pursued in our previous discussion of God and the mind, and I shall consider human actions and appetites just as if I were considering lines, planes, or bodies. (*Ethics* III, Preface)

The passage in the *Ethics* is repeated in Chapter 1, Section 4 of the *Tractatus Politicus*, but with a difference that is significant.

In order to investigate the topics pertaining to this branch of knowledge with the same objectivity as we generally show in mathematical inquiries, I have taken great care to understand human actions, and not to deride, deplore, or denounce them. I have therefore regarded human passions like love, hate, anger, envy, pride, pity, and the other feelings that agitate the mind, not as vices of human nature, but as properties which belong to it in the same way as heat, cold, storm, thunder and the like belong to the nature of the atmosphere. (Spinoza, 1958; Gebhardt III, p. 274)

Are the passions and actions to be studied as Euclid studied lines, planes, and solids, or are they to be studied empirically, as we study the capricious and inconstant agitations of the atmosphere? Does Spinoza fail to see the difference between the method of the geometer and the method of the meteorologist? Is it a careless slip, or has Spinoza changed his conception?

It is clear enough that Spinoza had not changed his views. Part of the first chapter and most of the second chapter of the *Tractatus Politicus* are a summary review of the central argument of the *Ethics*, and there is no suggestion of any shift in his views on geometry or the study of nature. Nor is Spinoza carelessly nodding. It is sufficiently evident, even although the two quoted passages (from the *Ethics* and from the *Tractatus Politicus*) are by different translators, that the *Tractatus Politicus* passage is carefully based upon the *Ethics*. A comparison of the Latin originals shows that Spinoza took care to use the same key words in the two passages.

A closer reading makes it obvious, however, that Spinoza saw no incompatibility between the geometrical method and the study of such natural agitations as the emotions and the weather. The *Ethics* passage refers to "the universal laws and rules of nature" according to which all things change. Later in *Ethics*, in a reference back to the Preface of Part III, he affirmed, "I consider human affects and their properties precisely as I consider other natural objects" (*Ethics* IV, Proposition LIX,

Scholium). Conversely, in the *Tractatus Politicus* passage Spinoza speaks of the objectivity (*animi libertate*) with which we pursue mathematical inquiries. How then does Spinoza conceive of the relation between mathematical and empirical methods in the study of nature and natural change? I will summarize under six headings what Spinoza saw as the positive value of the mathematical model, by which he generally understood the geometrical method of Euclid, for science. A seventh point, however, points to a major deficiency in the mathematical model.

(1) *Elimination of final causes*. Mathematics "does not deal with ends, but with the essences and properties of forms (and thus) places before us another standard of truth" (*Ethics* I, Appendix). When we try to explain things through final causes we are led by inadequate ideas to evaluate things as sound, putrid, corrupt, beautiful, deformed, harmonious, good, and evil. We deride, deplore, and denounce human actions. Mathematics shows us how to explain events without final causes, and thus to eliminate evaluations based upon inadequate ideas of what is to our private gain or hurt.

(2) *Necessity*. Just as it follows necessarily from the nature of a triangle that its three angles equal two rights angles, so infinite things in infinite ways follow necessarily from the nature of God (*Ethics* I, Proposition XVII, Scholium).

(3) *Demonstration*. "If men understood things, they would, as mathematics prove, at least be all alike convinced if they were not all alike attracted" (*Ethics* I, Appendix).

(4) *Order*. Geometry is a model of proper order, beginning with the clearest and simplest ideas, and deducing in correct order their more remote consequences. This is one of the points Spinoza makes several times in the *Tractatus de Intellectus Emendatione*.

(5) *Hermeneutic clarity*. "Euclid, who only wrote of matters very simple and easily understood, can easily be comprehended by anyone in any language" (*Tractatus Theologico-Politicus*, Spinoza, 1883, p. 113; Gebhardt III, p. 111). Knowledge of the special biographical and historical situation of the author is not needed for the understanding of clear and distinct premises and their necessary consequences.

(6) *Avoiding polemics*. Since all are alike convinced by rational demonstration (see point 3 above) mathematics makes for peaceful community rather than the divisive polemics of those who are led by the inadequate ideas of the imagination.

(7) *Abstraction*. In his last letter to von Tschirnhaus (*Letter LXXXIII*; Gebhardt IV, pp. 333—334), Spinoza says that geometrical figures are *entia rationis*, abstractions. This is a position he had consistently maintained from the beginning. An *ens rationis* is a mode of thinking which has "nothing corresponding to it in nature" (*Tractatus de Intellectus Emendatione*, Spinoza, 1930b, p. 27; Gebhardt II, p. 27). For this reason Spinoza wrote in the *Cogitata Metaphysica* I.1 concerning number and measure that "these modes of thinking are not ideas of things and cannot be classified as ideas" (Spinoza, 1963, p. 109; Gebhardt I, p. 234). This is an extreme prohibition which he soon abandoned. In the *Tractatus de Intellectus Emendatione* he says of the conception of a circle that it is a true idea (Spinoza, 1930b, p. 27; Gebhardt II, pp. 27—28). In 1663, about one year after he had stopped working on the *Tractatus de Intellectus Emendatione* (see *Letter VI*, Gebhardt IV, pp. 15—36), Spinoza gave up speaking of an idea of an *ens rationis* as true. In *Letter IX* to de Vries he wrote,

. . . a definition either explains a thing as it exists outside the understanding, and then it ought to be true . . . . Or else a definition explains a thing as it is conceived or can be conceived by us: and then, indeed, it differs from an axiom and a proposition because all that is required of it is merely that it should be conceived, and not, like an axiom, that it should be conceived as true. (Spinoza, 1928, pp. 106—107; Gebhardt IV, pp. 43—44)

In its dependence on abstractions mathematics is emphatically not a paradigm for philosophy and science. In the *Tractatus de Intellectus Emendatione* it is stated repeatedly that philosophy must never draw conclusions from abstractions but only from some particular affirmative essence, or from the definition (i.e. the correct analysis) of such a particular essence. Because Descartes failed to observe this vital distinction his "principles of natural things are useless, not to say absurd" (*Letter LXXXI* to von Tschirnhaus; Spinoza, 1928, p. 363; Gebhardt IV, p. 332). A sound physics must study extension as an active and dynamic expression of the free causal power of nature (*natura naturans*). This is not to deny that there exist in nature particular pyramids, domes, spheres, and rectangular structures from which the geometer abstracts his *entia rationis*. In the *Ethics*, when the point is relevant, Spinoza was careful to speak of the circle or triangle "existing in nature" (Part II, Proposition VII, Scholium). "The reason why a circle or triangle exists or does not exist is not drawn from their nature, but from the order of corporeal nature generally" (Part I, Proposition XI, Demonstration 2).

I conclude that while mathematics serves as a standard in the first six ways set out above, on Spinoza's view it can not serve as a model for the content of our knowledge of the singular things and processes of nature. The comparison of human actions and passions with lines, planes, and solids points to an analogy of method. In the *Ethics*, but not in the *Tractatus Politicus*, the geometrical order is followed. Beginning with what Spinoza considered to be the clearest, simplest, and most fundamental concepts and principles, the general configurations of the varieties of passions, desires, and actions are deduced *ordine geometrico*. The comparison in the *Tractatus Politicus* with storms and weather points to a similarity of content. Just as storm and tranquility are necessary properties of the atmosphere, so are passion and action necessary properties of human nature. "Man is necessarily always subject to passions" (*Ethics* IV, Proposition IV, Corollary). As the climate varies with geography and season, so too with the passions. One and the same object will arouse different passions in different people, while it stirs the same person to different passions at different times (*Ethics* III, Proposition LI). Indeed, "the affect of one person differs from the corresponding affect of another as much as the essence of the one person differs from that of the other" (*Ethics* III, Proposition LVII).

By what methods, then, are we to come to a scientific understanding of such finite physical modes as storms and nitre, and of such finite modes of thought as the passions, or the ideas of an historical text like the Bible? To answer this question I must review very briefly some familiar themes.

The fundamental activity of nature is the cause of each finite mode in two distinct ways. (Spinoza defines *natura naturans* as equivalent to *God* at *Ethics* I, Proposition XXIX, Scholium, and in the present context the term *nature* is the more appropriate.) First, the activity of nature, in the form of an active attribute, is the cause of each mode *absolutely* (Spinoza's term at *Ethics* I, Proposition XXVIII, Scholium), because without that free activity neither the essence nor the existence of the mode can be or be conceived. Second, the particular determinate existence and action of every finite mode is molded by the particular action upon it of other finite modes by which nature is affected (*Ethics* I, Proposition XXVIII; cf. II, Proposition IX). If we visualize the finite modes as limited areas on the surface of a vibrant organism, moving ceaselessly, continuously, and in all possible ways, then we may speak of nature's absolute causality as a radial axis. The second type of causality, nature as affected by particular and determinate modes of activity, might be regarded as a topographic axis.

Although each individual thing is determined by another individual thing to existence in a certain way, the force nevertheless by which each thing perseveres in its existence follows from the eternal necessity of the nature of God. (*Ethics* II, Proposition XLV, Scholium)

Each finite mode is an intersection of the two causal axes. It is one of Spinoza's first axioms (Axiom 3 in Part I of the *Ethics*) that "if no determinate cause be given, it is impossible that an effect can follow". Hence a body will continue in a uniform state of motion or rest until it is determined to a change of state by the determinate action of some other finite body (*Ethics* II, Proposition XIII, Lemma 3, Corollary). What is true of modal bodies is equally true of modal minds (*Ethics* III, Propositions VI and IX). To express that this continuing endurance is not passive or inert but is, rather, an affirmative activity Spinoza uses the term '*conatus*'.

Every extended mode — a storm, a specimen of nitre, a human body — is a direct radiation of the infinite and eternal activity of extension and the adequate idea of that mode must explain it through active extension and those features which are universal and necessary to that attribute. In Spinoza's terminology, these universal features are modes which flow from the absolute nature of the attribute and hence share the infinity and eternity of that attribute. In *Ethics* II, Proposition XIII, Spinoza mentions as features which are common to all bodies, even the simplest, motion, rest, variability in degree of motion, and causal interaction (Lemmas 1, 2 and 3). The ideas of what is common to all bodies must be adequate. Indicating that these adequate ideas have, since the Stoics, been called *common notions* ('*notionum, quae communes vocantur*', *Ethics* II, Proposition XL, Scholium 1), Spinoza speaks of them as "ideas or notions" (*Ethics* II, Proposition XXXVIII, Corollary). These adequate ideas or common notions are the foundations of all scientific reasoning, and Spinoza is incredulous

that the very learned Mr. Boyle had set before himself in his Treatise on Nitre no other end than merely to show that the puerile and trivial doctrine of Substantial Forms, Qualities, etc., rests on a weak foundation (*Letter XIII* to Oldenburg; Spinoza, 1928, p. 124; Gebhardt IV, p. 64).

So much is generally agreed. But Spinoza adds a further kind of common notion. In *Ethics* II, Proposition XIII, Spinoza had gone on, after outlining the features common to all bodies whatsoever, to sketch a hierarchical series of compound bodies of increasing degrees of structural complexity. What is proper to all instances of a particular complex structure of motion and rest clearly will not be common to all bodies whatsoever. At *Ethics* II, Proposition XXXIX, Spinoza draws the consequence for knowledge of the fact that a highly complex body — say, the human body — is a hierarchical organization of subordinate, but still complex, parts. The idea of what is proper to a particular complex structure and common to all its instances must be adequate, and a common notion.[4] The common notions, which are the foundations of reasoning, are common to all human minds, since they are the ideas of the overall structure proper to the human body and common to all human beings. Through this physical human community, all human beings are united by *feelings* of sympathy, approval, indignation, pity, and the desire to emulate others and to act benevolently toward them. As well, "under the guidance of reason (men) seek their own profit (and) desire nothing for themselves which they do not desire for other men" (*Ethics* IV, Proposition XVIII, Scholium).

Spinoza makes a further point, however, which is more relevant to the subject of this paper. If some aspect of the human body, essential to its overall structure (as are, for example, the fluid blood, the soft flesh, and the hard skeleton) is generally affected by external bodies through their common physical structure, there will be in the human mind an adequate idea of this affection of the human body (*Ethics* II, Proposition XXXIX, Demonstration). "The more things the body has in common with other bodies, the more things will the mind be adapted to perceive (adequately)" (*Ethics* II, Proposition XXXIX, Corollary). Indeed, it is one of Spinoza's axioms (*Ethics* II, Axiom 4) that "we sense (or feel — the Latin word is '*sentimus*') that a certain body is affected in many ways". The core of this bodily interaction is not only a set of adequate common notions but also, as *Ethics* II, Proposition XXXIX argues, a common sensing or feeling.[5]

Feeling or sensing is basic to whatever our body shares with the external bodies which affect it. Scientific knowledge develops on the basis of the body's sensing and acting, as well as it does on the common notions. I will mention briefly some corroborating examples. Spinoza defines an attribute as what the intellect perceives of substance, and he uses the word '*percipit*' because that word conveys "that the mind is passive in its relation to the object" (*Ethics* II, Definition 3, Explanation). We are passive to the extent that other ideas and other bodies act upon our minds and bodies. "When the human mind through the ideas of the affections of its body contemplates external bodies, we say that it then imagines" (*Ethics* II, Proposition XXVI, Corollary, Demonstration). We perceive the attribute of extension through each and every action of other bodies upon our body. In short, we perceive extension through our sensory imagination.[6]

The same point is made, this time explicitly concerning the common properties of extension (motion and rest, variability of speed, etc.) in the demonstration of Proposition VII in Part V.

> The affect which arises from reason is necessarily related to the common properties of things, which we always contemplate as present (for nothing can exist which excludes their present existence), and which we always imagine in the same way.

The reference to the common properties as always imagined in the same way is quite deliberate, since Spinoza directs the reader back to *Ethics* II, Proposition XXXVIII, whose demonstration parallels the passage just quoted from Part V, Proposition VII. The demonstration of *Ethics* II, Proposition XXXVIII, begins by speaking of a common property as something which can only be *conceived* adequately. After the demonstration introduces the idea of the affections of the human body, however, it concludes that the common property is necessarily *perceived* adequately.

Immediately after the demonstration of Proposition XIII of Part II, Spinoza states as a corollary that "the human body exists as we sense it (*sentimus*)". He then goes on to give a brief outline of the common properties of extension and concludes with six postulates concerning the composition of the human body and its interaction with other bodies. These postulates, Spinoza says, and the outline of the common properties of extension upon which they depend, rest upon our sensing of the existence of the human body. "No postulate which I have assumed contains anything which is not confirmed by an experience (*experientia*) that we can not mistrust, after we have proved the existence of the human body as we sense it" (*Ethics* II, Proposition XVII, Scholium).

We have sensory experience not only of the common properties of the body but also of the human mind. An affect is a *pathema* (General

Definition of the Affects), a sensory experience or imagination (*Ethics* IV, Proposition IX, Demonstration). Spinoza states that his analysis of the affects is given in general terms, through "the common properties of the mind and the affects" (*Ethics* III, Proposition LVI, Scholium).

The common properties do not constitute the essence of any individual thing (*Ethics* II, Proposition XXXVII), and hence their ideas can not, by themselves, give us adequate knowledge of the essence of any individual thing. Yet sensory experience is superior to reason in this respect. In a well known passage, Spinoza says that we sense not only the existence but also the essence of the human body and mind *sub specie aeternitatis*. "We feel and know by experience (*sentimus experimurque*) that we are eternal. For the mind is no less sensible (*sensit*) of those things which it conceives through intelligence than of those which it remembers ..." (*Ethics* V, Proposition XXIII, Scholium). This is further attested by the "common opinion of men ... (who) are indeed conscious of the eternity of their minds, but they confound it with duration . . ." (*Ethics* V, Proposition XXXIV, Scholium).

In summary, then, Spinoza does not separate scientific reasoning hermetically from common experience. Sensory experience might be compared to an unrefined melt out of which are purified and extracted the clear, distinct, adequate ideas which are the foundations of scientific reasoning. Since the common properties of extension and thought are present in all experience, no particular event or experiment, and no limited set of experiments, can be specified as necessary for the discovery of the common properties. Why, then, has it been so difficult for natural philosophy to reach the sure path of science?

The answer is that "although human bodies agree in many things, they differ in more, and therefore that which to one person is good will appear to another evil, that which to one is well arranged to another is confused . . ." (*Ethics* I, Appendix). Spinoza follows the philosophical tradition in ascribing to sensory experience a double character: the intermittent and vagrant experience of which he speaks in *Ethics* II, Proposition XL, Scholium 2, and the unwavering sense of what is common to the human body and external bodies. This double character makes it easy for careless minds to confuse eternity with duration, necessary causes with final causes, and quantitative extension with qualities and substantial forms. It was the example of mathematics that helped to rescue the foundations of science from these confusions. "In addition to mathematics, other causes also might be assigned, which it is superfluous here to enumerate" (*Ethics* I, Appendix), but he adds tantalizingly, he has set this subject, along with others, aside for another treatise (*Ethics* II, Proposition XL, Scholium 1).

One storm differs from another in innumerable ways. There are equally many differences between any two instances of fear, or anger, or love. Any emotion as it actually occurs is the response of a specific and peculiar individual to distinctive causes. Given a basic agreement in attribute, finite things limit one another through their differences (*Ethics* I, Definition 2), and "an effect differs from its cause precisely in that which it has from its cause" (*Ethics* I, Proposition XVII, Scholium). The absolute or radial causality of active nature does not, of itself, specify a given number of distinct individuals, nor does it determine their distinctive differences. This is the function of the axis of topographic causality. Every existing individual thing exists and acts in certain determinate ways because it is a member of a system of finite modes, each limiting the others by its differences from them, while at the same time all together are systematically coordinated and interlocked (*cordinationem et concatenationem* in *Tractatus Theologico-Politicus*, Chapter 4; Gebhardt III, pp. 57—68). Environmental causes may help or hinder any finite mode in its *conatus*, its activity of self maintenance. They may also exclude or destroy it. The major ways in which anything acts to maintain itself under the positive and negative pressures from its surroundings are sometimes called by Spinoza the 'actual' or 'given' essence (*Ethics* III, Proposition VII; and IV, Proposition IV).

Can we reach adequate knowledge of the immense variety and difference in nature? It would seem that we can not. Spinoza makes frequent remarks on the vast variety of natural phenomena of which we are almost totally ignorant.[7] More important, our bodies are only parts of larger natural processes, and we can know the ways in which external bodies differ from ours only through a fragmentary part of the total process — that part, namely, which is our body's response. The idea of that fragment must itself be a fractional portion of a larger whole. Spinoza calls such ideas of affections of the body *imaginations*. If such an idea is taken as representing the presence of the larger whole of which it is but a detached segment, the idea is inadequate. Inadequate ideas follow with the same necessity as adequate ideas (*Ethics* III, Proposition XXXVI), and it is impossible that all our ideas should be adequate (*Ethics* IV, Proposition IV). Clearly, Spinoza recognizes that for the philosopher of nature, the scientist who wishes to understand the permanent patterns of difference and the determinate forms of natural change, the model of mathematics, while it has an important value, is seriously deficient. Inadequate ideas "are like conclusions without premises" (*Ethics* II, Proposition XXVIII, Demonstration), and they are unavoidable.

However, Spinoza does not leave it at that. There are methods by which the scientist can amend inadequate ideas. At *Ethics* II, Proposition XXIX, Scholium, Spinoza contrasts the external with the internal determination of the mind. So long as the mind considers things only in the order in which they happen in everyday life, as if they turn up haphazardly or by coincidence, our ideas must be confused. However, if the mind is internally determined, itself initiating the consideration of several things at once, actively trying to understand their agreements, differences, and oppositions (*convenientias, differentias, et oppugnantias*),[8] it then contemplates things clearly and distinctly.

It is the positive content of our inadequate ideas that is the basis for several methods to which Spinoza alludes. It must be repeated here, however, that Spinoza's paramount interest in human salvation and blessedness diverted him from a focussed and detailed treatment of scientific method. In the *Ethics*, in particular, he was dealing with the agreement of things in their common properties rather than with their differences and oppositions. If we consider the positive content of even an inadequate idea it can not be said to be false (*Ethics* II, Proposition XXXIII). ". . . These imaginations of the mind, regarded by themselves, contain no error, and . . . the mind is not in error because it imagines . . ." (*Ethics* II, Proposition XVII, Scholium). This is a point he makes repeatedly (see *Ethics* II, Scholia to Propositions XXXV and XLIX; and IV, Proposition I). Optical illusions, dreams, and madness are not errors. They are simply ways — through mechanisms of which we are largely ignorant — in which our bodies respond positively to certain environmental conditions. The error in such cases arises from our ignorance of the true causes, and our urgent desire to fix upon some cause or other (*Ethics* I, Appendix; II, Proposition XXXV, Scholium; III, Proposition II, Scholium). Hence when we discover the actual cause the error is dissipated but the imaginations (illusions, dreams, etc.) remain (*Ethics* II, Proposition XXXV, Scholium; IV, Proposition I, Scholium). In fact, sensory imagination expresses the positive reactive power of the body and idea of the affection of the body (*Ethics* II, Proposition XVII, Scholium).

Although Spinoza will not call the inadequate ideas of the imagination false, neither will he call them true. This is a well known crux, much mooted by the commentators. To be consistent, it is argued, Spinoza ought to have called the ideas of the affections of the body true. After all, the idea and its physical correlate are one and the same in substance (*Ethics* II, Proposition VII, Scholium) and "a true idea must agree with that of which it is the idea" (*Ethics* I, Axiom 6). The objection overlooks two important points. First, the statement that a true idea must agree with its *ideatum* is not a definition of truth but an axiom. It states a necessary condition — *truth* refers to the extrinsic relation of thought to extension, whereas *adequacy* refers to the intrinsic relation of ideas to other ideas (*Ethics* II, Definition 4, Explanation; *Letter LX* to von Tschirnhaus, Gebhardt IV, pp. 270—271). Second, the objection disregards the close connection Spinoza sees between the identity of idea and extended correlate and the identity of their sequence of causes. The demonstration of *Ethics* II, Proposition VII rests explicitly on Axiom 4 of Part I that "The knowledge of an effect depends upon and involves the knowledge of the cause". *Ethics* II, Proposition VII, Scholium emphasizes that "whether we think of nature under the attribute of extension, or under the attribute of thought . . . we shall discover one and the same order, or one and the same connection of causes . . .". This is the additional sufficient condition of truth. "A true idea involves the highest certitude; to have a true idea signifying just this, to know a thing perfectly and as well as possible (*perfecte et optime*)" (*Ethics* II, Proposition XLIII, Scholium). A true idea must be perfect, that is to say, complete, including the idea of its cause. Spinoza had made the same point in the *Tractatus de Intellectus Emendatione*. A true idea "shows how and why something is or has been made . . . . This conclusion is identical with the saying of the ancients, that true science proceeds from cause to effect" (*Tractatus de Intellectus Emendatione*, Spinoza, 1930b, p. 34; Gebhardt II, p. 32). Without such knowledge of the cause the bare assertion that Peter exists "is false or if you choose (*si mavis*) not true" (*Tractatus de Intellectus Emendatione*, Spinoza, 1930b, p. 26; Gebhardt II, p. 26) whether or not Peter does in fact exist.

Although, then, Spinoza hesitates in the *Tractatus de Intellectus Emendatione* over whether incomplete ideas may not be called false, in the *Ethics* he is clear. The truth value which he ascribed to the inadequate ideas of the imagination is neither truth nor falsehood, but possibility.

I call these individual things (whose essence neither posits nor excludes their existence) possible, in so far as we are ignorant, whilst we attend to the causes from which they must be produced, whether these causes are determined to the production of these things (*Ethics* IV, Definition 4).

This designation of possibility as a third truth[9] value is quite deliberate, as can be seen from a passage in chapter 4 of the *Tractatus Theologico-Politicus*.

The actual system and interconnexion of things (*coordinationem et concatenationem*), i.e., the way in which things are really ordered and interconnected, is quite unknown to us; so for practical purposes it is better, indeed necessary, to regard things as possible (Spinoza, 1883, p. 58; Gebhardt II, p. 58).

This coordination and concatenation of determinate causes is, Spinoza says, a common property of individual things (*Ethics* II, Proposition XXXI, Demonstration). It is through their agreement in this common property of determinate causation that we can have clear and distinct ideas of the differences and oppositions of individual things, and of the necessary possibilities of the ideas of individual things and their affections.

It is important to notice that the distinction between contingency and possibility (Definitions 3 and 4 of *Ethics* IV) refers to the difference between the externally determined happenstance order of everyday life (*Ethics* II, Proposition XXXI) and the internally determined order in which we understand the existence and action of every finite mode to be necessarily embedded within an "infinite nexus of causes" (*Ethics* V, Proposition VI, Demonstration). To call something possible rather than contingent is to indicate that "in attending to the causes from which it must be produced" we have an adequate and scientific idea of that thing. At the same time, however, we have an inadequate idea of the particular existence and action of the possible thing, because "we are ignorant . . . whether these causes are determined to the production of (this thing)" (*Ethics* IV, Definition 4).

From the fact that the imagination, considered in itself, is not in error Spinoza drew an important methodological principle. It may be called *the principle of detachment*. "The power of the mind over the affects consists . . . in the separation by the mind of the imaginations from the thought of an external cause, which we imagine confusedly (*quam confuse imaginamur*)" (*Ethics* V, Proposition XX, Scholium; see also *Ethics* V, Proposition II). Since, as previously noted, the *Ethics* is primarily concerned with salvation and blessedness, Spinoza's concern is with the detachment of the affects, but he has already made it quite explicit that the "imagination is an affect" (*Ethics* IV, Proposition IX, Demonstration). The mind must take the initiative in detaching the imagination from the contingent chronological order of everyday life. We must assert not that we see a winged horse, but that a particular visual imagination is possible, as the necessary effect of causes which remain to be investigated. If we assert, not that we feel the painful shoulder, but that we feel pain in the area of the shoulder, investigation of the bodily correlate of that imagination may discover it to be a possible effect of a myocardial infarction.

It is a corollary of the principle of detachment that an exact description of the possible imagination is a prerequisite to the investigation of the necessary causes. In the opening chapter of the *Tractatus Theologico-Politicus*, Spinoza wrote,

We are not now inquiring into the causes of prophetic knowledge. We are only attempting ... to examine the Scriptural documents, and to draw our conclusions from them as from ultimate natural facts; the causes of the documents do not concern us (Spinoza, 1883, p. 25; Gebhardt III, p. 28).

In Chapter 7, after pointing out that it is most important to reach the clearest possible understanding of Scriptural passages, he wrote,

We are at work not on the truth of the passages, but solely on their meaning .... In order not to confound the meaning of a passage with its truth, we must examine it solely by means of the signification of the words, or by a reason acknowledging no foundation but Scripture (Spinoza, 1883, p. 101; Gebhardt III, p. 100).

It is the same principle, I believe, that led Spinoza to write to Oldenburg,

If of every liquid there were an account given as accurately as possible with the highest trustworthiness, I should consider it of the greatest service for the understanding of the special features which differentiate them: which is to be most earnestly desired by all Philosophers as something very necessary (*Letter VI*, Spinoza, 1928, p. 97; Gebhardt IV, p. 34).

Although Spinoza was not speaking directly of the imagination in his letter to Oldenburg, he was speaking of phenomena which, in the order of everyday life, are represented by the imagination.

Another methodological principle, which I will call *the principle of hypothetical explanation*, is closely related to the principle of detach-

ment. There are three parts to the principle of hypothetical explanation.
(a) There are alternative explanatory accounts of the particular causes,
or kinds of causes, which will determine a possible thing to exist and
act. (b) One or more of these explanatory accounts may be confirmed
to a certain limited extent, but here demonstrative certainty is impossible for the human mind. (c) Empirical evidence can weaken or
entirely destroy belief in the correctness of any explanatory account,
and the discovery of such empirical evidence can never be demonstrated to be impossible.

I expect that some readers will be sceptical that such a methodological principle, or triad of principles, is to be found in Spinoza. I
must, therefore, show that Spinoza argues for each of (a), (b), and (c),
and that he does in fact apply the principle of hypothetical explanation
to specific problems in the *Ethics*, the *Tractatus de Intellectus Emendatione*, the *Tractatus Theologico-Politicus*, and the correspondence with
Oldenburg.

According to Spinoza, we have an adequate idea of one aspect of
any thing we call possible, and an inadequate idea of another aspect of
that same thing. Since we recognize that we share with it the common
property of concatenation within a necessary causal nexus of unlimited
extent, we have an adequate idea of the thing. Since at the same time we
do not know which of several causes actually determines the existence
and characteristic actions of that thing, several alternative explanations
are open to us and, other things being equal, we will fluctuate and
hesitate in our belief (*Ethics* II, Proposition XLIV, Scholium; see also
III, Proposition XVII, Scholium). In short, we will doubt (*Ethics* III,
Proposition XVII, Scholium). As certainty is a mark of an idea that is
true and adequate (*Ethics* II, Proposition XLIII), doubt is a mark of an
inadequate idea. We know that one or more of the possible causes is
necessary, and so our idea is not false. But we do not know which of
the causes is actual and determinative, and hence our idea is not true. It
is inadequate. Indeed, Spinoza says that we *imagine* the particular
causes that posit the existence of a possible thing (*Ethics* IV, Proposition XII, Demonstration). We perceive these causes through the affections of the human body — that is to say, through our senses.

An inadequate idea can not be demonstrated to be true. That would
be a self-contradiction, since the inadequate idea would then be demonstrated to be adequate. An inadequate idea is not like a conclusion

which *happens* to lack premisses (*Ethics* II, Proposition XXVIII, Demonstration). In the human mind it *necessarily* lacks premisses, although those premisses are necessarily present in a perfect and detailed idea of the infinite whole of nature. Specific causal explanations of things that are possible are hypotheses, and that is what Spinoza calls them. In an important note in the *Tractatus de Intellectus Emendatione*, Spinoza wrote,

The same must be understood of hypotheses which are made to explain certain movements which are in harmony with heavenly phenomena, save that if these are applied to celestial movements, we conclude from them the nature of the heavens, which, however, can be quite different, especially as for the explanation of such movements many other causes can be conceived.[10]

In his *Letter XIII* to Oldenburg Spinoza referred to his explanation of the redintegration of nitre as "my hypothesis" (Spinoza, 1928, p. 125; Gebhardt IV, p. 65).

Inadequate ideas can be strengthened by additional experience, and our belief can thus be confirmed to the point where doubt and fluctuation cease. Of course, for Spinoza the absence of doubt is by no means the same as certainty (*Ethics* II, Proposition XLIX, Scholium). An aggregation of similar imaginations and inadequate ideas will fuse into a universal, but still inadequate, idea which Spinoza calls vagrant experience (*Ethics* II, Proposition XL, Scholia 1 and 2). It is through such experiential induction that we come to believe that oil feeds fire and that water extinguishes it. Such inductions can not yield certainty, and we are "like one who, having never seen any sheep except with short tails, is surprized at the sheep from Morocco which have long ones" (Spinoza, 1910a, Part II, Chapter 3; Gebhardt I, p. 56). In Chapter 7 of the *Tractatus Theologico-Politicus*, Spinoza is at pains to detail the "difficulties and shortcomings, which prevent our gaining a complete and certain (*certam*) knowledge of the Sacred Text" (Spinoza, 1883, p. 108; Gebhardt II, p. 106). In *Letter VI* to Oldenburg Spinoza recommends his hypothesis as the simplest (Spinoza, 1928, p. 86; Gebhardt IV, p. 17) and very easy (*Letter XIII*; Spinoza 1928, p. 125; Gebhardt IV, p. 65), but he does not claim that his experiments establish it. He says repeatedly, in both his letters on Boyle's work, that his own experiments confirm his hypothesis only "to some extent" (Spinoza, 1928, pp. 86, 126, 127; Gebhardt IV, pp. 17, 66, 67).

Our acceptance or belief in an hypothesis may be put into doubt by experience. It may be weakened to the point where it is destroyed. But adequate ideas cannot, by themselves, destroy a belief in an hypothetical causal explanation of a possible thing. Sense experience is essential. Spinoza's reasons are central to his philosophy. From an adequate idea and its adequate consequents we can, according to Spinoza, demonstrate the falsehood of such inadequate and false ideas as teleological explanation, free will, and other similar prejudices (*Ethics* I, Appendix). But that is because inadequate ideas of that superstitious kind *contradict* the adequate idea of the common property of necessary and nonteleological causation. On the other hand, the definition of a possible (*Ethics* IV, Definition 4) is formulated so as to bring to the fore that each of the alternative causes, and hence each of the alternative causal explanations, is entirely *consistent* with scientific causal necessity. Hence the common and adequate notions which are the foundations of science can not, by themselves, demonstrate the inconsistency or falsehood of a properly formulated explanation of causes. "An affect cannot be restrained nor removed unless by an opposed and stronger affect" (*Ethics* IV, Proposition VII). What Spinoza demonstrates concerning the affects in the first seven propositions of Part IV of the *Ethics* clearly hold also for all sensory and inadequate ideas which are consistent with the common and adequate notions of science. ". . . The human mind will contemplate (any) external body as actually existing or as present, until the human body be affected by an affect which excludes the existence or presence of the external body" (*Ethics* II, Proposition XVII). In other words, a particular imagination will endure indefinitely (*Ethics* II, Propositions XXX and XXXI), until it is destroyed and replaced by some other imagination. The same holds for general beliefs based upon inductive generalizations from particular experiences. Belief in an hypothetical causal explanation, since it is an inadequate idea, can not be weakened or eliminated except by some other belief based upon sensory experience.

It should be noted that I take 'belief' and 'believe' to come closer to Spinoza's word '*opinionem*' (*Ethics* II, Proposition XL, Scholium 2) than 'opinion' and 'opine'. For Spinoza it would be a contradiction to say, "I believe that *p* is true". A belief, however firmly held, can be put in doubt by future experience. A true idea can not possibly be doubted because it is known for certain to be complete (*Ethics* II, Proposition XLIII and Scholium). To believe that *p*, for Spinoza, is to accept *p* as more powerful in influencing our actions, whether of mind or body, than any present alternative belief.

Spinoza's use of the principle of hypothetical explanation is well exemplified in the Oldenburg correspondence, as well as in the *Ethics*. A brief summary of the methodological criticism of Boyle will show his use of the principle. Both men agree that the grounding concepts for the explanation of chemical phenomena are what Spinoza (1928, p. 93; Gebhardt IV, p. 28) calls notions — motion, rest, and the basic principles of mechanics. Their disagreement is on the particular causal explanations. While Boyle attempts to explain the phenomena of nitre on the hypothesis that nitre is a heterogeneous body, Spinoza maintains that the hypothesis that it is homogeneous is simpler, and better confirmed by experimental results. Nitre is clearly what Spinoza will later, in the *Ethics*, call a *possible* thing.

Spinoza is surprised that Boyle should think it possible or necessary to establish through special experiments that motion, rest, and their laws are the fundamental explanatory concepts for chemical phenomena. Special experiments are superfluous because the primary sensory experience of motion and rest is always present. Hence Spinoza (p. 91 and *passim*; IV, p. 25) mentions numerous examples of everyday observations which show sufficiently that the secondary qualities depend on motion and rest. However, no amount of observation can produce the certainty and conviction of a demonstration in the mathematical manner of the necessity of the fundamental concepts of natural science (p. 91; IV, p. 25). Because he is not mathematically rigorous Boyle sometimes confuses physical concepts like motion and rest with everyday concepts like visible, invisible, hot, and cold (p. 93; IV, p. 28). Likewise, he sometimes confusedly "seeks the cause in the purpose" (p. 96; IV, p. 32).

Spinoza writes repeatedly that his experiments confirm his hypothesis only "to a certain extent". While Boyle's experiments are excellent, they do not *demonstrate* his hypothesis of heterogeneity, and there are other common experiments which are not at all captured[11] by Boyle's proposed explanation (pp. 127—128; IV, p. 67). Spinoza recognized that experimentation could strengthen Boyle's hypothesis and weaken his own, but for this "yet another experiment is required by which it would be shown that the Spirit of Nitre is not really Nitre, and without salt of lye cannot be reduced to a state of coagulation, or be crystallized . . ." (p. 85; IV, pp. 16—17). Boyle seems to think, says Spinoza, that in his experiments he knows what nature contributes, and hence he thinks he knows with certainty what caused the heat observed in his experiment (p. 128; IV, p. 67). Spinoza is more sceptical than the sceptical chemist. What grounds did Boyle have, he asks, for thinking he knew *all* the determining factors? In his later terminology, our knowledge of specific causes is inadequate.

In the *Ethics*, at Part II, Proposition XVII and its Scholium, Spinoza again employs the principle of hypothetical explanation. In the corollary to that proposition Spinoza proposes an explanation of the physical mechanism correlative to the inadequate ideas of imagination and memory. This physical mechanism is, of course, a prime instance of an affection of the human body which is determined jointly by the human body and by other and different bodies. Of the imagination and memory Spinoza writes

This may indeed be produced by other causes, but I am satisfied with having here shown one cause through which I could explain it, just as if I had explained it through the true cause. I do not believe [12] however, that I am far from the truth, since no postulate which I have assumed contains anything which is not confirmed by an experience that we cannot mistrust after we have proved the existence of the human body as we sense it.

Although Spinoza does not use the word 'hypothesis' in this passage, it is clear that he recognizes that there are alternative explanations, that any scientific explanation must be in terms of the common properties which we both conceive adequately and perceive experientially, and that he does not claim demonstrative truth for the hypothesis he offers. He states only that he *believes* it to be the most acceptable explanation.

In addition to the principles of detachment and hypothetical explanation, a third methodological principle — I will call it *the modeling principle* — is developed by Spinoza in the Preface to Part IV of the *Ethics*. The principle is an explanation and generalization of the notion of a model (*exemplar*) of human perfection, presented earlier in the *Tractatus de Intellectus Emendatione* (Spinoza, 1930b, p. 5; Gebhardt

II, pp. 8—9). It is intended to apply to any particular class of things determined in their existence and action by determinate causes, and studied scientifically as sharing common properties. The principle may be formulated as follows. In order to make comparisons, identify differences, and order these differences serially, an abstract model may be constructed. *Entia rationis* and fictions are necessary for the construction and use of such a model. The model is a tool or instrument to which nothing positive in nature corresponds, and it comprises only that finite set of ranges in which comparisons of difference are to be made.

In the Preface to *Ethics* IV, Spinoza tries to show that something of value for human freedom and science can be rescued from the confused teleological thinking of everyday life. Ordinarily people use the notion of perfection in two contexts. First, in the processes of construction of some particular thing, the finished and completed product is called *perfect* if it fulfills the intention of the artisan or his employer. The product is imperfect when it is incomplete, before it is finished or after it begins to disintegrate. Second, this particular notion of perfection is generalized as people set up exemplars or types of perfection for various classes of products, each person according to his or her imagination and emotional preference. These notions of perfection and imperfection are then ascribed to natural processes in which human skill plays no part.

Spinoza does not propose to drop entirely the notion of perfection, as it applies to a type or exemplar of this or that classification of natural things. Granted, such an abstract model is only a mode of thought; nevertheless it is indispensable. Our minds are finite and our ignorance infinite. For both practical and theoretical purposes we must bring things together, compare them, identify their similarities and differences, and order these differences in some scale or ratio. Spinoza proposes a surprising paradox. The transcendental, *ens* (a being, entity), which he had earlier explained as formed through a fusion of all things (*Ethics* II, Proposition XL, Scholium 1), may be put into reverse to expose the fissures. If we take *ens* as our model we can identify perfection with reality, since to be a distinct entity is to act in affirming self-identity through a variety of circumstances. Active nature (*natura naturans*) is absolutely perfect and real, since it is self-caused (*Ethics* I,

Definition 1) and acts in an absolutely infinite variety of ways. Each specific genus we abstract — the examples cited at *Ethics* III, Proposition LVII, Scholium, are insects, fish, birds, and man — will have its place on the scale of reality or perfection according to its ability to maintain itself through action in some limited variety of situations. Indeed, at *Ethics* II, Proposition XIII, Scholium, Spinoza had already said that the degrees of perfection and reality are also grades (*gradibus*) of animation. The variety of kinds of interaction into which any thing can enter determines its degree of *perceptual* clarity or obscurity. The more autonomous a thing is in its actions, the more distinct is its *conceptual* understanding. Thus instead of a confusion of teleological archetypes based on personal bias Spinoza proposes a non-teleological transcendental model which generates an orderly series of graded models.

Further, the same general principle enables Spinoza to grade the individuals within the range of each exemplary model. So one individual may be more or less free or autonomous (*Ethics* IV, Propositions XX—XXXV), or more or less interactive (*Ethics* IV, Proposition XXXIX). Anything which we know to help an individual to come closer to its model we call good, and what hinders we call bad.

However, the models and their ancillary terms like perfection, degrees of reality, good and bad, are *entia rationis*, modes of thought to which nothing positive in things corresponds. In the *Cogitata Metaphysica* I.1 Spinoza had distinguished between fictions (*ficta*) and abstractions (*entia rationis*). A fiction is an actual entity to which we ascribe, in imagination, some counterfactual state. Examples of fictions are imagining that the sun circles the earth, that a candle burns in a vacuum, or that a horse has wings. *Entia rationis* are class concepts, formed by the intellect in order to bring actual things together for comparison of their similarities and differences. Examples of *entia rationis* are genus, species, time, and mathematical concepts — figure, number,[13] and measure. We need be in no fear of forming such constructions,[14] provided we first have a true and adequate idea of the first principles of science. The intellect may then freely and knowingly construct fictional entities and abstractions as aids, tools, or instruments to push its investigations further (*Cogitata Metaphysica* I.1; and *Tractatus de Intellectus Emendatione*, Spinoza 1930b, p. 11; Gebhardt II, pp. 13—14). So Spinoza points out Boyle did not appreciate the importance of exact and careful measurement, for he failed "to inquire whether the quantity of fixed salt which remains in the crucible is always found to the same from the same quantity of Nitre, and whether it is proportionate when there is more Nitre" (*Letter VI*, Spinoza, 1928, p. 85; Gebhardt IV, pp. 16—17). So too in politics Spinoza constructs abstract models of political organization (*Tractatus Politicus*, Chapters 6ff; Gebhardt III, p. 297ff). The various constitutions are ordered in accordance with the modeling principle — each exemplar must be able to maintain itself in a variety of circumstances, and they may be graded according to stability in the greatest variety of situations. But "to have shown what ought to be done is not enough; the main problem is to show how it can be done, i.e. how men, even when led by passion, may still have fixed and stable laws" (*Tractatus Politicus*, Spinoza, 1958, p. 335; Gebhardt III, p. 308). Spinoza believed that through a model of human nature the common passions of human beings could be better analyzed, and through the model constitutions it could be seen how these ordinary and determinate passions could be harnessed to the maintenance of a stable state. As his countrymen had studied the weather and its stormy winds, and found ways to make them turn the mill wheels, so Spinoza wishes to study the passions dispassionately, and to re-direct them so that they support social equilibrium and peace.

University of Toronto

NOTES

[1] *Philosophy* includes science, of course, in the usage of the time.

[2] See *Letter XXXIX* to Jelles, Gebhardt IV, pp. 193—195.

[3] Pfeiffer (1941, p. 46) writes, "The two founders of modern Biblical criticism are the Jewish philosopher, Baruch Spinoza, and Richard Simon, a French priest". Simon (1678) acknowledges the priority and importance of Spinoza's hermeneutic method.

[4] Gueroult's masterly work (1974) is particularly enlightening on *Ethics* II, Proposition XXXIX and the common notions. White's (Spinoza, 1930a) translation of *Ethics* II, Proposition XXXIX is misleading.

[5] In the *Tractatus de Intellectus Emendatione* (Spinoza, 1930b, p. 32; Gebhardt II, p. 31), Spinoza identifies the common sense with imagination.

[6] However, we *conceive* extension as an attribute expressing the activity of substance or nature (*Ethics* I, Propositon X, Demonstration).

[7] *Tractatus Theologico-Politicus*, Chapters 4 and 16 (Gebhardt III, pp. 57—68, 189—200); *Tractatus Politicus*, Chapter 2 (Gebhardt III, pp. 276—284); *Letter XXX* and *XXXII* (Gebhardt IV, p. 166 and 169—176); and *Ethics* I, Appendix and Part III, Proposition II, Scholium.

[8] Cf. *Tractatus de Intellectus Emendatione*, Gebhardt II, p. 12 (Spinoza, 1930b, p. 9). Gueroult (1974) believes that Spinoza is here indebted to the Stoics. I would suggest that a more immediate influence is Bacon (*Nov. Org.* II.11, on the inductive tables of Presence, Absence, and Degree).

[9] Spinoza is speaking of things, although it is ideas that are true (or false). However, "the order and connection of ideas is the same as the order and connection of things" (*Ethics* II, Proposition VII).

[10] This note is omitted in the Wild edition. I quote the Boyle translation (Spinoza, 1910b, pp. 244—245; Gebhardt II, p. 22).

[11] Spinoza wrote '*evincitur*', not '*sequitur*'.

[12] Spinoza wrote '*credo*', not '*cognitio*', since he is speaking of an hypothesis and not of an idea which the human mind can *know* to be true.

[13] "We imagine time because we imagine some bodies to move with a velocity less, or greater than, or equal to that of others" (*Ethics* II, Proposition XLIV, Scholium). As to numbers, *Letter L* (Gebhardt IV, pp. 238—241) makes clear that Spinoza considers them class concepts.

[14] "*Nullo ergo modo timendum erit, nos aliquid fingere*" (*Tractatus de Intellectus Emendatione*, Spinoza, 1930b, p. 23; Gebhardt II, p. 24).

REFERENCES

Geuroult, Martial: 1974, *Spinoza II: L'âme* (*Ethique, II*), Aubier-Montaigne, Paris.

Pfeiffer, R. H.: 1941, *Introduction to the New Testament*, Harper, New York.

Simon, Richard: 1678: *Histoire critique du Vieux Testament*, Billaine, Paris.

Spinoza, Baruch: 1883, *Tractatus Theologico-Politicus*, transl. by R. H. M. Elwes, Bohn edition, George Bell and Sons, London.

Spinoza, Baruch: 1910a, *Short Treatise on God, Man, and His Well-Being*, transl. and ed. by A. Wolf, A. and C. Black, London.

Spinoza, Baruch: 1910b, *Tractatus de Intellectus Emendatione*, transl. by A. Boyle, Everyman edition, Dent, London.

Spinoza, Baruch: 1925, *Spinoza Opera*, ed. by Carl Gebhardt, 4 vols., Carl Winter, Heidelberg. (Cited as 'Gebhardt' in the text.)

Spinoza, Baruch: 1928, *The Correspondence of Spinoza*, transl. By A. Wolf, Allen and Unwin, London.

Spinoza, Baruch: 1930a, *Ethics*, transl. by William Hale White, in John Wild (ed.), *Spinoza: Selections*, Charles Scribner's Sons, New York.

Spinoza, Baruch: 1930b: *Tractatus de Intellectus Emendatione*, transl. by R. H. M. Elwes, in John Wild (ed.), *Spinoza: Selections*, Charles Scribner's Sons, New York.
Spinoza, Baruch: 1958, *Tractatus Politicus*, transl. by A. G. Wernham, Clarendon, Oxford.
Spinoza, Baruch: 1963, *Earlier Philosophical Writings; The Cartesian Principles and Thoughts on Metaphysics*, transl. by Frank A. Hayes, Bobbs-Merrill, Indianapolis.

ALEXANDRE MATHERON

SPINOZA AND EUCLIDEAN ARITHMETIC:
THE EXAMPLE OF THE FOURTH PROPORTIONAL

Spinoza's example of the fourth proportional has often been mistreated: it has generally been considered either trivial or ill-chosen. My aim in this article is to show that it is perfectly germane. To do so it is enough to take seriouly his reference to Euclid's *Elements* on the assumption that Spinoza was both aware of its implications, and addressed himself to readers whom he expected to be equally aware; we shall then see that the example illustrates with extreme precision the distinction he establishes in the *Tractatus de Intellectus Emendatione* between the last two 'modes' of knowledge, if we grant that these are identical with the last two 'genres' of knowledge in the *Ethics*. Since the texts of the *Tractatus de Intellectus Emendatione* are, at one and the same time, those which are most in need of elucidation as well as those which best elucidate that example, we shall concentrate principally on that work, saving the *Ethics* for the end.

Knowledge of the third mode is defined in the *Tractatus de Intellectus Emendatione* as that "in which the essence of one thing is inferred from another thing, but not adequately" (Gebhardt II, p. 10). Two cases are then distinguished: the first, where one infers a cause from its effect and the second, in which a conclusion is drawn "from a certain universal which is always accompanied by a certain property" (II, p. 10). Since I have tried to show elsewhere that the first case is a particular form of the second, I shall confine myself to the latter. Knowledge of the fourth mode is knowledge in which "a thing is known by its essence alone or by its proximate cause" (II, p. 10). This means both that the essence of a thing is known here by itself or by its immanent proximate cause and that the properties of this thing are known by virtue of the essence of the thing alone or by the conjunction of that essence and an exterior proximate cause.

These two modes of knowledge — as well as the first two — can serve equally well for discovering the fourth proportional to three given numbers. Let us take the example of the three numbers 161, 9, 913 and 931.

125

Marjorie Grene and Debra Nails (eds.), Spinoza and the Sciences, 125—150.
© 1986 *by D. Reidel Publishing Company.*

1. The first three modes, in this very case, are alike in that they proceed by application of the 'Rule of Three'. But they differ in the way the validity of this rule was previously discovered. The mathematicians who practice the third mode know this validity 'through the force of Proposition 19 of Euclid, Book VII' (II, p. 12). The discovery, by his third mode, of the fourth proportional to our three given numbers comes about in two phases: first, one demonstrates, if that has not already been done long since, Proposition 19, and then one applies its conclusion to the particular case of the numbers 161, 9,913 and 931 by dividing the product of the last two by the first.

1.1. Spinoza presents this first phase, namely the demonstration which establishes the '*force*' of Proposition 19 in a form directly copied from that of the definition of the second kind of knowledge of the third mode given above. It consists in establishing that a certain universal U, the "nature of the proportion", is always accompanied by a certain property P: "its property consisting in the fact that the product of the first and the fourth numbers is equal to the product of the second and the third" (II, p. 12). Each of these two terms merits detailed examination.

1.1.1. Let us look first at the universal U, "the nature of the proportion". Definition 20 of Euclid, Book VII, tells us what this consists in: "numbers are proportional when the first is the same multiple, or the same part or the same parts of the second that the third is of the fourth". The general term, here, is clearly "the same parts". But what, exactly, does it mean? No definition is given, but its meaning is clarified by the context and, in particular, by the proof of Proposition 4 ("Any number is either a part or parts of any number, the less of the greater.") In fact it implicitly results from this proof that, two numbers a and b being given, a phrase such as 'a is m parts of b' always means: 'a is equal to m times the greatest common measure of a and of b'.[1] As a result, if this greatest common measure is itself contained n times in b, the expression 'part of b', in the context of this comparison between a and b, is synonymous with 'n-th part of b'. In these conditions — and this will be very important for understanding Spinoza's text — it is possible to define in *itself* the relation of a to b (their logos, their ratio) by the number (m) and the nature of the 'parts' (n-ths) of b whose sum is equal to a, in the precise sense which has just been given to the word 'parts.' To say that the relation of a to b is m/n (m and n clearly being relatively prime) is equivalent to saying, by definition, that the greatest common measure of *a* and *b* is contained *m* times in *a* and *n* times in *b*. Definition 20 then becomes perfectly clear: the numbers *a, b, c* and *d* are proportional when the relation of *a* to *b*, in this precise sense of the word 'relation', is the same as the relation of *c* to *d*, and this means that if the greatest common measure of *a* and *b* (e.g., *e*) is contained *m* times in *a* and *n* times in *b*, the greatest common measure of *c* and *d* (e.g., *f*) is also contained *m* times in *c* and *n* times in *d*, and conversely.[2]

Such is the essence of proportion in general. Spinoza, as we can see, could perfectly well consider the definition given in Book VII as a genetic definition: just as the sphere is engendered by the rotation of the semicircle as by its proximate cause (*TdIE*, Gebhardt II, p. 27), so, too, the proportion (that is, the ensemble of its four terms *a, b, c, d*) is engendered by the application of a ratio *m/n* to two 'greatest common measures' *e* and *f* ($a = e \times m, b = e \times n, c = f \times m, d = f \times n$). This is only possible, or course, because the essence of the ratio can itself be defined independently of the definition of the proportion and prior to it. And, in the same manner, simply by making this universal definition concrete, we can conceive by its proximate cause, no longer only the essence of the proportion in general, but also the essence of each particular species of proportion and even the essence of each individual proportion. We can engender as many specific essences of proportions as there are determinate numerical values given to *m* and *n*, on the sole condition that we choose these (*m* and *n*) to be relatively prime. Moreover, within each species defined in this way by a particular ratio *m/n* we can engender as many individual essences of proportions as there are determinate numerical values given to *e* and *f*.

If we proceed in the inverse direction, how can we find the greatest common measure of two given numbers and thus find their ratio as well? Propositions 1 and 2 of Book VII explain how, by defining the operation often called *anthyphairesis*. When two numbers are given such that $a < b$, this consists in subtracting *a* from *b* as many times as is necessary (*i* times, for instance) in order to find a remainder *r* less than *a*, then to begin over again with *a* and *r, et cetera*, until there is no longer any remainder. In other words, *anthyphairesis* is the application of the schema below:

$$b = (a \times i) + r, r < a$$
$$a = (r \times j) + s, s < r$$
$$r = (s \times k) + t, t < s$$
$$s = (t \times l)$$

We can then easily prove that t is the greatest common measure of a and b. (For moving from bottom to top in our schema, we immediately see that t is a sub-multiple common to a and to b; and moving from top to bottom, we immediately see that every common sub-multiple of a and b is also common to b and $(a \times i)$, hence of r as well, and thus of a and $(r \times j)$, *et cetera* . . . and therefore of t.) This allows us, when we move from bottom to top, to calculate the two numbers m and n as functions of the quotients l, k, j and i, such that $a = t \times m$ and $b = t \times n$.

If a and b, instead of being integers, are magnitudes of whatever kind, lines, for example, the operation would retain its [*un*] sense, but it could well continue indefinitely, always leaving some remainder. In this case, as Euclid shows in Proposition 2 of Book X, a and b would be incommensurable. It would, however, still be possible to define the ratio of a to b, although the Euclidean books do not do so; but in that case we would have to reverse the perspective and define this ratio genetically by the series of quotients i, j, k, l . . . , whether this series is finite or infinite, instead of considering those quotients as simple means for calculating a previously defined ratio (similarly, it is necessary, according to Spinoza, to define the circle by the rotation of its radius instead of considering this rotation as a simple means for constructing a circle already defined elsewhere (*TdIE*, Gebhardt II, p. 35)). When this series is finite, we would drop back to the preceding case; when it is unlimited, on the contrary, the ratio of a to b would take the form of a continued fraction, perfectly intelligible in itself, but irreducible to the form m/n, with integral m and n.[3] This generalization starts to be sketched out in the seventeenth century, at least "in a practical state": Huygens, for example, uses continued fractions. What could Spinoza have thought of this? Nothing, perhaps, since for him by definition there are no other numbers than whole numbers. But he clearly knew of the existence of irrationals, even if he did not class them under the category of number. Even in the case of whole numbers, moreover, he was able to interest himself in researches tending to give a more genetic character to the concept of ratio. What is certain, at all events, is that the recourse to *anthyphairesis*, generalized or not, allowed him to say that a ratio, considered in itself and in abstraction from any relation to other ratios, *has an essence*.

Now this is by no means evident. Spinoza had to make a choice. Everyone knows, in fact, that there is another definition of proportion in Euclid: Definition 5 of Book V which can be applied just as it stands to incommensurable magnitudes since it allows us to avoid the difficulties raised by the passage *ad infinitum.* According to this definition, to say that *a* and *b* have the same ratio as *c* and *d* amounts to saying that, for every pair of whole numbers *m* and *n, ma > nb* if and only if *mc > nd, ma = nb* if and only if *mc = nd, ma < nb,* if and only if *mc < nd.* A generalization of 'genius', it is often rightly said. But, taken as an epistemological model it does not allow us to illustrate what Spinoza wants to illustrate. For in this case the proportion is not defined in terms of ratio; on the contrary, it is the ratio which is defined in terms of proportion (or, in terms of 'same ratio'): a ratio is defined only by its equality with other ratios. One could say, if one likes, that the ratio of *a* to *b* has as its 'essence' the set of all other ratios to which it is equal in the sense of Definition 5, but it is impossible to ask what it is *in itself*; the question lacks sense. According to Book VII, on the contrary, it does make sense: we can first determine that the ratios *a/b* and *c/d* are taken separately and then conclude from this that they are equal because their *anthyphaireses* are equal and because the numbers *m* and *n* are the same in both cases. But, of course, Book VII also allows us to arrive at that same conclusion by other routes and this is precisely what the property P is going to show.

1.1.2. This property P, which is deduced from the 'nature of the proportion' and is announced in Proposition 19 of Book VII, is the equality of the product of the extremes and the product of the means. If we trace Proposition 19 back to those on which its proof is based and then to those from which the latter are deduced, *et cetera . . .* , we get the following schema:

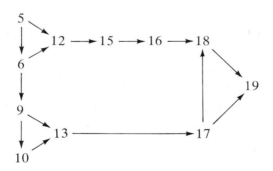

Let us summarize briefly (using symbols which are not in Euclid, notably '$a/b = c/d$' for 'a is to b as c is to d '), by beginning with the linear sequence leading from Propositions 5 and 6 to Proposition 16:

PROPOSITION 5. If a is the n-th 'part' of b and if c is the n-th 'part' of d, then $(a + c)$ is the n-th 'part' of $(b + d)$.

Proof. pair each of the n parts of b equal to a with one of the n parts of d equal to c; then there are n pairs obtained in this way, all the sums $(a + c)$ are equal to one another and their sum is equal to $(b + d)$; therefore, etc. . . .

PROPOSITION 6. If a is m n-th parts of b and c is m n-th parts of d, then $(a + c)$ is m n-th parts of $(b + d)$.

Proof. by hypothesis, the integers (a/m) and (b/n) are equal, and so the intergers (c/m) and (d/n) are equal; but, it follows from Proposition 5 that $(b/n) + (d/n) = (b + d)/n$ and that $(a/m) + (c/m) = (a + c)/m$; therefore, *etc*. . . .

PROPOSITION 12. If $a/b = c/d = e/f = \ldots$, then $a/b = (a + c + e + \ldots)/(b + d + f + \ldots)$. This is merely a rewriting of Propositions 5 and 6 in the language of proportions, by applying Definition 20; since Proposition 6 is itself only the generalization of Propositon 5 we can treat the group 5—6—12 as a single proposition.

PROPOSITION 15. If $1/c = b/d$, then $1/b = c/d$.

Proof. by hypothesis and Definition 20, $d = b \times c$; but it follows from Proposition 12 (when we replace all the numerators with 1, all the denominators with b and treat a number c as the fraction $1/b$,) that $1/b = c/(b \times c)$; therefore, *etc*. . . .

PROPOSITION 16. $b \times c = c \times b$.

Proof. According to Definition 20, $1/c = b/(b \times c)$; therefore, by Proposition 15 (when we replace d with $(b \times c)$), $1/b = c/(b \times c)$; but, according to Definition 20, $1/b = c/(c \times b)$; therefore, *etc*. . . .

The result of the first sequence is, therefore, to establish the commutativity of multiplication. Let us pass on to the second linear sequence, from Propositons 9 and 10 to Propositon 17:

PROPOSITION 9. If a is the n-th part ot b, c is the n-th part of d and a is the $p \cdot q$-th parts of b, then c is $p \cdot q$-th parts of d.

[*Proof.* It follows from Proposition 6 (when we replace m with p, n with q, c with a, b and d with c and we add the a's n times and the c's n times), that $(a \cdot n)$ is $p \cdot q$-th parts of $(c \cdot n)$; therefore, etc. . . .]

PROPOSITION 10. If a is $m \cdot n$-th parts of b, c is $m \cdot n$-th parts of d and a is $p \cdot q$-th parts of c, then b is $p \cdot q$-th parts of d.

[*Proof.* it follows from Proposition 9, when we make the appropriate substitutions, that a is 'the same parts' of c as $(a \cdot m)$ is of (c/m); therefore, (a/m) is $p \cdot q$-th parts of

(c/m); but, by hypothesis, (a/m) is the n-th part of b and (c/m) is the n-th part of d. Therefore, it follows from Proposition 9 that, etc. . . .]

PROPOSITION 13. If a/b, $= c/d$, then $a/c = b/d$. [This is simply a rewriting of Propositions 9 and 10 in the language of proportions, with the use of Definition 20. Since Proposition 10 is itself merely the generalization of Proposition 9, we can therefore treat the group 9—10—13 as forming a single proposition.]

PROPOSITION 17. $a/b = (a \cdot c)/(b \cdot c)$.
   [*Proof.* By Definition 20, $a/(a \cdot c) = b/(b \cdot c)$; therefore, it follows from Proposition 13 that, etc. . . .]

Finally, the two linear sequences meet:

PROPOSITION 18. $a/b = (c \cdot a)/(c \cdot b)$. This is obvious from Propositions 16 and 17.

PROPOSITION 19. If $a/b = c/d$, then $d \cdot a = c \cdot b$ and *vice versa*.
   [*Proof.* By Proposition 17, with the appropriate substitutions, $c/d = (c \cdot a)/(d \cdot a)$; therefore, etc. . . . The proof of the converse proceeds inversely, using the same two Proportions 17 and 18.]

If we grant that the two groups 5—6—12 and 9—10—13 form in reality just two proportions, the proof of Proposition 19 is based exclusively on only six anterior propositions. We could reduce that number even further, for instance, by observing that the commutativity of multiplication, established in Proposition 16, can be proved more easily; or that Proposition 15 is merely a particular case of Proposition 13. But let us leave this aside and take Euclid as he is. The whole deductive sequence just summarized can very readily be made 'intuitive' in either the Cartesian or the ordinary sense of the word. With a little training, anyone can grasp it with a single glance. But, even though it would not be intuitive in the Cartesian sense, it would be so at any rate in the Spinozistic sense — by means, it is true, of a precision which will be given later. For Proposition 19, connected with all those on which it is based, makes us know *solely by the essence* of proportion in general a property of proportion in general; and Definition 20 causes us to know the essence of proportion in general by way of its proximate cause, that is to say, by the essence of ratio in general. Thus, apart from one reservation to which I shall return, the knowledge given to us in this first phase (the deduction of the necessary connection of P to U) is not, in itself, of the third mode; it belongs to the fourth mode, in the sense defined in *Tractatus de Intellectus Emendatione,* with respect to a proportion in general. It is only the second phase that is going to render it of the third mode with regard to the given individual proportion.

1.2. This second phase, therefore, is the application of the universal truth demonstrated in Book VII to the particular case of our three given numbers. However, to be completely precise, we would have to say that this application itself contains two moments.

The first moment is a very simple syllogism in Barbara. The *Major*: every set U of four numbers a, b, c and d, such that $a/b = c/d$ (in the sense of Definition 20) has the property P of being such that $d \times a = c \times b$. The *Minor*: the set formed by the four numbers 161, 9, 913, 931 and x (x unknown) is, by hypothesis, a set U such that $161/9{,}913 = 931/x$. *Conclusion*: the set formed by these four numbers is such that $x \times 161 = 931 \times 9{,}913$, which amounts to saying that $x = (931 \times 9{,}913)/161$.

The second moment then consists in carrying out the indicated multiplication and division. Certainly, it is always possible, at least in principle, to add, one by one, 9,913 numbers each equal to 931, then to subtract the number 161, one by one, from their sum, as many times as necessary so that there is no remainder, and then to conclude that the number of these subtractions gives us the value of x. But, in practice, we do not go about it this way; in order to go more quickly we apply the canonical rules of 'multiplication' and 'division' as they are called. But then, on pain of falling back into knowledge of the first or second mode, we must have previously demonstrated the validity of these rules. Thus we shall discover, within this second moment, a reduced model of the two phases characteristic of knowledge of the third mode. Then we can prove (a proof which, in itself, is of the fourth mode) that, given any three numbers whatever, a, b, c, the number obtained by multiplying c by b according to the canonical rules (the universal U) has the property P of being the a-th part of a number equal to b times c. Afterwards, by means of a syllogism, we can apply this universal truth to the particular case of the three given numbers: the major will be a recalling of that universal truth; the minor will be the carrying out of the operations which can be put into words as follows: "57,323 is the number obtained by using the two procedures defined in the major when $c = 931$, $b = 9{,}913$ and $a = 161$"; and the conclusion will be: "$57{,}323 = (931 \times 9{,}913)/161$". Finally, by comparing the conclusion of the second moment with that of the first, we shall deduce the final conclusion: "$x = 57,323$" and the fourth proportional will be found.

The two moments within this second phase thus both come down to treating the universal truths demonstrated in the first phase as general rules applied from outside to resolve particular problems: the conclusions of the proofs in the first phase serve as a 'program' for the operations in the second phase. Hence Spinoza quite deliberately uses the expression 'perform an operation' in this regard. This expression is self-evident insofar as the second moment is concerned. But the first moment, too, is an operation, for it follows exactly the same schema: the application of the 'rule of three' to the three given numbers allows us to put in place the calculations which the application of the rules of multiplication and division to these same numbers will then allow us to carry out.

One comment, however. The second moment, taken in itself, is not irremediably of the third mode: it can be brought back to the fourth if, at each step in the multiplication and division, we ourselves redo the proof establishing the validity of our procedure with regard to *the particular example* of the numbers on which we are in the course of operating. No one will ever proceed in this way, because it is useless and tedious, but it is possible in principle. On the contrary, as we shall see, the first moment in the second phase belongs *irremediably* to the third mode.

1.3. What does the inadequacy of this third mode actually consist of? In the fact that the mathematicians who proceed in this way "do not see the adequate proportionality of the given numbers" (*TdIE,* Gebhardt II, p. 12). The given numbers are obviously the individual numbers 161, 9,913, 931 and 57,323, and not the four numbers *a, b, c,* and *d* in general. As for the 'adequate proportionality', this phrase can be interpreted in two ways, which are in no way mutually exclusive. 'To see the adequate proportionality' can mean 'to see the proportionality adequately'. However, we can also take the word 'adequate' in its etymological and its ordinary sense (which, besides, is the source of its sense in Spinoza): 'equal to', 'adapted to', 'adjusted to', 'corresponding exactly to'. In this last case 'the adequate proportionality of the given numbers' would mean 'the proportionality which is quite exactly and precisely that of those four numbers'; not *the fact that they are* proportional, but the very thing in which their proportionality *to one another exactly con-*

sists; that is, if one wants, the singular relation by which we conceive the singular essence of the singular proportion which they form together. In fact, the two senses are complementary; Spinoza means that, although we comprehend adequately *that* our four numbers are proportional, we do not comprehend adequately *in what* they are so, because our conclusion with respect to the fourth term has not been deduced from knowledge of the relation between the first two terms, the relation which the fourth itself ought to have with the third.

This clarifies the literal meaning of the general definition of the third mode of knowledge: knowledge "in which the essence of a thing is concluded from another thing, but not adequately" (II, p. 12). This in no way means that this knowledge is not adequate *in itself*; it is adequate through and through, in the second phase as much as in the first — what Proposition 19 states is adequate, and its applicaton to our three given numbers makes us comprehend adequately that their fourth proportional is 57,323. This conclusion clearly bears on something which *belongs to the essence* of the singular proportion determined by these three numbers, since, when the latter are given, there is a proportion if and only if the fourth term is the former.[4] Thus, in coming to this conclusion we conclude at the same time that we could deduce it from the essence *if* we knew that essence, the discovery of which would have to pass through that of the ratio of 161 to 9,913. Hence we can in a fashion say that the 'essence has been concluded', at least in this sense that something linked necessarily to it has been concluded about it. But this conclusion concerning the essence of a singular proportion has not been deduced *from that essence itself*: we arrived there without knowing anything of the relation of 161 to 9,913; we deduced it, not from this relation, but 'from another thing', that is, from the nature of proportion in general and from the equality between the product of the extremes and the product of the means which stems from that nature. This is so much the case that once we have obtained that conclusion we still do not know what 161 is to 9,913; consequently, we also do not know what 931 is to 57,323: we know, with complete certainty, that these two ratios are identical but we do not know what ratios they are. Thus, in fact, we do not see in what the proportionality of *these very numbers precisely* consists. This amounts to saying that the essence of *their* singular proportion 'has not been adequately concluded': all that we know of it is that it *has to be such* that we can deduce from it what in fact we deduced adequately from something else.

We can now see why it was important for Spinoza to refer to Book VII of Euclid rather than to Book V. According to Book V, in fact, the equality of the product of the extremes and the means is not a *consequence* of the equality of the two ratios, of which each, taken on its own, could be defined in itself: not only is a ratio defined only by its equality with other ratios, but, in the case of whole numbers, the equality of the two ratios (in the sense of Definition 5) is *almost defined* by the equality of the product of the extremes and the product of the means: each of these two equalities is immediately inferred from the other.[5] So much is this the case that when we know that $57{,}323 \times 161 = 931 \times 9{,}913$, we recognize *ipso facto* everything which is to be known of the 'essence' of the proportion formed by our four numbers, assuming that this notion still has a meaning (which is doubtful); in any case, there is no 'hidden essence' behind that equality. Book VII, on the contrary, offers Spinoza an epistemological model capable of illustrating the difference between two ways of knowing adequately the same truth about the same thing: either by the essence of that thing, or by a universal character which it has in common with other things.

2. In these conditions how can the fourth proportional of our three given numbers be discovered by the fourth mode of knowledge? The answer is now clear: we must first search for the ratio of the first two numbers and then deduce from it what the fourth number has to be if the third is to have to it the same ratio. This implies the two conditions are realized; there is nothing exorbitant in the first of these, but the second renders the fourth mode of knowledge *in mathematics* useless and tedious in the majority of cases (except in very simple ones), if not *de jure,* at least *de facto.*

2.1. The first condition is easily recognized. The process which comes closest to the fourth mode of knowledge has two phases:

In the first phase, we contruct genetically the essence of the ratio of 161 to 9,913, proceeding by *anthyphairesis,* on the model of Proposition 2 of Euclid, Book VII:

$$9{,}913 = (161 \times 61) + 92$$
$$161 = (92 \times 1) + 69$$
$$92 = (69 \times 1) + 23$$
$$69 = 23 \times 3$$

The greatest common measure of our two numbers is therefore 23. From this we deduce, using the quotients 3, 1, 1 and 61, that this greatest common measure is contained 7 times in 161 and 431 times in 9,913. The *essence of the ratio* of 161 to 9,913 is therefore 7/431. It is true that if we proceed in this way, content to *apply* the rule of *anthyphairesis,* we would remain at the third mode of knowledge. But there is nothing irremediable in this situation: instead of giving first the general proof of VII, Proposition 2 and thus applying its conclusion to our two numbers, nothing prevents us from re-thinking this proof directly *for this particular example* of our two numbers; this is made all the easier inasmuch as that proof is almost instantaneous.

Then, the second phase consists in determining the value of the number x by establishing what it must be in order to show with 931 a greatest common measure which is contained 7 times in 931 and 431 times in x. In order for that greatest common measure to be contained 7 times in 931 it must be equal to 133. We conclude from this that $x = 133 \times 431 = 57{,}323$.

This time, therefore, without using the rule of three, we have deduced the value of the fourth proportional *from the essence* of our singular proportion, itself genetically conceived through the combination of the ratio 7/431 and the two 'greatest common measures' 23 and 133. It is true that, from a purely mathematical point of view, this approach scarcely pays off: whereas the application of the rule of three consisted uniquely in a multiplication followed by a division, here we had first to simplify the fraction 161/9,913 and then proceed to a division followed by a multiplication. But, it is also true that, if numbers were *real physical entities,* this knowledge would be ontologically more perfect than the former knowledge. And, even in mathematics if it could be put to work in a pure state, it would be epistemologically quite superior to the application of the rule of three: it would be *intuitive,* in the quite precise sense that it would allow us to reach the *vision of an essence*; and this would in no way prevent it from being at the same time deductive, since something would have been concluded from the essence, namely the value of x.

2.2. This is where the second condition intervenes. In the case of our four given numbers this process, *as a matter of fact,* cannot be carried out in a pure state, although *de jure* it can always be so. For if we have not applied the rule of three it remains no less the case that we have carried out divisions and multiplications; thus, in spite of everything, we have proceeded by the application of general rules to one particular case. Does this mean that this knowledge, in the final account, reduces to the third mode of knowledge? No, because it makes us know its object through its essence and that essence through its proximate cause. What now has to be made precise is that there are *degrees of purity* in the fourth mode of knowledge: it can be more or less intuitive. For it to be *purely* intuitive, all the stages through which it passes must be equally intuitive, that is, that we do not call upon the third mode of knowledge as an auxiliary procedure. The fourth mode of knowledge remains, on the contrary, *in an impure state* when, in the procedure which leads us from knowledge of the essence to knowledge of its consequences, we have recourse, if only for convenience, to intermediate steps which come from the third mode.

In the case of our three numbers, therefore, the deduction of their fourth proportional would be purely intuitive only if we carried out our divisions and multiplications, not by mechanically applying rules whose justification would have been understood once and for all, but by rethinking each of the steps involved under the form of an addition or a subtraction unit by unit, or decade by decade, *et cetera* . . . by rethinking, for example, 23×3 under the form:

$$(3 \times 3) + (20 \times 3) = (3 + 3 + 3) + (20 + 20 + 20)$$
$$= [(1 + 1 + 1) + \cdots] + [(10 + 10) \cdots], \textit{ et cetera.}$$

To be sure this is nearly impracticable in fact, although it is never impossible in principle — if, at least, as Spinoza requires, it belongs to the nature of number to be finite.[6]

2.3. This is why, in fact, a fourth proportional can be found by the fourth mode of knowledge only when *the three given numbers are very simple* (*TdIE,* Gebhardt II, p. 12). For in that case we no longer need any operations. As soon as there are operations to be performed, not only are more required than when we apply the rule of three, but we do not even attain, *per contra,* a purely intuitive knowledge; thus, this does not offer any interest and we never proceed in this way. On the other hand, in the case of very simple numbers, the procedure of the fourth mode is obligatory. To take once again the example given here by Spinoza, if we are looking for the fourth proportional of the numbers 2, 4 and 3, there is no division or multiplication needed for us to understand, in the first phase, that the greatest common measure of 2 and 4 is 2, that it is contained once in 2 and twice in 4 and that the ratio of these two numbers is therefore $1/2$; and there is no division or multiplication needed. For us to understand, in the second phase, that the number contained once in 3 is 3, that the greatest common measure of 3 and x is therefore 3, that the number which contains 3 two times is 6 and that, consequently, $x = 6$. We see it "intuitively, without performing any operation" (II, p. 12): while impurely intuitive knowledge only eliminates one operation, that called 'the rule of three', pure intuitive knowledge eliminates them all without exception.

What risks deceiving us in this last example is clearly its very simplicity. Since it is a matter here of a quasi-instantaneous step, we are tempted to think that it belongs to the nature of the fourth mode of knowledge generally to be *immediate* in every sense of the word — and not simply in the sense that it concludes something from *the thing itself,* without passing through mediation of universal rules applied from the outside. If we take one more step, then we are tempted to imagine that the fourth mode of knowledge *is not deductive,* whereas this is already false in the case of the numbers 2, 3, and 4: 'intuitively' does indeed mean 'without performing any operation', but, if every operation is a deduction, not every deduction is an operation; there are intuitive and non-intuitive deductions, just as there are deductive and non-deductive intuitions. In reality, instantaneity is only contingently linked to the fourth mode of knowledge: if purely intuitive knowledge can be used to discover a fourth proportional only in the case of very simple numbers, this is only a limitation in fact. *De jure,* we could proceed in the same manner in every case, although at enormous expense in time, and the illusion of non-discursivity would disappear. Put simply, this would not be worth the effort, for this kind of calculation is never an end in itself, not even for the mathematicians. But *it would be worth the effort* when it is a question of knowing the essence of a real physical entity, especially *our own essence.*

3. We could, of course, take many other, seemingly less 'trivial', mathematical examples to illustrate the difference between the fourth mode of knowledge and the third. But, it must be noticed that in the majority of cases these examples would not be as relevant as the one Spinoza chose.

3.1. Let us consider from this point of view the *resolution of an equation*. In an initial phase we have deduced once and for all the validity of a general formula: for instance, every equation of the form $ax^2 + bx + c = 0$ (universal U) has the property P of being such that $x = \ldots$, *et cetera*. In itself this deduction is of the fourth mode in relation to the second-degree equation in general. Then, in a second phase, each time we want to resolve a particular second-degree equation, we apply the formula; this causes us to know, through the third mode of knowledge, the particular value of x in that equation. If, however, instead of simply applying the formula, we go through the entire deduction of the first phase again, conserving its structure exactly, but replacing *from the start* (and not only at the end) the coefficients *a, b,* and *c* by the value they have in this particular equation, then the knowledge we would obtain of the value of x would be purely intuitive.

We could also take the example of the *algebraic resolution of a geometrical problem*. After having expressed the problem in equations, we resolve these equations in a deduction which can be of the fourth mode relative to them *if* we have proceeded as I have just indicated. When we retranslate our final result into geometrical terms we reach a knowledge which is of the third mode relatively to the figure considered: we conducted our reasoning on something other than that figure, namely on certain symbols and we have applied the conclusions of our deduction to it from the outside. But if we had made the same deduction while keeping in mind from start to finish the geometrical equivalent of each of its steps (and this, according to Descartes's *Géométrie*, ought always to be possible *de jure*, although horribly tedious), then this knowledge in the third mode would have been elevated to the fourth mode without any change in its structure.

Let me add, finally, that even in the case of the geometrical demonstration of a geometrical theorem or of an arithmetical demonstration of an arithmetical theorem we most often use the third mode of knowledge in the course of the intermediate steps; someone who expounds that demonstration orally or in writing is obliged to proceed in that way. But, nothing prevents someone who thinks it through again on his own account from making the proof purely intuitive.

3.2. Let me show this using the same example, the deduction of

Proposition 19 of Euclid, Book VII. When I said earlier that this deduction was of the fourth mode, I expressed some reservation. For, if it is of this mode, it is not so in a pure state. It does indeed establish a property of proportion in general from the essence of proportion in general; but, in passing from the latter to the former it has recourse at several points to steps belonging to the third mode: the proof of a proposition Q from a proposition P frequently amounts to giving a particular value, adapted to Q, to the general terms figuring in the enunciation of P; for instance, Proposition 12 is deduced from Proposition 15 by applying the enunciation of Proposition 12, where the numerators and denominators are any numbers whatever, to the particular case in which the numerators are equal to 1 and all the denominators are equal to one another. In a treatise of arithmetic (or, for that matter, of metaphysics) this recourse to intermediate steps of the third mode is practically inevitable, both for reasons of economy and for pedagogical reasons. The *reasons of economy* (economy of thought, of course) are obvious; it pays better to demonstrate once and for all a proposition in its general form than to prove it in each of the particular forms one would have to give it in order to use it in later proofs. As for *pedagogical reasons,* they are of two kinds. On the one hand, it is more commodious and more striking to prove a proposition by basing oneself exclusively on the *enunciations* of earlier propositions, abstracting from anything which can figure in their proofs (and which the reader can forget without losing the thread); this leads to a rather curious result: Proposition 16 is intuitively deduced from the equality '$1/b = c/(b \times c)$', but since this equality, which has been deduced from Proposition 12 in the course of proving Proposition 15, has not gained the status of a number proposition, the proof of Proposition 16 can only utilize it by re-proving it from Proposition 15, applying the latter to the apparently particular case where d has the explicit value $(b \times c)$. On the other hand, it is more commodious when proving a proposition to rely on an earlier proposition closer to, rather than farther from, the one in question. This also leads to a result analogous to the preceding: Proposition 17 is in reality intuitively deducible from Proposition 9, which it merely reexpresses in the language of proportions; but, since Proposition 10 is the generalization of Proposition 19, while Proposition 13 is the translation of Proposition 10 into the language of proportions and Proposition 13 is closer to Proposition 17 than Proposition 9 is, the demonstration of Proposition 17 takes the form of an application of Proposition 13 to the particular case where *a* and *b* are sub-multiples of *c* and *d.*

However, this impurity of the fourth mode of knowledge is by no means irremediable here. To make the deduction of Proposition 19 purely intuitive it would be enough to base all the proofs on a single one, using the same symbols all along without ever making any substitutions: Proposition 5 would have to be enunciated in such a way that the *a, b, c, d* which figure in it are *the very same terms* that we find again in Proposition 19. This implies, on the one hand, the reestablishment of everything which the third mode allows us to dispense with: each proposition would have to be written in each of the particular forms in which it is going to be used subsequently. But this implies, on the other hand, the elimination of all the propositions invoked in the deduction of Proposition 19 for pedagogical reasons alone: Propositions 15, 10 and 13 would be short-circuited, since 16 can be directly deduced from Propositions 12, 17 and 9. Finally, since Proposition 9 is precisely the particular form in which Proposition 6 would have to be rewritten, Proposition 9 would disappear as a separate proposition and Proposition 17, at last, would be deduced directly from Proposition 6.

Here is the scheme of what a purely intuitive deduction of Book VII, Proposition 19 would be:

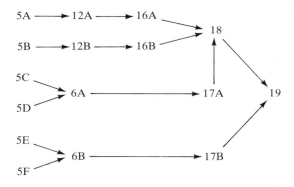

PROPOSITION 5A. Since 1 is the *a*-th part of *a*, it follows that $(1 + 1 + 1 \ldots)$ is the *a*-th part of $(a + a + a + \ldots)$; that is, if the 1's and the a's are added together *c* times, then *c* is the *a*-th part of $(a \cdot c)$. [The same proof holds as in the general case.]

PROPOSITION 12A. $1/a = c/(a \cdot c)$. [A simple rewriting of Proposition 5A in the language of proportions.]

PROPOSITION 16A. $a \cdot c = c \cdot a$ [For, by definition 20, $1/a = c/(c \cdot a)$; therefore, by Proposition 12A, etc. . . .]

PROPOSITION 5B, 12B and 16B. The same as the preceding, with b in place of a. [Same proof as for Proposition 5A.]

PROPOSITION 5D. For the same reasons, since the integer b/n is the n-th part of b, $(b/n) \cdot c$ is the n-th part of $(b \cdot c)$.

PROPOSITION 6A. If a is the $m \cdot n$-th parts of b, $(a \cdot c$ is the $m \cdot n$-parts of $(b \cdot c)$. [For then the integers (a/m) and (b/n) are equal. Therefore, by Propositions 5C and 5D, etc. . . .]

PROPOSITION 17A. $a/b = (a \cdot c)/(b \cdot c)$. [Translation of Proposition 6A into the language of proportions.]

PROPOSITIONS 5E, 5F, 6B and 17B. The same as the preceding, with c in place of a, d in place of b and a in place of c. [Same proofs.]

PROPOSITION 18. $a/b = (c \cdot a)/(c \cdot b)$. [Obvious by Propositions 16A, 16B and 17A.]

PROPOSITION 19. If $a/b = c/d$, then $d \cdot a = c \cdot b$ and *vice versa*. [Obvious by Propositions 17B and 18.]

In this way we would achieve a *perfect* knowledge of the truth enunciated in Proposition 19. This is the way God knows real physical entities, without ever going through any universals and this is also the way it will be important for us to know them as much as we possibly can.[7]

3.3. It is true that it makes little difference whether we comprehend mathematical truths as perfectly as this, since knowledge of them is never an end in itself and thus it is worth more to us to reach them by the most economical means, provided simply that these are rigorous: access to perfection of that sort would be of interest only if it did not require any effort on our part — whereas in ontological matters, on the contrary, the effort is worth the pain. This is why Spinoza, before working out the *Ethics,* declares that "up till now" he has understood only "a very few things" in this perfect manner: "$3 + 2 = 5$" a truth deducible without any operation from the definition of the three numbers (*TdIE*, Gebhardt II, p. 11); "two straight lines parallel to a third are parallel to one another", a theorem immediately deducible from Euclid's Fifth Postulate (II, p. 11). For among the Euclidean proofs only the shortest are purely 'intuitive'; and since we need not elevate the others to the same degree of perfection, Spinoza himself never committed himself to so elevating them; not having done so, he could not present them as examples of perfect knowledge.

In principle, however, their imperfection is by no means irremediable. It very often happens in Euclidean mathematics that we can, *de jure*, proceed in the same way as in our second schema; that is, when we find ourselves in the presence of the third mode of knowledge, we can resort to a stop which has the same structure as the deduction carried out in its first phase, but which concretizes it by thinking it through again entirely in regard to the particular example of the object studied. In these cases the difference between the fourth mode of knowledge and the third is not very palpable; this a difference of degree, not of nature, and this third mode, remediable as it is, in reality is the fourth mode in an impure state.

However, this is not always the case. And the *relevance of the case chosen by Spinoza* derives precisely from the fact that, in the case of the discovery of the fourth proportional, the fourth mode of knowledge *is not* a simple concretization of the third: their respective structures are entirely different and thus it is not enough to particularize one in order to rediscover the other. For if the Euclidean deduction of Proposition 19 is in itself intuitive (and can be rendered purely intuitive), this is because it follows in its entirety from the definition of proportion taken at a certain level of generality — that of four indeterminate numbers a, b, c, and d — and is situated in its entirety *at precisely that level of generality,* thus issuing in a conclusion relative to these same undetermined numbers. The reason why it can play the role of the first stage in a cognition in the third mode, with respect to our four given numbers, is precisely that a change of level takes place in the course of the process. Thus, if we wanted to particularize this proof while still conserving its structure even though rethinking it wholly in terms of the numbers 161, 9,913, 931 and x, we would have to begin by particularizing, *at the level of these very numbers,* the definition of proportion itself, in other words, we would have to start by defining, not proportion in general, but the proportion of *these four individual numbers.*

Now, in order to define proportion in its singularity we obviously must execute the *anthyphairesis* of the first two numbers; this will give us, at one and the same time, their greatest common measure and their ratio. Then we have to deduce from these what has to be the greatest common measure of the last two numbers. But then the fourth proportional is immediately discovered: there is no need to appeal to Proposition 19. If we kept to it there is no doubt that we could go on to particularize at the same level the whole deductive chain, including Proposition 19, which stems from the initial definition; to do so, however, will not help us to discover the fourth proportional, since we would know it already. Consequently if we do need Proposition 19 in order to discover it, this is because we did not particularize the initial definition at the start; and, in this case, any particularization which we can subsequently carry out will depend *irremediably* on the third mode of knowledge: in order to know a property of a thing at a level of generality inferior to that at which one knows the essence of that thing one must necessarily apply a general rule to a particular case. Here, consequently, the difference between the two modes of knowledge is striking: no longer is there any homology or continuity between them. This is no doubt why the example of the fourth proportional appeared so suggestive to Spinoza that he retained it even in the *Ethics*.

4. In fact, in the *Ethics,* in the second scholium to Book II, Proposition XL, this example reappears in just this form, with three apparent differences and one real difference requiring explanation.

4.1 Spinoza treats the first kind of knowledge (i.e., the first and second modes of *TdIE*) much more rapidly than in *TdIE*, but he says exactly the same thing. As for the second kind of knowledge (the third mode of *TdIE*), Spinoza, instead of saying that Proposition 19 of Euclid, Book VII concerns "the nature of the proportion and its property", says that it concerns "the property common to proportional numbers"; but this comes to the same thing: a proportion being a set of four proportional numbers and the "nature of the proportion" mentioned in the *TdIE* being that of proportional in general, "its property" does designate a property common to all sets of four proportional numbers, i.e., to all proportional numbers. As for the third kind of knowledge (the fourth mode in the *TdIE*), Spinoza is more precise: instead of simply saying that we see the proportionality "intuitively, without performing any operation", he expressly analyzes the two steps in the procedure: first we see *uno intuiti* the ratio of the first two numbers, then we "conclude" from this what the fourth has to be in order to have the same ratio to the third; but we have seen that the text of the *TdIE,* if we take seriously the reference to Euclid, must necessarily be interpreted in the same way. As for the fourth difference, it could seem insignificant: instead of taking as his example the numbers 2, 4, 3, and 6, Spinoza chooses the numbers 1, 2, 3, and 6. In reality, however, this modification does have a meaning: it is intended to adapt the arithmetical example to the more complete form given to the doctrine of the kinds of knowledge in the *Ethics.*

4.2. We saw that the *TdIE* characterizes the modes of knowledge only in terms of their respective points of arrival, and its goal was precisely to guide us towards the discovery of *that from which we have to start* in order to attain to the most perfect knowledge. Now this discovery has been made, and using it as our basis, we can genetically reconstruct everything, including knowledge itself; that is why the second scholium defines each kind of knowledge by its point of departure and its point of arrival *simultaneously.*[8] The point of arrival of the second kind of knowledge corresponds quite precisely to the third mode in the *TdIE*; but the second scholium discloses to us the first principles on which the necessary link between such and such a 'universal' and such and such a 'property' rests. These first principles ground the adequate character of this deduction: they are (1) the *universal common notions* whose genesis was the theme of Proposition XXXVIII of Book II (adequate ideas of the properties common to all things and which are equally in the whole and the part of each thing) and (2) the *proper common notions,* whose genesis was the theme of Proposition XXXIX (adequate ideas of the properties common to the human body and to certain external bodies by which it is commonly affected, and which are equally in the whole and in the part of each of these bodies). Likewise, the point of arrival of the third kind of knowledge corresponds quite precisely to the fourth mode of the *TdIE,* but the second scholium indicates to us whence we must "proceed" in order to comprehend "the essence of things": we must proceed "from adequate knowledge of the formal essence of certain attributes of God"; for God, himself known through his essence alone, is the true proximate cause of all real entities.

We can then ask, however, what relevance our arithmetical example has now. Since numbers, according to Spinoza, do not express any real property of things, what relation can they have to the common notions? For an even stronger reason, what relation can there be between the discovery of a fourth proportional and knowledge which proceeds from an attribute of God?

4.2.1. That numbers do not express any real property of things does *not* mean that arithmetic is not true.

We know the genesis of number from *Letter XII*. In the first place, we necessarily imagine singular things by separating them from the substance of which they are the modes, for substance cannot be imagined: we imagine them, therefore, as so many substances independent of one another (Gebhardt IV, p. 57). In the second place, inasmuch as we cannot recall all the details of each of these things taken separately, in order to imagine them in the easiest way possible we group them into classes whose elements, already separated from one another for the preceding reason, now appear to us as so many discrete and homogeneous units.[9] Finally, in the third place, we feel the need, if only for practical reasons, of 'determining' these different classes, that is, of delimiting them by comparing them to one another (IV, p. 57).[10] This gives birth to number, the "auxiliary of the imagination" (IV, p. 57): In order to compare these classes we put them into one-to-one correspondence, discrete unit by discrete unit, by means of manipulative operations which can at first be materially performed and then simply imagined. Finally, the images of these operations can be associated with symbols: '1', '2', '3', *etc.* Hence it is certain that numbers, contrary to geometrical entities, are nothing in things themselves. Whereas a square table really has the property of being square, two tables do not really have the property of being two: it is we who bestow this property on them.

However, it remains no less the case that the operation by which we bestow this property on them is, itself, perfectly real. And these operations, which are always directly or indirectly material, are subject to necessary laws: it is eternally true that, if a human body performs in succession the operation '2' and the operation '3', it will obtain the same result as it would in performing the operation '5'. Thus, the adequate knowledge of arithmetical truths, contrary to knowledge of geometrical truths, is not connected to the *universal common notions* of

Proposition XXXVIII, it is in fact connected, it seems, to one of the *proper common notions* of Proposition XXXIX: namely, to the adequate idea of that common property possessed by all human bodies (which customarily affect one another), of being able to perform the material operations of enumeration (operations which, to the extent they can imply global displacements of our body, concern it as a whole at the same time as concerning each of its parts).

4.2.2. But then, what relation with the third kind of knowledge is involved? To understand this relation a single remark will be enough. Knowledge whose point of departure belongs to the third kind can indeed be, in respect to a given object, of the second kind as far as its point of arrival is concerned. Spinoza brings this out in the Scholium of Proposition XXXVI in Book V: Book I, beginning from the idea of God, demonstrated that "all things depend on God both for their essence and their existence"; this conclusion, in itself, belonged to the third kind of knowledge relatively to the notion of 'thing' in general; but at the same time, once it was applied from the outside to the particular case of the human mind, it gave us a knowledge belonging to the second kind in respect to the human mind and in regard to its point of arrival: "our mind, which is a thing among others (universal U) has the property (P) of depending . . . , *etc.*" It is only in Book V that knowledge of the essence of our mind springing from God will make this same truth intuitive. In these conditions couldn't we conceive of a knowledge, springing from common notions and leading to knowledge of the essence of a thing, as of the second kind in terms of its point of departure and of the third kind in terms of its point of arrival?

In the case of real physical entities this possibility is clearly excluded: the best physicist will never understand anything of the essence of bodies if he does not know that they are modes of an attribute of God. In the case of mathematical entities, however, such a transition from the second to the third kind of knowledge is perfectly possible. Mathematical entities are precisely *not* real physical entities; they *are* common properties. Considered in themselves, in abstraction from the things of which they are the properties, they will therefore be known in the most perfect way when we have reconstructed them genetically from those common notions (whether universal or proper) which are ultimately implied in their concept. And if, in accord with convention, we decide to raise them to the status of distinct objects by treating them as if they were real physical entities, we will be able to consider this reconstruction as giving us knowledge of their essence. No doubt, this knowledge will not proceed from the idea of God, but there is no need for it to do so; in fact it would be a mistake to make it do so: the circle, as such, is not a mode of God and number is even less so. The best way of understanding their essence will therefore be to confer on them as "naturans" the equivalent in their case of what God is in regard to real entities.

4.2.3. Something of the sort is perhaps the reason why Spinoza replaced the numbers 2, 4, 3, and 6 with the numbers 1, 2, 3 and 6. In the case of the proportion $2/4 = 3/6$ the two steps in the deduction of the fourth proportional were on exactly the same level. The first step consisted in the discovery of the 'proximate cause' of the fraction $2/4$: the ratio $1/2$ joined to the greatest common measure of 2. The second step consisted in deducing from this that the 'proximate cause' of the fraction $3/x$ was the same ratio $1/2$ joined to the greatest common measure of 3. From one fraction to the next, therefore, the proof proceeded by the mediation of $1/2$: it was "$1/2$ insofar as affected by the modification resulting from its application to 2" which produced "$1/2$ insofar as affected by the modification resulting from its application to 3", just as "God, insofar as affected by a finite modification", is the cause of "God insofar as affected by another finite modification".[11] This is no longer the case with the proportion $1/2 = 3/6$: on the contrary, $1/2$ has its proximate cause the ratio $1/2$ joined to the greatest common measure 1, that is to say, $1/2$: put differently, $1/2$ is the *cause of itself.* And just as the second step in the deduction of the number 6 remains unaltered, "$1/2$ considered in its absolute nature" (viz. $1/2$ *naturans*) is the cause of "$1/2$ insofar as affected by the modification resulting from its application to 3" ($1/2$ *naturatum*) *in the very same way* that it is its own cause. This example, recast in this way, thus constitutes the best equivalent within the arithmetical theory of proportions of a knowledge moving from God to the essence of real entities.

4.3. We should certainly not conclude from this that Spinoza claimed to reduce mathematics to this elementary model: he is surely not unaware that the mathematics of his day tends to distance itself more and more from it and he certainly does not condemn this development.

He does not have the ambitions of Hobbes in this domain. To repeat: according to Spinoza mathematics have no interest in themselves: they are only instruments; those who do mathematics are therefore completely free to resort to the most economical procedures without worrying about their greater or lesser perfection.[12] On the contrary, when we are demanding from the discipline a model for *ontological* knowledge of real entities, we must in actual fact look for it in regions where mathematicians employ the third kind of knowledge in a pure state, whatever may be the intramathematical importance of those regions. Spinoza, through this example, simply wants to say to his readers "follow me all the way to the end and you will come to know your nature starting from the nature of God *in the same way* and *just as well as* you understand the equality $1/2 = 3/6$".

But, to follow Spinoza all the way to the end does not consist simply in reading the *Ethics* nor even in comprehending it formally. His formal exposition, for the same reasons as that of the Euclidean books, depends, and could only depend, on the third kind of knowledge in an impure state. It falls to us, therefore, to render this exposition purely intuitive by rethinking it on our own account: we shall draw from it "the highest possible satisfaction of mind" (*Ethics* V, Proposition XXVII).

École Normale Supérieure de Saint-Cloud

NOTES

[1] Cf. Itard (1961, p. 91). On this point we follow completely Itard's interpretation.

[2] Cf. Itard (1961, p. 91). It seems to me impossible that Spinoza did not understand Euclid in just this way. If he had understood him in a less rigorous way, his example would be quite trivial or ill-chosen and the importance he attaches to it would no longer make sense.

[3] According to the schema of *anthyphairesis*, we see that:

$$a/b = I$$
$$i + I$$
$$j + I$$
$$k + \ldots$$

If we can prove that the same quotients $i, j, k, \ldots$ renew periodically, this amounts to a proof of irrationality.

[4] Cf. *Ethics* II, Definition 2.

[5] For in the case of whole numbers, if we assume the commutativity of multiplication, it is sufficient to take Definition 5 and to set $m + d$ and $n + c$, in order to deduce that $da = cb$. Conversely, from $da = cb$, it follows that $ndma = mcnb$, from which we can immediately deduce that a, b, c and d satisfy Definition 5.

[6] Cf. *Letter XII*, Gebhardt IV, pp. 58—59.
[7] Cf. *Tractatus de Intellectus Emendatione*, Gebhardt II, p. 36.
[8] Cf. Gueroult (1974, p. 606).
[9] Cf. also *Letter L*, Gebhardt IV, p. 239.
[10] Cf. also *Cogitata Metaphysica*, Gebhardt I, p. 234.
[11] Cf. *Ethics* I, Proposition XXVIII.
[12] Cf. *Tractatus de Intellectus Emendatione*, Gebhardt II, p. 35: Spinoza, on one and the same page, declares that the way one defines mathematical entities has no importance, and takes as an example of a perfect definition the genetic definition of the circle.

REFERENCES

Euclid: 1952, *Elements*, transl. by Sir Thomas L. Heath, Chicago, University of Chicago Press.
Itard, Jean: 1961, *Les livres arithmétiques d'Euclide*, Herman, Paris.
Gueroult, Martial: 1974, *Spinoza II: L'âme*, Aubier-Montaigne, Paris.
Spinoza, Baruch: 1925, *Spinoza Opera*, ed. by Carl Gebhardt, 4 vols., Carl Winter, Heidelberg. (Cited as 'Gebhardt' in the text.)

Translated by
DAVID LACHTERMAN

PART III

SPINOZA AND THE HUMAN SCIENCES:
POLITICS AND HERMENEUTICS

JOSEPH AGASSI

TOWARDS A CANONIC VERSION OF CLASSICAL POLITICAL THEORY

0. INTRODUCTION AND ABSTRACT

This study relates to Spinoza in an obvious manner: he is presented here as the leading contributor to what should count as the classical political theory. The way this connects with the sciences is dual: first, it is assumed that the classical theory is scientific, especially since in its later variant it includes classical economic theory; and, second, it is presented here as canonic in the sense in which classical physics is canonic since Einstein.

Classical physics achieved a canonic version with Einstein's 1921 *The Meaning of Relativity* (1956). Classical political theory has not, and for diverse reasons. First, much of it is still contested — yet it has been superseded since even those who lean towards it hold modified versions of it. Second, much of it is couched within the traditional metaphysical framework within which it evolved. But just as classical physical science is now presented with minimal metaphysical material, so classical political science can and should be stated. This will also relieve us of some traditional metaphysical confusions which may simply drop out of the picture. Finally, just as there are different versions of classical physics — action-at-a-distance and field theories — which historically competed but in the canonic presentation are put side by side, so one has to detach the two traditional liberal political theories, the realist and the Utopian strains. The difference here is that whereas physicists who worked with both action-at-a-distance and fields were aware of the conflict and openly attempted to reconcile them, in political theory we have an attempt to confuse the two variants of classical political theory which won respectability — an attempt which was successful until very recently.

1. THE TWO REVOLUTIONS: IN PHYSICS AND IN POLITICS

As tradition has it, the scientific revolution consisted chiefly of the rise of mathematics and mathematical physics. The life sciences, we are

Marjorie Grene and Debra Nails (eds.), Spinoza and the Sciences, 153—170.

told, lagged behind; sociology and political science are the latest arrivals, or perhaps simply they are still in their embryonic state. The tradition in question has some variants, to be sure. Whether psychology or economics are scientific, in particular, is not quite universally decided. Nor is the place of scientific history, or even the possibility of scientific history. And the same could be said of scientific ethics. Despite these discrepancies, however, the traditional view has stayed virtually unchallenged, and the ease with which its thesis is proven is truly astonishing. Indeed, the argument is so overwhelming in favor of the traditional view, that it defies analysis. Yet, once analyzed, it crumbles.

One intuitive measure of an idea, good or bad, is the lengths to which one has to go in order to acquire it. When one invests enormous efforts following the footsteps of any trailblazer, and only at the end of the road one is able to see that the desired destination is reached, then, and perhaps only then, is one able to appreciate the heroic pioneering deed of the trailblazer. There is much truth in all this, though counter-examples also come to mind fairly easily. We all remember moments of great enlightenment, of great transition, of the veil falling and vision all of a sudden greatly improved. At times such experiences arise out of great labors — when all the many hooks are finally pried loose the veil falls altogether. But at times this is not so at all: at times the veil lifts with ease as if by magic: at times the fog and mist vanish at once as the sun's rays break through the cloud. People are no longer alive who experience this when reading Einstein or Bohr, but some of them have told their stories this way. And, no doubt, the Copernican revolution was something of the kind too: a vision which seizes one immediately.

Yet even for cases where efforts are the measure of the pioneer's foresight and of the length of the road they traverse, even then we may easily misjudge matters in retrospect. Thus, the more advanced our education system is, the more we develop a society familiar with some elementary kinematic theory or simple calculus, the more we may find it easy to underestimate not only Copernicus but also Kepler or Newton. This is, quite possibly, why Bertrand Russell has appealed — subjunctively — to the abilities of the best ancient Greeks: the first modern thinker they could not comprehend, said Russell, was Sir Isaac Newton. He meant, presumably, that they would need quite a lot of preparatory training in order to understand Newton, but would understand Copernicus right away.

In what follows the political theory of Machiavelli will be presented as a Copernican revolution in politics. That revolution, too, is in a sense a revival of some ancient ideas. It is the application of the ancient idea of the autonomy of the individual and of comprehensive utilitarianism — Epicureanism, for short — an application of Epicureanism to politics. The application had a surprisingly coherent outcome, including the resultant theory of checks and balances, of separation of state and church, of constitutionality, and of tolerance. This composite theory was ancient in parts, but the parts belonged to ancient political practice, not to any articulated theory. The resultant comprehensive theory is democratic in thrust, yet applicable, and intentionally so, also to constitutional monarchy, and thus to all sorts of regimes on the ancient scale with monarchy and democracy as its two extremes. Indeed, Machiavelli declared the separation of powers a synthesis between monarchy, oligarchy and democracy. Spinoza's unfinished *Political Treatise* seems to agree with Machiavelli, unlike his *Politico-Theological Treatise*, which is democratic. In any way, I contend, the theory looks to us fairly obvious, but I dare say, contrary to Russell, an ancient would find it surprising and barely comprehensible.

Another measure of the novelty and import of an idea, good or bad, is how much it alters the lives of those who learn it. Historians can praise Copernicus or Newton by reference to the increased predictive power of their theories as compared with earlier ones. Why, however, do we care about predictive power? Everyone knows that Einstein's theory has a higher predictive power than Newton's, and is therefore given much praise. Yet nobody will dare express the same contempt for Newton that is so often meted Ptolemy and more so Aristotle. Something is suspect here. Newton's predictive power is today good enough for many a purpose, we are informed. Was not Ptolemy's predictive power good then — or even now — for some purposes? Of course, a theory which wins a crucial test should win. But the importance of its winning is judged otherwise. Galileo had a suggestion: he said, in order to accommodate Copernicanism we have to alter our views not only of the heavenly bodies — the earth too has become heavenly — but also of mechanics; and we must limit the authority of the Church to the exclusion of physics. This measure makes immediate sense. It also makes political science much more important than physical science — or it makes physics important because of its political implications.

There is still one more sense by which to measure success, particu-

larly important for a transition period. In times of change it is far from
clear who is going to win, or who has won. But at times a magic wand is
waved and at once victory is obvious. Anyone who has examined the
introduction of new medical techniques, say innoculation, with some
empathy, knows how hard it was to decide the efficiency of the new
techniques, the superiority of its advantages over its disadvantages.
When antibiotics entered the market, there was no doubt: they looked
like veritable magic. (The verdict was an exaggeration, it turned out
later; but this is another matter.) One reason Galileo was so very
important, is that he showed that the moon has seas and mountains, no
crystalline smooth surface (he was mistaken: there is no water on the
moon), that the inner planets have phases like the moon, that Jupiter
has moons like Earth, that Aristole's theory of gravity is no match to
Archimedes', that Aristotle's theory of free fall is inconsistent, and
more. Galileo simply closed the debate.

The political debate has, in a sense, never been closed. Promises to
increase the application of government brute force so as to reduce the
crime rate still get individuals elected to office, even to high office, and
even in some of the most enlightened countries in the world. And this is
impossible without some thinkers siding with these individuals. Hence
Machiavelli's idea, which was central to all liberal philosophy, is still not
fully endorsed or exhausted. People still refuse to accept the view that
private vices can be public virtues, and that we should strive to make
this the case whenever possible. Nevertheless, when we remember that
Machiavelli's given name became, in vernacular English, the name of
the devil — old Nick — and when we remember how much hatred
Spinoza drew — which I shall illustrate later on — then we can say that
Spinoza did win in the manner in which Galileo did. The Glorious
Revolution in England in 1688 was the beginning of an age of toleration throughout Western Europe, which was unconceivable decades
earlier. We need not forget Voltaire's protest against the legal murder
of a heretic in the early eighteenth century. Yet comparing this with
similar experiences but a century earlier shows enormous progress
though, of course, the battle for toleration is still not over — not by a
long chalk. In a sense, then, but only in a limited sense, we can say, the
Glorious Revolution closed a debate. And in no small part this is due to
the tradition of Machiavelli, Hobbes, and Spinoza.

So much for what I could come up with in my search for an analysis
and a justification of the traditional view which sees in Kepler, Galileo,

and Newton the crowning glory of the scientific revolution and the trailblazers of the modern scientific world. I find the analysis wanting.

There are, of course, other virtues traditionally recognized, like tenacity and like courage. These, of course, are not specific to scientific ventures. And there are scientific ventures intellectually not so challenging yet morally more taxing and physically more taxing than some of the greatest. And surely Diaz, Columbus and Magellan are so very admirable explorers on account of these qualities as were Captains Cook and Scott. And even Darwin's much less taxing voyage on the Beagle was more courageous than his publication of his later books, which are scientifically, of course, much more impressive than his Beagle diary. But we need not examine all the dimensions of the scientific endeavour to see the point of — and to contest respectfully — the traditional view of physics as the true and leading pioneer science.

One must concede one point, however, to the traditional view: it is no doubt the case that Galileo has made it quite obvious that the new philosophy, as it was called, had won over the old. All the rest does not hold. The revolution in politics was much more profound, both in its depth and in the effects it had on our outlook on life. Moreover, it made possible victories which were not as clear-cut as that of Galileo and Newton. Furthermore, the two revolutions, the one in physics and the one in politics, were not unrelated. Not only did Galileo succeed in his effort to limit the authority of the Church of Rome. Without his challenge to religious authority, the scientific revolution could not have any success. And *vice versa*: one reason Locke's political philosophy won so much acclaim, was that it was triumphant variant of the militant politics of Machiavelli and Spinoza. Another reason for Locke's popularity was his personal friendship with Newton.

2. THE POLITICAL REVOLUTION: THE ADVENT OF THE NEW EPICUREANISM

Renaissance political theory may be divided into two main streams, Utopian and realist, with Machiavelli, Hobbes, and Spinoza belonging to the realist school as its leading representatives. Their fundamental thesis was Epicureanism, which roughly amounts to naturalistic ethics or to the preaching of enlightened self-interest. They were viewed as atheists because they denied the divine supervision over human affairs known as divine providence and because they denied it at the times when dominant opinion had it that the abolition of the fear of God will lead to the increase of crime rate to the point of collapse of civil society. What was new in this revial of Epicureanism? Courage aside, one needs not much intelligence to repeat or revive an ancient doctrine. Was there also some innovation in the Epicurean revival?

The idea that Machiavelli or Hobbes could count as an ideologist of the liberal domocratic camp is rather puzzling, since the one is the notorious author of *The Prince* and the other advocated absolute monarchy. But at least they are praised for their realism and for their parenting the application of scientific method to politics. This is absurd: any application of scientific method produces scientific laws. Did they discover any scientific law of politics?

This is not to deny their courage. It is well-known that Hobbes is reported (Aubrey, 1972) to have expressed admiration for Spinoza, whose courage, he said, permitted him to go further than Hobbes himself dared go. Quite possibly, Hobbes meant nothing more than that his intention was to offer as bold a Bible criticism, but only Spinoza pioneered the field by explicit statement. This may sound surprising. It is one thing to get priority for an idea and quite another for the courage of stating it. Who was the first to have the idea that the Bible is not divine? This hardly matters. The idea was always commonplace, as it is an ancient heresy. What mattered, however, quite apart from courage, was arguing from internal inconsistencies that the Bible is fallible. This, however, is not as easy as it looks today: the hunting of inconsistencies — in the Bible, in Aristotle, in any authoritative text — is a game not in the least confined to heretics: believers do that too, from the ancient Rabbis to Althusser. As long as the inner inconsistencies are problems for the sincere believer, tasks for the faithful hermeneutic scholar to settle, this is not cause for too much concern; and some cause for some concern is always recommendable. Even the question, did Spinoza finally settle matters and prove the Bible fallible, even that is still debated among Bible scholars. It is really the very declaration that Aristotle is fallible, and more so that the Bible is fallible, that mattered. And this was a matter not only of courage but of philosophy. But this is not yet political science — at most political philosophy, and perhaps simply of philosophy in general.

It is, of course, obvious what this philosophy in general was: Epicureanism, the idea of autonomous enlightened self-interest, which defies, as a matter of course, all authority, political, philosophical, or divine. We seem to be inching towards a political science. We have the fundamental traditional Judeo-Christian political philosophy, which appoints divine providence policeman in charge of the public order. Less facetiously, it appoints to that role of policing the individual's fear of God. Thus, in situations in which modern rabble-rousers clamour for law-and-order, for tightening of the law and of the pressure of the police and the courts on lawbreakers, in these situations some centuries ago rabble-rousers would put pressure on the public to go to their churches or to their synagogues and demand more hell fire and brimstone from preachers. Savonarolla. And this is why the Epicureans were deemed the public enemy: they were the demoralizers, the ones to deflate the major means of the maintenance of the public order. But how does this make Epicureanism into an alternative political philosophy?

The question thus posed is vague, and so it is unwise to answer it prior to some clarification and to nailing down the terms of the possible answers to it. The question may be historical and it may be an abstract, logical or intellectual exercise. The two questions can diverge: we may have history choose one alternative and abstract thinking another, for example. In such cases it might be advisable to cast a glance at the history of the case, at least so as to know how close to history we are. Yet in the present case of Epicureanism history is of little help. It is especially confusing since influential writers found it impolitic to admit that Machiavelli, Hobbes, or Spinoza served them as sources. This was license to some authors, such as Montesquieu, to lift passages from Machiavelli (E. Levi-Malvano, 1912).

The question, how does enlightened self-interest work as a political principle, is answered by a set of theories which are often labelled theories of the natural goodness of Man. Here we come up with an added source of confusion. The label is a bit misleading, since Epicureanism is utilitarian or hedonistic in the sense that it denies the good or the evil of any impulse, not that it affirms that all impulse is good. Utilitarianism or hedonsim is a theory which does recognize good and evil; it even attempts to explain them by reference to natural impulse and to reason. And explaining good and evil this way would be futile if Man's natural impulses were deemed good by a supposition. Yet the theory is known as the theory of natural goodness in the sense that the expression 'goodness by nature' was understood in mediaeval and Renaissance literature (including one of Moses Maimonides, eight introductory chapters to his commentary on the *Mishnah, Aboth,* and an essay by Sir Francis Bacon, 1906): the good by nature has the natural inclination to do the right deed whereas the merely good is one who has to overcome his nature in order to do the right deed. The theory of the natural goodness of man, then, is the theory hostile to sermonizing and moralizing. Now, how can we maintain civil society without sermons? This seems to the modern audiences a downright stupid question — and we can take this fact as a major piece of relevant evidence — yet at the time the importance of the preacher was deemed by the general public as obvious as that of the policeman today. So how can society function without the preacher?

The very existence of many theories of politics based on enlightened self-interest makes all of them highly suspect. It is customary to ridicule their logic this way, with the prime target as Spinoza and for his claim of basing his politics on his axioms, for his having argued *in modo geometrico.* This is partly unfair in its oversight of logical errors in Galileo and in Newton, but there is still a difference between physics and politics: the logic of physics was corrected and that of politics was not. Nor is this a small matter. The correction of the logic of physics called for new foundations of the calculus to overcome Berkeley's critique of Newton, as J. O. Wisdom (1939—41) showed just before World War II. The attempts were made by the leading mathematicians McLaurin, Lagrange Cauchy, and Weierstrass. Hardly anything like that has happened to politics. Even in economics, which might count as political to some extent, where the logic has been corrected sufficiently, the net profit from the corrections is very small, at least by comparison.

Perhaps logical inadequacies are generally neither here nor there, yet the correction of the logic of physics has led to the establishment of a canonic classical physical science, but not a canonic classical political science. Why? This partly has to do with the classical view of science as authoritative and obligatory. But it is really tiring to remind ourselves that canonic classical physics commands no authority any longer. The canonic version of classical physics is, indeed, due to Albert Einstein, and its principles are best expressed in Einstein's *The Meaning of Relativity* (1956), first chapter. Indeed, one can generalize the point and say, and I think in truth, that only refuted theories can receive truly canonic formulation. Moreover, the canonic formulation incorporates both the viewpoint that supersedes the theory to be canonically formulated and some considerations of historical content to be incorporated.

And when the two conflict we may compromise. Hence, the very best canonic wording is far from being smoothest and most economic. Thus, the canonic version of classical physical science will remain that of Einstein, despite the existence of a better variant due to Elie Cartan. There may be an attempt to make Cartan's version canonic, and by recasting the history of the classical theories to fit it; what will result, I think, would be too much of a distortion of history.

3. THE RISE OF UTOPIAN LIBERALISM

If we are after a canonic wording of classical politics, we must notice the two strains in it, the Utopian and the realist. The idea of a civil society of enlightened self-motivated individuals is, most obviously, Utopian: it is always wiser and more profitable to all sides of a dispute to settle it by consensus and compromise rather than by force, it seems. If this thesis is true, if force is always contrary to self-interest, then there is no need for force ever, and its use always indicates the lack of enlightenment. And a society governed with no force is the liberal's Utopia. It is not a Utopia in the sense of Sir Thomas More, who, after all, invented the word, nor in the sense of the Utopians discussed in Friedrich Engels's *Socialism: From Utopia to Science*. It is a Utopia in the sense of being liberal to the utmost, and so we should judge as Utopian in this sense all Epicureans, including Machiavelli, Hobbes and Spinoza, as well as Locke, Hume, Smith, Bentham, and Marx — not to mention the fathers of the American Revolution and the French Revolution. There was never a good war or a bad peace, said Ben Franklin. Is this true? Of course it is. But the question remains, assuming the utilitarian canons, is this still true?

Clearly, the neighbors who have their property to protect from each other are better off if they can trust each other to live and let live. What about the penniless who has nothing to lose? Is he not better off taking other's property without the fear of his property being taken since he has none? The penniless, thus, refutes the basing of the liberal Utopia on Epicureanism. Not so, said Spinoza and Bentham: it is in the interest of the community to support the penniless — to give him enough property, indeed, so as to make it in his interest to live and let live. Is it not immoral to bribe a potential criminal? Not if it makes peace the treasure of all citizens, said the Epicureans. And will not the economy suffer thereby? No, said Spinoza, since once he has property the state

can levy high taxes from him, and he will be happy to pay them, as the
Dutch experience shows. No, said Bentham's teacher, Adam Smith,
since the settled stranger will become an entrepreneur and, provided he
is not taxed highly, he will increase the wealth of the nation. Will there
be no lapses from rationality? Probably there will, but not enough to
upset the established order, said Smith's friend and colleague David
Hume. And what should be done to the deviant and how? If he does
not agree to live and let live, he may be punished, and for that purpose
we need the state as a night-watchman, they all said. Does this not
refute the Epicurean view? No, said Rousseau, since putting a person in
jail under this condition is no less than forcing him to be free. This, of
course, is absurd. No, said Bentham, if jail will prepare him for civil
society. But what if he prefers to remain as he is?

Despite the absurdity, the liberal Utopians were the leaders, since
they developed a metaphysics, a psychology, a political science, a
political economy, a theory of legal reform and penology all of a piece.
This system of thought has approached its canonic presentation — from
the outside — in Elie Halévy's superb 1901—4 *The Growth of
Philosophical Radicalism* (1955), but has still not reached its final
canonic version, and largely due to the fact that some contemporary
political debates lean heavily on the writings of these classical liberals
— the radical right presumably on Smith and the radical left presumably on Marx — so that they find it hard to approach matters from the
outside. Yet we are sufficiently outside classical radicalism or classical
liberal Utopianism to draw a few conclusions.

First, Spinoza was in error to dismiss the Utopians on account of
their irrealism. Their ideas were very interesting, they had enormous
immediate political impact, and they could and did quite often temper
their Utopianism with *ad hoc* measures of realism. The most obvious
example is the realism with which the fathers of the American
Revolution tempered their Utopianism. Not only did they accept the
Machiavellian realist proposals of separation of state and church, the
separation of powers, and of checks and balances. They even instituted
the constitutional provision of constitutional amendments, and explicitly
on the ground of human fallibility, no less. Moreover, the application of
ideas from Utopian theory required that current society is a near-
Utopia. And this was done regularly, with a mixture of an unbelievably
naîve and tremendous optimism with an admirable sense of proportion.
With this optimism Condorcet saw the brilliant near future in the

darkness of his sorry state as a person wrongly persecuted by the Revolution which he wholeheartedly supported to the very end. With this sense of proportion liberal David Hume viewed his own country as a republic — and at the very same period when conservative Samuel Johnson saw it as tyranny.

The most important contribution of the liberal Utopians to political life was the boosting of the rise of the autonomous individuals. This was part and parcel of their Epicureanism. Yet it may be doubted whether this success were at all possible but for the realist branch of the Epicurean movement, namely Machiavelli, Hobbes and Spinoza. So we have to move a step back, from the eighteenth century, to its immediate past.

4. THE RISE OF REALIST LIBERALISM

How can one view the authors of *The Prince* and *Leviathan* as Liberals? Soon after Machiavelli's death a pamphlet was published, allegedly by himself, in England and in English, declaring his book an ironic and satirical work. This view was echoed by Rousseau (in his *Social Contract*). Spinoza was more cautious. He left open the question, why *The Prince* was written, and noted that in his other writings Machiavelli advised the people not to trust a prince. There is no doubt that Machiavelli did express very liberal, very democratic, very republican ideas in his other writings, ideas which had an enormous influence on the fathers of the American Revolution. With so much important liberal and democratic influence, how could he also publish such a subversive work as *The Prince*? The problem has worried the author of the essay on Machiavelli in the prestigious *Encyclopedia of Philosophy* (Gilbert, 1967). Not that the author articulated this problem; but he did answer it: Machiavelli wrote for ulterior motives, and he who paid him called the tune. I mention this to illustrate the depth of intellectual misery that is still to be found amongst leaders of political thinking five centuries after Machiavelli's death. One can imagine how frustrated he must have felt in the company of the leaders and employees and scholars of Florence of his days. What was one to do under the circumstances? How does one feel as an Epicurean amongst people who show so much ineptness in the art of thinking and in the art of living?

It is hard to fathom today the depth of this question. And it is not

only a political matter. Contrary to what one might presume, political
life was varied and dynamic at the time, democracy was a familiar
form of government, and even in trade guilds elections and voting
were common. But the sense of impotence which pervaded medieval
Europe like a miasma was still extremely detrimental. About half a
century before Machiavelli organized a national militia Fillipo Lappi
Brunelleschi built the dome of the Florence cathedral in an ancient
Roman way and caused great excitement. Machiavelli got the idea of a
militia from Antiquity, as he did with other political ideas of his. The
point is that he created a militia — and it worked. The Renaissance was
just that, the rebirth of Antiquity and the overcoming of the sense of
helplessness of the Middle Ages. This is part and parcel of the revival
of the Epicurean spirit which made the rebirth of science possible too.
Not only is the rise of political science in the Renaissance equal to the
rise of the physical science. It was a precondition for it, an incentive for
it, and an ally to it.

But let us return to our question. How does one render Epicureanism a political view with no one enlightened enough to share the idea of
enlightened self-interest and to create a community of enlightened
citizens? The ancient Epicureans (and Stoics) shrugged their shoulders.
Like Confucius, they said, if people behaved properly politics would be
unproblematic. But this is not good enough. The trend had to start
somewhere. It is obvious that Machiavelli, Hobbes and Spinoza had to
precede Hume and Smith and Bentham and Marx. But how did they
begin the trend?

Unless one is fully cognizant of this question, or of this quest, one is
bound to misread Machiavelli, and to a lesser extent but in the same
manner, Hobbes and Spinoza. Consider Bertrand Russell, whose view
(*History of Western Philosophy*) is the best. He had an admiration for
Machiavelli for his intellectual honesty; he notices that Machiavelli's
Prince reminds the reader of his having written of republics elsewhere,
thus making the *Prince* and the *Discourses* two parts of one book; and
he notices that the *Discourses* propose a liberal democratic philosophy.
Nevertheless, Russell says Machiavelli would have approved of Hitler's
burning of the Reichstag and of other tactics which, one might add,
Hitler may well have learned from Machiavelli by one route or another.

Would Machiavelli — or Spinoza — approve of Hitler? This is the
central problem cutting through metaphysics to politics. Epicureans do
not disapprove, and Spinoza even declared everybody a part of God

and thus everybody's act an act of God. Evils were often blamed not on the evil doer but on the good people who permitted it, and this sounds perverse. To take a recent example, since Israel is blamed for having permitted a massacre to occur, whereas the brutes who have committed the massacre are dismissed with no more than a disgusted glance, and their government's involvement in the massacre is totally ignored, many Israelis deem this a great injustice. Now, Israel is, of course, complimented by this outcry, and if things do go downhill further, then Israel too, like Lebanon, will deserve no more than a disgusted glance. Machiavelli's admiration for Cesare Borgia includes both a realistic dismissal of all other politicians and an assessment of the special talent of Borgia: he could, despite all his corruption, do what the equally corrupt competitors he was facing could not do: in brief Machiavelli thought that only brute force could unite Italy in her corrupt state and return it its old dignity and enable Italy to arise as a strong democracy. Hence Russell seriously misjudged Machiavelli's subjunctive judgement of a Hitler. Were Machiavelli alive he would not have defended or attacked Hitler but he would have denounced Hitler's opponents, in Germany and outside it, especially the British, for having been so foolishly unprepared. On this, doubtless, he would have agreed fully with Winston Churchill. Let us remember that Spinoza criticized the leaders of the Dutch Republic, his lynched friend de Witt, that is, in the same manner. Spinoza's critique was a harsh self-criticism.

But suppose a tyrant would unite Italy; how will this liberalize the unified country?

It is this part of *The Prince*, let alone the *Discourses*, which is so important, both by internal logic and historically, and which is so very often simply overlooked by historians and commentators of all sorts — the prince who has won power should do all to maintain it, and for this Machiavelli has a simple recipe: make your servants fear you but make the nation love you. This is Epicureanism. Jeremy Bentham declared himself the Newton of the human sciences since, he said, he discovered the principle of universal gravitation in the human sphere: men need friendship, and they acquire it by mutuality, as in Newton's third law. Nor is the analogue with Newton a thing of the distant past: in his very last work in 1942, R. G. Collingwood announced his three laws of politics as a framework for (a modified version of) Bentham's universal gravity. Of course, what Bentham staked a claim for was not that people are gregarious, but that penology can be founded on this general fact. And, of course, his logic was wanting and diverse attempts to fill gaps, from Mill through Pareto and Hayek, were and still are offered.

Machiavelli's idea that the prince's enlightened self-interest should make him serve the people is the beginning of modern political science. By this I do not mean to say I agree with Machiavelli. It is hard for me to say. Were it shown that Hitler could not maintain the peace by terror, there still is Machiavelli's other monstrous disciple, Stalin, to take account of. Stalin won the love of millions by practicing terror! Hence, it seems, Machiavelli's theory is false. But I cannot say. What I do positively propose is that this is the starting point of the future canonic version of the traditional classical liberal philosophy. Before offering an outline of a proposal for a canonic version, let me return to the problem of transition. For, clearly, leaving matters in the hands of a prince is not enough, since the population under a prince may learn to be so very submissive as to make tyranny the status quo. Nor was Machiavelli insensitive to this problem, which he discussed at some length in his *Discourses,* though it was neither he nor Hobbes but Spinoza who attended to it seriously. Nor is the question easy to solve when put in concrete political terms.

The most obvious and important concrete political case is religion and religious tolerance. Machiavelli, Hobbes, and Spinoza — all three recommended both state religion and religious tolerance. Bertrand Russell confessed inability to comprehend this point. Let us try to see whether the imposition of intolerance can be graded so that it can be reduced in stages. In a sense this can be done. After the Glorious Revolution of 1688, which granted religious freedom in England to Catholic and protestant alike, John Locke advocated toleration of Jews and Mohammedans too, and even to heathens, but not to atheists, meaning not to people who belong to no church — like Spinoza. In England today atheism is tolerated too. But this grading is obviously unsatisfactory. Rather, England has also graded toleration differently. It still has a state religion though of almost no import — it was politically more important a century ago and still more important two centuries ago. How does this gradation come about? Through public education, pretty much as envisaged by Spinoza.

5. AN OUTLINE OF THE HISTORY OF CLASSICAL LIBERALISM

Machiavelli proposed that a tyrant who comes to destroy corruption by corrupt means is admirable, especially since his single-mindedness may guarantee that he will not engage in petty corruption and that when his interest calls for proper conduct he will conduct himself properly. And, Machiavelli said, the best interest of a successful tyrant is to be loved and the best means for this is to be good. It was David Hume, let us remember, who has ascribed to Machiavelli the liberal political principle: private vices can be public virtues. The prince is vicious but should do good all the same. Hobbes turned this end point of Machiavelli's to his central axiom: a ruler is permitted all on two and only two conditions: first, that he maintain the peace, and second that what does not pertain to the peace is the private affairs of the citizens and should be treated as such in good faith. Not in vain did Leo Strauss view Hobbes's politics as thoroughly capitalistic. But what one need observe is its liberal thrust within politics.

Lest one might see here an apologetic attitude to Hobbes, let me briefly state, first that I do not deem Hobbes's politics correct, and second that the King and his Restoration entourage were not deceived by Hobbes's permission to the tyrant to rule his nation as he deems fit and they explicitly denounced Hobbes.

The liberal aspect of Hobbes can be seen clearly in his idea of a state religion. There is no doubt he was Machiavellian about it: he both abhorred it and recommended it: following Spinoza's lead we may simply suppose that Hobbes recommended state religion for political purposes on the understanding that when its services are no longer required it be rescinded, and on the understanding that only the ignorant, not the enlightened, need it, but that the enlightened can always wisely pay it the required lip-service.

With Spinoza this became the cornerstone; interestingly Russell ascribes this view more to Spinoza than to Hobbes, and he finds it hard to follow even in its Spinozist garb. Now, the Epicurean principle means that we live in a democracy no matter what regime we live in. This principle was reaffirmed by both Hume and Rousseau: no government can survive without some consent. Further, there is never any moral blame. What deserves any blame is either legal and so a practical affair, or, more deeply, it is rooted in error. This is Socratic eudaimony: we never knowingly act against ourselves, and our own interest (by definition) prescribes goodness. Hence all sin is but error. Hence the enlightenment of the individual guarantees his liberty. Hence a state religion is but for the unenlightened. Shlomo Pines has shown that

Spinoza was serious in his advocacy of a state religion, and Pines is right. Yet this should not for a moment bring us to overlook the profound liberalism of Spinoza.

This is the crux of the matter, and, in the history of politics, the major discovery as presented here. Epicureanism seems to hardly apply to politics, or else to apply in a liberal Utopian manner. Spinoza was anti-Utopian, as he clearly declares in the very opening of his *Political Treatise*. He argues geometrically, i.e., deductively from his principles. But , he promises, he will only deduce non-Utopian corollaries from his axioms. That is like saying, geometry applies both to perfect circles and to squares, but since most people are squares, I will apply geometry only to squares. This is not clear enough, since, in my reading, Spinoza wants the squares to become circles.

He puts it like that: the educated needs no laws. The purpose of laws is both to direct the uneducated and to educate him. The end of law, then, is a state where law is no longer required. This is Bentham's whole legal philosophy, and it is wholly Spinozist, and through Bentham we look at Spinoza and see his views as trite. Yet, when we compare Spinoza's view of the law with the views of his predecessors, we can see the contempt for the common man expressed in demands for harsh laws which cannot educate, and which repress even the most enlightened.

The transition from medieval views to Renaissance views was the transition to Epicureanism. Machiavelli's Epicureanism was realistic but not visionary, Spinoza was visionary; his vision became common sense when it was taken to be a matter of course. This process was the transition from realistic liberalism to Utopian.

The transition from realist liberalism to Utopian was Locke's Second Essay. It is, indeed, an inconsistent book which contains allusions to Machiavelli, but none to Hobbes and Spinoza whom he is reported to have denounced. (He makes a generous acknowledgement to Richard Hooker, which the latter's biographer in the *Encyclopedia of Philosophy* (Munz, 1967) declares unfounded.) Clearly, the theory of separation of powers and of checks and balances, which is often ascribed to Locke with a reasonable measure of justice, this theory has no room within Utopian liberalism; it was borrowed from realist liberalism. This inconsistency was diminished by Hume and Smith who declared the real very near Utopia. It was solved by Marx who claimed that private property made the real ever further from Utopia, yet the socialist

revolution etc. What is common to all liberals is the idea, invented by
Machiavelli, that private vices can be public virtues. Liberalism is the
attempt to make them virtues. The line of descent from Machiavelli to
Marx is thus very clear with powerful inner logic. The central figure on
the process clearly was Spinoza, who democratized liberalism. The last
version of Spinozism is Marx's mature political theory. In it he refuses
to blame the capitalists — they are as much caught in the net, he said,
as the workers. In it he refused to blame the workers for being content
with wage increase and he advised his followers to join workers in their
struggle as the workers understood it and help workers develop a better
understanding of the situation and of the need for the proletarian
revolution. Yet, unlike Spinoza, Marx could not refer to the law as
educational, but referred instead to the logic of the situation.

There is, finally, the question, why was Spinoza not the acknowledged master I claim he was? The answer is quite simply political:
people were cowards. Let me mention one example: the fact that
Immanuel Kant praises Descartes and Leibniz, yet notices that they left
the crowning achievement of their philosophy to another to attain, who
is the one to have united epistemology and ethics. We shall never know
who that other was (Cf. Agassi, 1969, pp. 501—502). Yet, let me
mention also the important idea of Kant's ethics: we have to be moved
by a sense of duty, he said, though in principle, were we all as
enlightened as we should be, enlightened self-interest would be both
sufficient and preferable. This became the cornerstone of the ethics
of Schopenhauer. This should suffice to show how popular this
unmentionable fellow really was who, I conjecture, is none else than
Spinoza. It is hard for me to decide how much Spinoza owed to either
Machiavelli or Hobbes. It is clear, however, that his unrestrained
expression of views which were clearer and more logically streamlined
than anything that went before, made him rightly the object of both so
much admiration and so much contempt. Today we take so much of his
teaching for granted that we overlook it. Paradoxically, our very
rejection of his views, which we should express with more intellectnal
honesty and civil courage, will help us appreciate them — from the
outside — all the more. And, not paradoxically, the defects in his logic
should bring us to attempts to update his views — as was done for
example by R. G. Collingwood in his *New Leviathan* (1942), in which
he only grudgingly recognized Spinoza (in a footnote), but without
endorsing his philosophy; or as was done, at about the same time,

1933, by Alexandre Kojève (1969) in his commentary on Hegel's *Phenomenology*, in which Hegel is presented as a follower of Hobbes's Epricureanism — yet without endorsing the corrected versions of Spinoza's views. It is a sad fact that Spinoza's radical ideas are the cornerstone of today's neo-conservatism and neo-radicalism. I do not think he would have felt flattered by such a compliment.

Tel-Aviv University, Israel, and
York University, Canada

REFERENCES

Agassi, Joseph: 1969, 'Unity and Diversity in Science', in R. S. Cohen and M. W. Wartofsky (eds.), *Boston Studies in the Philosophy of Science*, Vol. 4, Reidel, Dordrecht, 463—522. Reprinted in Joseph Agassi, *Science in Flux*, Reidel, Dordrecht, 1975, pp. 404—468.
Aubrey, John: 1972, *Brief Lives*, Penguin Book, Harmondswork, England.
Bacon, Francis: 1906, 'Of Goodness and Goodness of Nature', in *Essays,* Everyman's Library, New York.
Collingwood, Robin G.: 1942, *The New Leviathan,* Clarendon, Oxford.
Einstein, Albert: 1956, *The Meaning of Relativity,* transl. by E. P. Adams, 5th edition, Princeton University Press, Princeton.
Gilbert, Felix: 1967, 'Machiavelli, Niccolò', in Paul Edwards (ed.), *Encyclopedia of Philosophy,* Vol. 5, Macmillan, New York, pp. 119—121.
Halévy, Elie: 1955, *The Growth of Philosophic Radicalism,* transl. by Mary Morris, Beacon, Boston.
Kojève, Alexandre: 1969, *Introduction to the Reading of Hegel; Lectures on the Phenomenology of the Spirit,* assembled by Raymond Queneau, ed. by Allan Bloom, transl. by James H. Nichols, Basic Books, New York.
Levi-Malvano, E.: 1912, *Montesquieu e Machiavelli,* Université de Grenoble and Institut Français de Florence, Grenoble.
Munz, Peter: 1967, 'Hooker, Richard', in Paul Edwards (ed.), *Encyclopedia of Philosophy,* Vol. 4, Macmillan, New York, pp. 63—66.
Wisdom, J. O.: 1939—41, 'The *Analyst* Controversy: Berkeley's Influence on the Development of Mathematics', *Hermathene* **29**, 3—29; 'The *Analyst* Controversy: Berkeley as a Mathematician', *Hermathene* **53**; 'The Compensation of Error in the Method of Fluxions', *Hermathene* **57**, 49—81.

RICHARD H. POPKIN

SOME NEW LIGHT ON THE ROOTS OF SPINOZA'S SCIENCE OF BIBLE STUDY

In previous writings about Spinoza's Biblical criticism, I have stressed the connection of some of Spinoza's central points with those presented by Isaac La Peyrère in his *Prae-Adamitae* of 1655 (Popkin, 1974, 1976, 1979, 1973, 1978). La Peyrère had denied the Mosaic authorship of the Pentateuch, the accuracy of the existing Biblical texts, and the Biblical claim that everybody is descended from Adam. La Peyrère's book was published in Amsterdam in five editions in 1655, was denounced far and wide and its author arrested in 1656 in Brussels.

Spinoza and his associates of the time, Juan de Prado and Daniel Ribera seem to have adopted some of La Peyrère's heretical views, and these ideas apparently played a role in the excommunication charges against the three rebels. Spinoza owned a copy of La Peyrère's *Prae-Adamitae*, and passages from it are used in the *Tractatus-Theologico Politicus.*[1]

More recently, while rummaging through some recondite material in the William Andrews Clark Library in Los Angeles, and by examining some of the rcords of the Quaker missionaries in Amsterdam in the period right after Spinoza's excommunication, I have found two other critiques of the Bible which also seem to have played an important role in the development of Spinoza's analysis of the Bible. One is the epistemological-historical attack on the claim that the Bible, the written word, the Scripture, is the Word of God offered by the leading Quaker polemicist, Samuel Fisher, 1605—1665, and the other is the embryonic formulation of the horrendous thesis of *Les Trois Imposteurs,* the contention that Moses, Jesus and Mohammed were political imposters, using their claim to divine revelation as a means of establishing political orders.

In the last couple of years I have examined evidence that Spinoza came in contact with the Quakers soon after his excommunication, that he probably translated a pamphlet of theirs, by Margaret Fell, from Dutch into Hebrew, that he probably knew Samuel Fisher personally, who before he went off to try to convert the Pope and the Sultan was in

171

Marjorie Grene and Debra Nails (eds.), Spinoza and the Sciences, 171—188.

Amsterdam in 1656—57 (and was the only Quaker in the Amsterdam group who knew Hebrew). Fisher has been called the most radical Bible critic of the period by Christopher Hill (1972, pp. 208—215). As we shall see, a careful comparison of some of his arguments (in his 760 folio page opus) with those of Spinoza is most striking.

After Spinoza's excommunication on July 27, 1656, he went to live with the Collegiants, a group of radical millenarian Christians. The leader of the collegiants (insofar as they had any organization) was the prominent Dutch Hebraist, Adam Boreel, a good friend of Rabbi Menasseh ben Israel. In April, 1656, Boreel received a letter from Henry Oldenburg (who was to become Spinoza's most important foreign correspondent) informing him that somebody was claiming that the whole story of Genesis was composed in order for Moses to gain political control of the Jews, that Jesus tried to gain political power more prudently by enticing people wth the hope of eternal rewards, and Mohammed attempted this even more so. (This is the basis thesis of *Les Trois Imposteurs*, Chapter 3 and some of the same sentences appear in Oldenburg's letter and in the eighteenth century manuscript and printed texts.) Oldenburg beseeched Boreel to save Christianity by refuting these claims. Further correspondence shows that Boreel was working on his refutation for the rest of his life (until 1665), and that while he was dying Oldenburg, while corresponding regularly with Spinoza, was paying Spinoza's patron, Peter Serrarius, to have a copy of Boreel's work made. Robert Boyle paid the substantial bill, and Boreel's manuscript is among the Boyle papers in the Royal Society. (Henry More also had a copy of Boreel's work that he got from Francis Mercurius van Helmont.)

Thus when Spinoza moved in with the Collegiants, after his excommunication from the Synagogue, its leader Adam Boreel, was working to refute the central claim of the *Trois Inposteurs*. When one realizes Spinoza had probably heard of this claim, either from Boreel, or later from Oldenburg, one can see how Spinoza's interpretation of Biblical history is a way of dealing with this radical challenge, and a way that is quite different from Boreel's.

Lastly, in the spring of 1657, Rabbi Nathan Shapira of Jerusalem visited Amsterdam, and astounded the Dutch and English Millenarians with his philo-Christian interpretation of Scripture, offering what I have called a Jewish interpretation of Christianity rather than a Jewish rejection of it. Rabbi Shapira's views, as we know of them through the

Millenarian accounts, also seem to have been taken up in Spinoza's *Tractatus*.

Hence, in addition to the critical material and outlook Spinoza gained from Isaac La Peyrère, I think he was also greatly influenced by the anti-Scripturalism of Samuel Fisher, the discussion of political interpretations of the roles of Moses and Jesus, and the evaluation of Christianity of Rabbi Shapira. I will examine the possible contribution each of these could have made to Spinoza's Biblical criticism.

Samuel Fisher, 1605—1665, was an Oxford graduate, who first became a Baptist minister. In 1654, two Quakers, John Stubbs and William Caton, persuaded Fisher to join their movement. He tried to witness his faith during a meeting of Parliament, and was ejected. He decided to convert the Pope and the Sultan.[2] En route he went to Amsterdam with Stubbs, and worked with Caton in attempting to get the Jews to see the light. During 1656—57, he attended the Amsterdam synagogue, and discussed religious issues at length with members of the Jewish community.[3] He apparently was involved with Spinoza in arranging the translation of two of Margaret Fell's pamphlets into Hebrew. He appended a Hebrew exhortation of his own to Spinoza's translation of Margaret Fell's *A Loving Salutation to the Seed of Abraham among the Jews*.[4] In 1658, he went to Rome and Constantinople, without succeeding in converting either Pope Alexander VII or the Sultan. In 1660, he was back in England debating various Puritan theologians. His massive answer to them, especially to John Owen, the vice-chancellor of Oxford, entitled *The Rustics Alarm to the Rabbis*, was published in 1660.[5] It is in this work that he presented many arguments that seem extremely close to what Spinoza developed in the *Tractatus-Theologico-Politicus*, Chapter 7—12.

For Fisher the overall problem to be faced by those who claim that Scripture is the Word of God, is to establish what is the real Divine Writing, rather than a man-made set of written letters. This human effort, the written work, called 'Scripture' is the result of a series of historical events that have occurred from early Jewish history down to the present. The actual Word of God is independent of any attempts to put words on paper, and God's Word existed from all time, and exists in all places. The Word existed long before Moses wrote down anything. From Abel, from the period of the Patriarchs, to the time of Moses, there was the Word of God, which was what the faithful lived by.[6]

Fisher contended that the Protestants have confused the latter with the Word. They have made Scripture, the man-made letters, the rule of faith. But they could actually only tell if Scripture was God's Word, if they knew the Word independently in order to compare it with what has been written down.[7]

Historically the writing down process seems to have begun with Moses. Since then people have been transcribing and translating what was written by Moses and others after him. (Fisher, like Spinoza and La Peyrère, denied that Moses was the sole and complete author of the *Pentateuch*.[8]) Human beings have written down and copied even the words attributed to God in the Scripture. No original holograph copies by God, by Moses, by early scribes exist. Scripture itself tells us that the texts that existed at the time of Ezra had become corrupt, and that Ezra had to correct and restore the text.[9] Thereafter, it was copied by one copyist after another. The point Spinoza stressed, the addition of Hebrew vowels sometime from Ezra's day or later, was made by Fisher. The vowels were not in the original, pre-Ezra, documents, and so another crucial human addition to the text book place.[10] Something like this happened with the text of the New Testament, when the iota subscripts were added to the written documents.[11]

Further, Fisher pointed out, it is not stated in Scripture that only the books presently bound together, and no other ones, constitute the Bible. The decision concerning what is canonical was made by a group of rabbis at a later date than the writing of the books. Many books, whose titles we know of, were left out, presumably for human reasons. Similarly some early Christians decided that the present collection of documents, the four Gospels, the Book of Acts of the Apostles, the epistles, and the Revelation of Saint John, were the only writings to be taken as Scripture. Other documents, including letters of St. Paul, were excluded from the canon by human beings for human reasons.[12]

If the Protestants try to rebut Fisher's point, and seek to justify the selection of what constitutes Scripture by appealing to the Jewish and Christian tradition, as Jews and Catholics do, then they are relying on fallible Jews and Christians of the past to determine what is and what is not the Word of God.[13] At present we have no original copies of the Scriptures. It is possible that in the transcription process from ancient times to the present that all kinds of errors have occurred in the course of copyists' copying the documents that came down to them. The late addition of Hebrew vowels and Greek iota subscripts have allowed for all sorts of variants. At present, Fisher said, there are about seven thousand known variants amongst the surviving manuscripts.[14]

If this is the case, then how can we now be sure what constitutes the right text? The text itself does not tell us. We can only appeal to the actual Word of God, or to the Light or the Spirit. From such a basis we can judge which man-written text properly expresses the Word of God: the Letter, the Word Writ, may not be the Word of God. We can only determine whether it is, if independently we know the Word of God.[15]

The Protestants who worship a book are as idolatrous as the Jews who worship the Bible or Torah Scrolls. Any existent Bible or Torah Scroll is a man-made object that has come to be what it is through a series of human activities like transcribing, editing or printing. These activities per se could not guarantee that the resulting documents state the Word of God, or are correct copies of written documents that did contain the Word of God. One could only determine the accuracy and exclusiveness of the text, if one knew independently what constituted the Word of God. This was known, at least to the Patriarchs, before there were any written documents, and it is still being revealed to the faithful through the Light or the Spirit.[16]

Fisher was willing to agree with his opponents that the original Biblical authors could have been divinely inspired, which would explain how they managed to contain the Word of God in their writings. But was there any reason at all to believe that the many misguided Jewish and Catholic copyists throughout the ages, or the greedy printers from Guttenberg onward, have been divinely inspired?[17] The Divine Message is eternal. However, there is no reason to believe that any presently existing written documents contain or constitute this message, unless we find this to be the case by the Light or the Spirit, which Scripture is supposed to reveal.

Fisher only published his series of arguments against taking the Letter for the Word in England in English in 1660, in a huge work of over 900 pages.[18] If, as sems to be the case, he knew and worked with Spinoza in Amsterdam in 1657, he could have told him some or all of his points in person (or he could have learned some of them from Spinoza). It is the case that many of the same points, and even some examples, appear in Spinoza's *Tractatus-Theologico-Politicus*, Chapters 7—12, as appear in Fisher's rambling text, when they discuss the history of the Scriptural texts, the loss of the originals, the transcription problems, the problems resulting from the introduction of Hebrew vowels, the state of present copies, and the human character of the existing text.[19] Spinoza himself said almost the same thing as Fisher did when he declared,

> They that look upon the Bible, however it be, as a Letter sent from Heaven by God to Man, will certainly claim and say, I am guilty of the sin against the Holy Ghost, in maintaining that the Word of God is faulty, marred, adulterated and contradictory to itself: that we have but fragments of it, and that the Original Writing of the Covenant which God made with the Jews perished.

Spinoza attacked those who worship paper and ink as if they constituted the Word of God (Spinoza, 1951, pp. 165—166, Chapter 12; Gebhardt III, pp. 158, 160—161).

The striking similarities between Spinoza's views and Fisher's about the nature of the written Bible need to be examined in detail (as well as the similarities of Fisher's case to that of Richard Simon). In view of the vastness of Fisher's work, a patient study of it will have to be put off for the present, except to note that much of Spinoza's Biblical criticism about the epistemological status of the present Scripture as a statement of the Word of God appears in Fisher's work, which was published ten years before Spinoza's *Tractatus* and which could have been transmitted from Fisher to Spinoza by personal contact.[20]

Before going on to consider other possible sources of Spinoza's views on the Bible, a few words should be said about whether Spinoza could have known the contents of works written in English, like Fisher's *The Rustics Alarm to the Rabbis*. What we know about Spinoza indicates that he knew Spanish, Portuguese, Dutch, Latin and Hebrew and perhaps some French and Greek. (The Quaker documents relating to him stress that he did not know English.[21]) He cited no works in English, used no English terms, and had no English books in his library. His fifteen year correspondence with Henry Oldenburg, who was in England, shows no sign that Spinoza picked up any English. However, a paper presented at the Amsterdam commemoration of Spinoza's 350th birthday by Sarah Hutton shows there is an amazing similarity between Spinoza's chapter on prophecy in the *Tractatus* and an essay on the subject by the Cambridge Platonist, John Smith, that was only published in English in 1652 (Hutton, 1985). Smith gave his citations from authors like Maimonides in English, Latin and Hebrew. Cambridge Platonist authors were in contact with various theologians in Holland, and their works were read by many of them. Spinoza could have seen what source materials Smith had used, and could easily have asked people he knew who knew English to tell him what Smith had said about these texts. Spinoza knew many people who knew English. Two of them, who are important for the rest of the material in this paper, were Adam Boreel, the leader of the Amsterdam Collegiants, into whose lodgings Spinoza moved after his excommunication, and Peter Serrarius, the 'dean' of the dissident Milleniarian theologians in Amsterdam, who seems to have been Spinoza's patron. Boreel was a graduate of Oxford, class of '45, and was often in England. He was in London the fall of 1655, entertained Menassah ben Israel in his home there, and interfered in the discussions about whether to readmit the Jews to England.[22] Serrarius is usually misidentified as a "Belgian chiliast, born in 1636", and is portrayed as the wildest religious lunatic of the time. Actually he was a French Protestant, born in 1600 in London, and raised there until he went to study theology in Leiden. He gradually became a theologian-at-large, belonging to no church. His home was the center for Millenarian theological discussion. Jan Comenius, John Dury, and many others stayed with him. Serrarius regularly communicated with England and sent messages, books and manuscripts to Samuel Hartlib, Henry Oldenburg, and Robert Boyle, and received materials back for Boreel, and others including Spinoza. He was Spinoza's actual link with England. He also was apparently the person who introduced Spinoza to the English Quaker mission.[23] Either Boreel or Serrarius could have told Spinoza what was said in a work written in English. So, Spinoza was not limited to his own personal linguistic abilities. (The catalogue of Serrarius's books that I located shows that he possessed a great many works in English.[24])

Adam Boreel received a letter in April, 1656, from Henry Oldenburg, then at Oxford, reporting about two problems that had recently been raised. (It is not made clear if they were stated in a book, or told to Oldenburg.) The first was that it claimed that the whole story of Creation was invented by Moses in order to introduce the Sabbath, which Moses used in order to establish political control over the Jewish people. The second problem was that Moses, Jesus and Mohammed, each in his own way, became leaders by promising people certain rewards.[25] Oldenburg's description of these problems constitutes an outline of what usually appears as Chapter 3, "Qu'est-que c'est la religion?" in the clandestine work, *Les Trois Imposteurs*.[26] Some of the sentences in Oldenburg's letter (in Latin) are exactly the same, word for word, as the French text that was published in the eighteenth century. In this text Hobbes' *Leviathan*, which appeared in 1651, is also cited.[27] So, Oldenburg had apparently been confronted by a recently composed or stated version of some of the heretical theses of *Les Trois Imposteurs*. He was shocked, and urged his friend, Boreel, whom he greatly admired as a theologian, to refute these irreligious claims, and thereby to save Christianity.[28] Oldenburg's letters from January 24, 1657, until 1665, when Boreel was dying report that Boreel was working on a refutation in which he would show "that the origin of religion is truly divine and that God appoints no one but himself to be the legislator for the whole human race", thus refuting a human political interpretation of Judaism, Christianity and Islam.[29] In 1665 Robert Boyle got Oldenburg to pay Serrarius a large sum of money to copy Boreel's manuscript on Jesus Christ the Divine Legislator.[30] At least two copies existed. Boyle had one that is now in the Royal Society archives, and Henry More had another that he discussed in some of his theological works.[31]

When Spinoza was excommunicated from the Synagogue in July 1656, he entered the world of the Collegiants at just the moment when its leader, Adam Boreel, was commencing his defense of Christianity against the political interpretation of the religion offered in *Les Trois Imposteurs*. While Spinoza mingled with the Collegiants, and presumably discussed religious matters with them, it was of paramount importance to the group in Holland and to their friends in England to have Boreel produce evidence that Christianity was not just a political fraud or imposture. When Oldenburg visited Holland in 1661, and made a trip to Rijnsburg to meet Spinoza personally, and to have a long talk with him about religious and scientific matters, Oldenburg was greatly concerned about Boreel's efforts to defend Christianity.[32] A couple of years later, we learn through Oldenburg's correspondence that Spinoza was writing the *Tractatus* at exactly the same time as Boreel was composing his defense of Christianity.[33]

I have not yet seen Boreel's work (which never seems to have been published), but from what is said about it in the Oldenburg correspondence, Boreel and Spinoza were apparently dealing with the same themes raised by this early version of *Les Trois Imposteurs*. Spinoza, in the *Tractatus*, explained how Moses gained political power over the Jews, how the Hebrew theocracy was established, and why Jesus should not be considered as a political leader, either human or divine (which is contrary to Boreel's thesis that Jesus was and is The Divine Legislator).[34] In one of Spinoza's letters, he made clear that he, like most Europeans, was sure that Mohammed was an imposter.[35]

In Spinoza's account of how Moses came to power, and how he became the lawgiver for the Jews, Spinoza pointed out that after the exodus from Egypt the Israelites were no longer subject to the laws of any nation. They were in the state of nature, an insupportable condition. At this point, Moses induced them "to transfer their power upon no mortal man, but upon God only". Moses, therefore, by his virtue and the Divine command, introduced a religion, so that the people might do their duty from devotion, rather than from fear. He bound them over by benefits, and prophesied many advantages for the Isrelites in the future (Spinoza, 1951, pp. 218—219; Gebhardt III, pp. 205—206).

So, in Spinoza's interpretation of this part of the Biblical account, Moses benignly rescued the Jews from their lapse into the state of nature. They, as most people in the state of nature, were unfit and unable to frame a wise legal code. Moses did this for them, and gave it sanction by putting it in religious terms. This was a benign fraud rather than a malicious one. The Mosaic framed legal system was fine for its time and circumstances, but, Spinoza contended, has no moral force for later societies.[36]

Having explained Moses's role as a political legislator using religion for good purpose, Spinoza offered a radically different account of Jesus's role. Jesus was not a political lawgiver at all. "Christ, as I have said, was sent into the world not to preserve the state nor to lay down laws, but solely to teach the universal moral law".[37] Moses was, for Spinoza, a legal benefactor, not an imposter in the bad sense. Considering the state of affairs after the Exodus, he did what was fine in that time a circumstance. Jesus, much later on, was teaching universal morality, not law, so was not a political imposter either.

Boreel attempted to answer the charges of *Les Trois Imposteurs* by describing Jesus as the Divine Legislator. Spinoza offered another solution — make Moses a great legislator for his time and Jesus no legislator at all. Seeing Spinoza as constructing an answer to *Les Trois Imposteurs*, Chapter 3, helps clarify much of what he said about the roles or Moses and Jesus in human society. And, it is touching, nonetheless, that most of the manuscripts of *Les Trois Imposteurs* we now know of, written in the late seventeenth or early eighteenth centuries, are entitled *Les Trois Imposteurs, ou l'esprit de M. Spinoza.*[38] These texts use some Spinozistic terminology about the nature of God and substance, and seem to have reached their final form through the efforts of some people in Spinoza's circle or people immediately influenced by him.[39]

If Spinoza was influenced concerning his interpretation of the role of Moses and Jesus by the embryonic form of *Les Trois Imposteurs* that he and Boreel were answering, his views also seem to reflect another development in the Amsterdam theological scene shortly after his excommunication. In 1657, Rabbi Nathan Shapira of Jerusalem came to Amsterdam to raise funds for his starving brethren in the Holy Land. Rabbi Shapira was originally from Cracow, and then had studied the Lurianic Cabbala at Safed in Palestine under the great Cabbalist, Hayyim Vital. He taught, and was the teacher of Nathan of Gaza, the Elijah of Sabbatai Zevi's Messianic movement. The rabbi regularly went to Poland to raise money for the Jews of Palestine. The Swedish invasion of Poland prevented this from 1655 onward. Rabbi Shapira had told Menassah ben Israel of the plight of his brethren now that they were cut off from funds from Poland, and Menassah presented a letter from the Rabbi to Oliver Cromwell on the subject.[40] In 1657, Rabbi Shapira came to Amsterdam seeking help for the Jews of Palestine, and was rejected by the Spanish-Portuguese community. However, he met with Serrarius and other Christian Millenarian leaders who were most helpful.[41]

Serrarius was amazed by the philo-Christian views of the rabbi, and communicated them to his friends, John Dury and Henry Jessey, who published his views and raised lots of money for his cause. Rabbi Shapira's views, as reported by Serrarius and the English Millenarians, were an exciting sign that the Jews were about to convert. The noise the Millenarians made about this must have come to Spinoza's attention. And, as we shall see, Spinoza offered an interpretation of Christianity that was extremely close to that of Rabbi Shapira.

According to Serrarius, Rabbi Shapira, unlike leading Amsterdam Jewish leaders (such as Haham Saul Levi Morteira and Isaac Orobio de Castro)[42] did not hold that *Isaiah* 53 about the suffering servant did not apply to the Messiah. Christians had brought up this passage over and over again to show the Jews that the Messiah had come in the person of Jesus of Nazareth. There is a long series of Jewish answers disputing this (Neubauer, 1876). Rabbi Shapira instead offered the view that *Isaiah* 53 was about the Messiah, and that the Messiah was reborn in every generation, but could not stay because human beings did not repent and reform. The Spirit of the Messiah had preceded Adam and had appeared in such figures as Heziekiah, Habbakuk and Jesus (Dury, 1658, pp. 12—13).

Another of Rabbi Shapira's views was enunciated when he joined Serrarius and his friends in studying the Gospel according to Matthew. The Rabbi offered as his opinion that the Sermon on the Mount was the fount of all wisdom, and its teachings had appeared in the sayings of the purest rabbis and in those of Jesus of Nazareth (Dury, 1659, p. 13). Further Rabbi Shapira told Serrarius that the Millenarians he had met in Amsterdam were so pure of heart that if ten like them were to pray in Jerusalem for the coming of the Messiah he would surely come (Dury, 1658, pp. 13—14).

Serrarius was overwhelmed by the rabbi's views, and was certain that they indicated the beginning of the process of the conversion of the Jews. He wrote a glowing account of his meetings with Rabbi Shapira in a letter to John Dury, in which he said,

When I heard these things, my bowels were inwardly stirred within me, and it seemed to me, that I did not hear a Jew, but a Christian, and a Christian of no mean understanding, who did relish the things of the spirit, and was admitted to the inward mysteries of our religion (Dury, 1658, p. 13).

Dury published Serrarius's letter in a pamphlet entitled, *An Information concerning the Present State of the Jewish Nation in Europe and Judea Wherein the Footsteps of Providence Preparing a Way for their Conversion to Christ, and for their Deliverance from Captivity Are Discovered.*[43] The Dutch and English Millenarians raised the money the rabbi needed, and sent it to him in Jerusalem along with a plan for the conversion of the rabbi and his brethren, and a request that Rabbi Shapira translate the New Testament into Hebrew.[44]

Rabbi Shapira's positive Jewish interpretation of Christianity, in contrast to the very negative ones being expressed by the leaders of the Amsterdam synagogue, is much like the philo-Christian views of Spinoza. The latter made the message of the Sermon on the Mount *the* central moral teaching. Rabbi Shapira is reported to have said that *Matthew 5—7* is "The Head of all Wisdom: and whoever walk according to it are more just than we" (Dury, 1658, p. 13).

In Chapter 12 of the *Tractatus*, Spinoza shared many of Rabbi Shapira's philo-Christian views. Both Spinoza and the rabbi rated Jesus extremely high, although both of them avoided claiming that he had any supernatural properties. Spinoza further asserted that the central message of the Scriptures, no matter how corrupt or faulty the text might be, was uncorrupted "even though the original wording may have been more often changed than we suppose". This divine Message

is an assertion which admits of no dispute. For from the Bible itself we learn without the smallest difficulty or ambiguity, that its cardinal precept is: To love God above all things, and one's neighbor as one's self. . . . This is the conerstone of religion, without which the whole fabric would fall headlong to the ground (Spinoza, 1951, p. 172; Gebhardt III, p. 165).

Later on, in Chapter 14, Spinoza said that "The Bible teaches very clearly in a great many passages what everyone ought to do in order to obey God: The whole duty is summed up in love to one's neighbor" (Spinoza, 1951, p. 183; Gebhardt III, p. 174). Spinoza reasserted this as point 5 of his list of dogmas of universal religion (Spinoza, 1951, p. 187; Gebhardt III, p. 178). Spinoza concluded his summation of this with a statement that is very much what Serrarius said about Rabbi Shapira. Spinoza asserted,

He who firmly believes that God, out of mercy and grace with which He directs all things, forgives the sins of men, and who feels his love of God kindled thereby, he, I say, does really know Christ according to the spirit, and Christ is in him (Spinoza, 1951, p. 187; Gebhardt III, p. 178).

Serrarius had told Dury regarding Rabbi Nathan Shapira,

is it to be believed that Christ is far distant from a soul thus constituted? or that any such thing can be formed without Christ in a man; for my own part, I confess I think, I see Christ in his Spirit, and I cannot but love him (Dury, 1658, p. 16).

As yet I have not found any data showing that Spinoza knew Rabbi Shapira personally. We have practically no information about Spinoza's activities at this time, except that he was in the ambiance of the Collegiants, the Quakers and others in Serrarius' world. Rabbi Shapira's startling views must surely have been known amongst these people, who were so concerned with bringing the Jews to a union with the Christians. The rabbi had, it appears, prepared the ground for Spinoza in proposing a Jewish version of Christianity, rather than a Jewish denial of it. Jesus's message was the sum of all wisdom. Jesus himself was part of the human historical drama, although he rose somewhat above it in incorporating the spirit of the Messiah. Nonetheless, he did not become a supernatural being. Spinoza held to this position throughout his career. In his final correspondence with Oldenburg, he was willing to accept the entire Gospel story except for the resurrection. Oldenburg told him that this just destroys "the whole truth of the Gospel history" (Spinoza, 1928, p. 361; Gebhardt IV, p. 330). Spinoza was unimpressed in view of the fact that the aim of Scripture was just to inculcate obedience to the Golden Rule, not to impart knowledge.[45]

In sum, Spinoza's evaluation of the Bible and his way of interpreting it seem to have included at least three additional aspects beyond what he had picked up from La Peyrère's challenge to the Mosaic authorship, the accuracy of our existing texts, and the universality of Biblical history. Much of Samuel Fisher's stock of epistemological and historical arguments undermining acceptance of the present text of Scripture as The Word of God appear in Spinoza's *Tractatus*. Since Spinoza probably knew Fisher personally, he could have been influenced by the Quaker polemicist, or they could have mutually influenced each other. Spinoza's concern with interpreting positively Moses's role as a politician, using religion to achieve political goals and denying Jesus any such role, seems to stem from the concern of the Collegiants, and especially their leader, Adam Boreel, to answer a portion of the irreligious claims of *Les Trois Imposteurs*. When one realizes that there is strong reason to believe Spinoza was party to the discussions about Chapter 3 of this horrendous text, some of Spinoza's concern about the political role of Moses and the non-political role of Jesus become clearer. And, finally, when Spinoza's surprisingly pro-Christian views are compared with those expressed by Rabbi Nathan Shapira in Amsterdam in 1657, Spinoza seems to have been following out the positive Jewish interpretation of Christianity of the Jerusalem rabbi.

All of this suggests, I think, that Spinoza's views about the Bible underwent quite a development from his rebellious attitude at the time of his excommunication, to a more restrained and positive view, at least about the moral message of Scripture, and role of Moses and Jesus, in the years in which he was in intimate contact with the Quakers, the Collegiants, and other Millenarians like Serrarius and Oldenburg. Further researches into the published and unpublished materials of these people may throw more light on Spinoza's development and his message in the context of his time. Spinoza's Biblical criticism may well represent a distillation of some of the radical views of his Millenarian friends, and a rational reformulation of their theory of mystical redemption through the internal purification of the spirit.

Washington University

NOTES

[1] See the list of Spinoza's books in Freudenthal (1899), Item 54. On Juan de Prado and Daniel Ribera, see Révah (1959, 1964).

[2] On Fisher's career, see the article on him in the *Dictionary of National Biography* and Braithwaite (1955), esp. pp. 288—294 and 426—428.

[3] See William Caton's letter to Margaret Fell, March 15, 1658 in Caton (n.d.), fol. 507; and Caton (1689), p. 40.

[4] On Spinoza's relations with the Quakers and his role in translating Margaret Fell's pamphlet, see Popkin (1984*c, b, d*).

Fisher's exhortation appears as the last two pages. The Hebrew translation was published in 1658. Two copies of it are in the Friends' Library in London. Professor Michael Signer of Hebrew Union College and I have prepared an edition of it for publication.

[5] The 1660 edition of Fisher is over 900 pages. It was republished in the posthumous collection of his writings (Fisher, 1679) where it is 768 folio pages long.

[6] Fisher (1679), pp. 50, 131, 196—201, 240—241, 264, 442—445, 460, 540 and 619. On p. 214 he said that Scripture is "meer a Graven Image as that is with Ink and Pen on Paper, or Skin of Parchment, a dead Letter".

[7] *Ibid.*, pp. 52, 128, 175, 296—299, 311, 336, 338, 424, 442, 449, 454, 476, 487, 541—542 and 680—681, where he said, "Truth is where it was before the Letter was, and will be in the Light and Spirit".

[8] *Ibid.*, p. 265. Did Moses "*write of his own Death and Burial and of Israels Mourning for him, after he was dead?*"

[9] Fisher's (1679) text is very repetitious, and the same points are made over and over again. These items are dealt with on pp. 275—276, 295, 325—327, 338—340, 354 and 414—418.

[10] *Ibid.*, pp. 194—195, 241 and 300—315.

[11] *Ibid.*, pp. 350, 378 and 444.

[12] *Ibid.*, pp. 265—295 and 470.

[13] *Ibid.*, pp. 218 and 287.

[14] *Ibid.*, pp. 300—311, 351, 354—360, 364—365, 371, 378, 395, 418, 433, 444, 459, 510.

[15] *Ibid.*, pp. 52, 128, 175, 296—299, 311, 336, 424, 449, 454, 476, 487, 541—542 and 680—681.

[16] *Ibid.*, pp. 199, 239—241, 242, 327—328, 418, 420—433, 459, 510, 521—523 and 530. On p. 523 Fisher accused his opponents of deifying "adored dead Transcript, Text and Corps of Greek and Hebrew Copies of the Scriptures".

[17] *Ibid.*, pp. 238, 263, 330—336, 338—339, 344—345, 350, 365—367, 372, 420, 433, 521—522 and 566.

[18] See Note 5 for details. Fisher, who knew Latin, deliberately argued in English to show his scorn for the 'rabbis' or 'professors' who tried to hide their ignorance in Latin, Greek and Hebrew.

[19] A detailed comparison of Spinoza's arguments and examples with those of Fisher needs to be made. I hope to do this in the not too distant future.

[20] They could easily have met in 1657 when Spinoza was interested in learning about Quaker doctrines, and was working on translating Margaret Fell's pamphlets. We are told that Fisher took Spinoza's translation of Margaret Fell's letter to Menasseh ben Israel with him when he left for Rome and Constantinople. See William Caton's letter to Margaret Fell, March 15, 1658, in Caton (n.d.), fol. 507.

[21] The letter cited in note 20 states that "the Jew that is to translate it unto Hebrew, could not translate it out of English". A later letter, William Ames to George Fox, October 14, 1658, *Barclay Mss.*, also in the Friends' Library in London, fol. 19, says that the translator "Who hath been a Jew" "cannot understand English".

[22] Henry Oldenburg's letter to Menasseh ben Israel, July 25, 1657, in Hall and Hall (1965), Vol. 1, pp. 124—125, describes a reception at Boreel's domicile where Oldenburg met Menasseh. For reasons unknown, the editors say that this cannot have been Adam Boreel. But he was in London in 1655. In December, 1655, he is reported by his friend, Samuel Hartlib, to have made a proposal concerning the negotiations going on about readmitting the Jews to England. His proposal was to readmit the Caraites, a heretical Jewish sect, as well. Hartlib endorsed the proposal. Cf. Samuel Hartlib's letter to John Worthington, December 12, 1655, in Worthington (n.d.), pp. 78—79.

[23] See van den Berg (1977), pp. 186—193. Ernestine van der Wall of Leiden is preparing a doctoral dissertation on Serrarius under the direction of Professor van den Berg.

In the French edition of K. O. Meinsma, *Spinoza et son cercle*, 1984, I have added a long note trying to correct the misinformation about Serrarius.

[24] There is a copy of this catalogue of his books and manuscripts that were sold after his death at the Herzog August Biblioteek. Drs. van der Wall and I plan to edit it in the near future.

[25] See Henry Oldenburg's letter to Boreel, April 1656, in Hall and Hall (1965), Vol. 1, pp. 89—92.

[26] See Rétat (1973). *Les Trois Imposteurs* is quite different from *De Tribus Impostoribus* and has quite a different history. Professor Margaret Jacob and I hope to untangle the history of the French text.

[27] Compare what is said in the Oldenburg letter with the text published by Rétat (1973), pp. 46, 47, 58 and 74. Hobbes is cited on pp. 32, 36, 38 and 51.

[28] It may be that Henry Stubbe was the person who told or showed Oldenburg the *Trois Imposteurs* material. Stubbe was at Oxford at the time, and was working on a Latin translation of Hobbes's *Leviathan*. See Jacob (1983), p. 18.

Oldenburg's plea to Boreel to save Christianity is in the letter of April, 1656 (Hall and Hall, 1965, Vol. 1, pp. 90—91).

[29] Oldenburg to Boreel, January 24, 1657 (Hall and Hall, 1965, Vol. 1, pp. 115—116).

[30] In the letters of Oldenburg to and from Boyle in 1665, and 1666 there is a lot of discussion about the cost of Serrarius's having Boreel's manuscript copied. The price

was 67 guilders and 10 stuivers. See Hall and Hall (1965), Vols. 2 and 3. The cost figures are given in Vol. 3, p. 18.

[31] Boyle's copy is listed in the Boyle papers at the Royal Society. Henry More (1708, preface) mentioned some of Boreel's points in his treatise *Universi humani Generis Legislator.* More reported that he got his copy from Francis Mercurius van Helmont (pp. iv—v).

[32] The matter comes up in several of Oldenburg's letter of the time to Boyle, Hartlib and Boreel.

[33] See Oldenburg's correspondence with Spinoza, 1663—1665.

[34] See Spinoza (1951), Ch. 17, pp. 218ff, Ch. 19, p. 247 and Ch. 20, p. 257; Gebhardt III, 205ff, p. 239, and 250.

[35] See Spinoza's *Letter XLIII* to Jacob Ostens, February, 1671, in Spinoza (1928), p. 259; Gebhardt IV, p. 226.

[36] Spinoza (1951), Ch. 5, especially the end of the chapter; Gebhardt III, pp. 69—80.

[37] *Ibid.*, Ch. 5, p. 70; Gebhardt III, pp. 70—71.

[38] Most of the manuscripts in England, France, The Netherlands, Germany and the United States have this title. It sometimes appears with Lucas's *La Vie de M. Spinoza*, and the works were apparently printed together in 1719. Spinoza, however, is not mentioned in *Les Trois Imposteurs.*

[39] Professor Margaret Jacob and I are trying to trace the development of the text. Information about Jean Lucas, supposed author of *La Vie de M. Spinoza*, indicates he was associated with some of Spinoza's followers. The 1719 text was put together by Rousset de Missy out of materials available in The Hague at the time. On his role, see Jacob (1981), Ch. 7, and my forthcoming review of this work in the *Journal of the History of Philosophy.*

[40] On Rabbi Nathan Shapira, see Scholem (1973), the references to him; and Ya'ari (1977). I am grateful to Professor Amos Funkenstein of UCLA for translating this material for me.

Jessey (1656) mentioned that Menasseh showed Oliver Cromwell letters from Rabbi Shapira about the condition of the Jews in Jerusalem.

[41] Serrarius (1657, p. 37) mentioned meeting the rabbi. He described the rabbi's stay in Amsterdam in a letter published in Dury (1658).

[42] Chief Rabbi Saul Levi Morteira and the Spanish trained theologian, Orobio de Castro, disputed the Christian interpretation of *Isaiah* 53 in various unpublished works that were widely distributed in Europe at the time.

[43] Katz (1982) attributed the work to Henry Jessey. However, I think the internal content indicates that John Dury was the author. The bulk of the work is a letter of Peter Serrarius to Dury.

[44] The fund raising efforts are explained at the end of the pamphlet. Much more detail about this appears in anon. (1671), pp. 69—77. The negative Jewish reaction is discussed in Ya'ari (1977).

More detail about Rabbi Shapira's visit to Amsterdam appears in my (1985a).

[45] Besides the constant statement of this throughout the *Tractatus*, see also Spinoza's *Letters LXXV* and *LXXVIII* to Oldenburg of December, 1675, and February 7, 1675, respectively (Spinoza, 1928, pp. 346—350 and 357—359; Gebhardt IV, 311—314 and 326—329).

REFERENCES

Anonymous: 1671, *The Life and Death of Mr. Henry Jessey, Late Preacher of the Gospel of Christ in London*, n.p.

Braithwaite, William C.: 1955, *The Beginnings of Quakerism*, 2nd ed., revised by H. J. Cadbury, Cambridge.

Caton, William: 1689, *A Journal of the Life of that Faithful Servant and Minister of Jesus Christ, Will, Caton*, London.

Caton, William: (n.d.), *William Caton Manuscripts*, Friends' Library, London.

Dury, John: 1658, *An Information concerning the Present State of the Jewish Nation in Europe and Judea. Wherein the Footsteps of Providence Preparing a way for their Conversion to Christ, and for their Deliverance from Captivity are Discovered*, London.

Fisher, Samuel: 1600, *Rusticus ad Academicos in Exercitationibus Apologeticus Quatuor. The Rustick's Alarm to the Rabbies, Or the Country correcting the University and Clergy and (not without good Cause) Contesting for the Truth, Against Nursing Mothers and their Children*, n.p.

Fisher, Samuel: 1679, *Rusticus ad Academicus in Exercitationibus Apologeticus Quatour. The Rustick's Alarm to the Rabbies, Or the Country correcting the University and Clergy and (not without good Cause) Contesting for the Truth, against Nursing Mothers and their Children*, in *The Testimony of that Worthy Man, Good Scribe, and Truthful Minister of Jesus Christ*, n.p.

Freudenthal, Jacob: 1899, *Die Lebensgeschichte Spinoza's*, Leipzig.

Hall, A. Rupert and Marie Boas Hall (eds.): 1965, *The Correspondence of Henry Oldenburg*, Madison Milwaukee and London.

Hill, Christopher: 1972, *The World Turned Upside Down*, London.

Hutton, Sarah: 1985, 'The Prophetic Imagination: a Comparative Study of Spinoza and the Cambridge Platonist, John Smith', *Proceedings of the 350th Anniversary Commemoration of the Birth of Spinoza*, November 1982, Amsterdam.

Jacob, James R.: 1983, *Henry Stubbe, Radical Protestantism and the Early Enlightenment*, Cambridge.

Jacob, Margaret: 1981, *The Radical Enlightenment, Pantheists, Freemasons and Republicans*, London.

Jessey, Henry: 1656, *a Narrative of the Late Proceedings at Whitehall Concerning the Jews*, London.

Katz, David S.: 1982, *Philo-Semitism and the Readmission of the Jews to England 1603—1655*, Oxford.

More, Henry: 1708, *The Theological Works of the Most Pious and Learned Henry More*, London.

Neubauer, A.: 1876, *The 53rd Chapter of Isaiah*, Oxford.

Popkin, Richard H.: 1973, 'The Marrano Theology of Isaac La Peyrère', *Studi Internazionali di Filosofia* **5**, 97—126.

Popkin, Richard H.: 1974, 'Bible Criticism and Social Science', in *Methodological and Historical Essays in the Natural and the Social Sciences; Proceedings of the Boston Colloquium for the Philosophy of Science, 1969—1972*, Boston Studies in the Philosophy of Science, Vol. XIV, D. Reidel, Dordrecht and Boston, pp. 339—360.

Popkin, Richard H.: 1976, 'The Development of Religious Scepticism and the Influence of Isaac La Peyrère: Pre-Adamism and Biblical Criticism', in R. R. Bolgar (ed.), *Classical Influences on European Culture*, Cambridge, pp. 271—280.

Popkin, Richard H.: 1978, 'La Peyrère and Spinoza', in R. Shahan and J. Biro (eds.), *Spinoza: New Perspectives*, University of Oklahoma press, Norman, pp. 177—195.

Popkin, Richard H.: 1979, *The History of Scepticism from Erasmus to Spinoza*, Berkeley and Los Angeles, Chapters 11 and 12.

Popkin, Richard H.: 1984a, 'Rabbi Nathan Shapira's Visit to Amsterdam in 1657', *Proceedings of the Symposium on Dutch Jewish History*, Tel Aviv and Jerusalem, December, 1982, pp. 185—205.

Popkin, Richard H.: 1984b, 'Spinoza and the Conversion of the Jews', *Proceedings of the 350th Anniversary Commemoration of the Birth of Spinoza*, November 1982, Amsterdam, pp. 171—181.

Popkin, Richard H.: 1984c, 'Spinoza, the Quakers and the Millenarians, 1656—1658', *Manuscrito*, VI, pp. 113—133.

Popkin, Richard H.: 1984d, 'Spinoza's Relations with the Quakers in Amsterdam', *Quaker History*, LXXIII, pp. 14—28.

Rétat, Pierre (ed.): 1973, *Traité des Trois Imposteurs*, Images et temoins de l'âge classique, Université de la Region Rhone-Alpes.

Révah, I. S.: 1959, *Spinoza et Juan de Prado*, Paris and The Hague.

Révah, I. S.: 1964, 'Aux Origines de la rupture spinozienne: Nouveaux documents sur l'incroyance dans la communauté Judeo-Portugaise d'Amsterdam à l'époque de l'excommunication de Spinoza', *Revue des Études juives*, Tome 3 (123), 357—431.

Scholem, Gershon: 1973, *Sabbatai Zevi, The Mystical Messiah*, Princeton.

Serrarius, Petrus: 1657, *Assertion du Regne de Mille Ans*, Amsterdam.

Spinoza, Baruch: 1925, *Spinoza Opera*, ed. by Carl Gebhardt, 4 vols., Carl Winter, Heidelberg. (Cited as 'Gebhardt' in the text.)

Spinoza, Baruch: 1928, *Correspondence of Spinoza*, ed. by Abraham Wolf, London, George Allen and Unwin.

Spinoza, Baruch: 1951, 1955, *Chief Works of Spinoza*, transl. by R. H. M. Elwes, 2vols., Dover, New York.

van den Berg, Jan: 1977, 'Quaker and Chiliast: The Contrary Thoughts of William Ames and Petrus Serrarius', in R. Buick Knox (ed.), *Reformation, Conformity and Dissent. Essays in honour of Geoffroy Nuthall*, London.

Worthington, John: (n.d.), *The Diary and Correspondence of Dr. John Worthington*, ed. by James Crossley, Vol. 1, in Cheltam Society, *Remains Historical and Literary, Connected with the Palatine Counties of Lancaster and Chester*, Vol. 3.

Ya'ari, Abraham: 1977, *Schluche Erets Israel* (in Hebrew), Jerusalem.

PART IV

SCIENTIFIC-METAPHYSICAL REFLECTIONS

J. THOMAS COOK

SELF-KNOWLEDGE AS SELF-PRESERVATION?*

In this century much attention has been given to what might be called the naturalistic or scientific side of Spinoza's thought. This attention is appropriate, I think, for it reflects a recognition of the seriousness with which our author claimed, in the Preface to Part III of the *Ethics,* that

... nature's laws and ordinances, whereby all things come to pass and change from one form to another, are everywhere and always the same; so that there should be one and the same method of understanding the nature of all things whatsoever, namely, through nature's universal laws and rules (Spinoza, 1955, p. 129).

The uncompromising sweep of this methodological pronouncement is impressive, and recent commentators such as Hampshire (1951), Curley (1969) and Matson (1977) have responded appropriately by emphasizing Spinoza's naturalism, seeing his system as an attempt to lay the metaphysical foundations for the new, developing "natural philosophy" of his time — that natural philosophy which was in many ways the progenitor of our own natural science.

The recent emphasis upon the naturalistic reading of Spinoza seems fundamentally right to me, but it is hard to escape the sense that in reading him this way we fail to do justice to the breadth of his vision. Specifically, it is difficult for many of us to understand or take seriously Spinoza's claims about the ethically salutary, emotionally satisfying effects of understanding when we construe understanding in anything like a natural-scientific way. If we think of rational understanding on a scientific model, what do we make of Spinoza's conviction that the person who rationally understands nature and him/herself as a part thereof is *eo ipso* a wise, good, strong, free and pious person? And if Spinoza's vision is a vision of nature understood scientifically, what are we prosaic and secular-minded twentieth-century types supposed to do with that element of affective religiosity — bordering on the ecstatic — which is so undeniably present in Spinoza's view?

These are the broad interpretive problems which underlie my own interest in the topic of this paper. Though I shall not be offering a full-scale natural-scientific interpretation of Spinoza's system, I hope to

191

Marjorie Grene and Debra Nails (eds.), Spinoza and the Sciences, 191—210.
© *1986 by D. Reidel Publishing Company.*

further the cause of such an interpretive approach by focusing upon a specific ethical claim within the system. I hope to render that claim intelligible in a way which does justice to Spinoza while revealing the relevance of the claim to the concerns of contemporary natural-scientifically minded philosophers.

The specific tenet of Spinoza's thought upon which I shall be focusing is his conception of the importance of self-knowledge for self-preservation. Spinoza held that every individual is characterized by an endeavor to persevere in being. This endeavor, we are told, is the source of all virtue, and the self-preservation to which it tends is the good which we all, always, seek. Spinoza also held that knowledge, and more specifically self-knowledge is a necessary and ultimately a sufficient condition for the achievement of self-preservation. Now the notion that self-knowledge can be of aid in attaining the human good was not original with Spinoza — one thinks immediately of the Platonic Socrates, perhaps Aristotle, certainly the Stoics. But Spinoza went further in this direction than any of these, his predecessors. For him, self-knowledge is not related to the good to be achieved as means is related to end. Rather, means and end ultimately coincide in the activity of self-understanding, for an increase in understanding, we are told, *is* an increase in expressed power of self-preservation, and perfect self-knowledge — intuitive knowledge of one's essence and of oneself as that essence — *is* eternal life. To know oneself is actively to preserve oneself; to know oneself fully is to participate directly and eternally in the life divine.

These are powerful (some say extravagant) claims. On a naturalistic interpretation, Spinoza is claiming that self-preservation is first and foremost a matter of understanding oneself through "nature's universal laws". A number of philosophers have wondered, lately, whether it is possible for us to develop a fully articulated "scientific image of man",[1] and what might be the consequences of one's understanding man or oneself in terms of that image. Spinoza seems to be suggesting that such a self-understanding on the part of an individual would be the pinnacle of human virtue — indeed, that such a self-understanding would be eternal life. It is this suggestion which I want to consider here.

In the *Ethics*, Spinoza first addresses the issue at hand in Propositions XXVI and XXVII of Part IV. These are important propositions, for they initiate the shift from metaphysical and psychological description to the morally prescriptive account of the wise and free individual in the later propositions of Part IV. In Part IV, Proposition XXVI, we are told that "Whatever we endeavor according to reason is nothing else but to understand . . .". Proposition XXVII reads "We know nothing to be certainly good or evil except what is really conducive to understanding or what can hinder understanding" (Spinoza, 1982, pp. 168—169). With these two propositions Spinoza establishes that the endeavor to persevere in one's being, when seen aright, is identical with one's endeavor to understand. Spinoza does not explicitly mention *self*-understanding here, but I shall be arguing that for him *all* understanding proceeds, as it were, by way of self-understanding. For the moment, let us focus upon the explicit claim that the endeavor for self-preservation, rightly viewed, is an endeavor to understand.

This pivotal claim in the *Ethics* has been often, understandably, called into question. Critics have charged Spinoza with a variety of offenses here. Bidney, for example, holds that Spinoza is simply equivocating on the term "self-preservation". "It appears", he writes,

that by self-preservation Spinoza means two different things. In the first place, he means preservation of the individual considered as a psycho-physical organism. In the second instance, he means preservation of the intellect which is the real or true self and principle of being and activity (Bidney, 1940, p. 347).

Bidney traces the first meaning of 'self-preservation' to Spinoza's interest in Hobbes, the second meaning to Plato and Aristotle, and concludes that Spinoza's attempt at synthesis has led him into equivocation and inconsistency.

Even as sympathetic a reader as Santayana found Propositions XXVI and XXVII of Part IV too much to swallow. In his own personal copy of a German translation of the *Ethics* (now in the Houghton Library),[2] he pencilled in the margin next to Proposition XXVI "a great *non sequitur* of the system". Now if Bidney and Santayana are correct, if Spinoza's identification of the effort for self-preservation with the effort to understand is an inference fallaciously drawn, much of the specifically ethical content of the system is lost, or at least evidentially severed from the metaphysical and physical doctrines from which it is supposed to follow. If we would save the continuity between the metaphysical and ethical views, a way must be found to remove the appearance of fallacy in Spinoza's argument for the identification of the endeavor for self-preservation with the endeavor to understand.

I am quite sympathetic with those commentators who have had

difficulty with Propositions XXVI and XXVII of Part IV. The proofs of these propositions make formal sense, given the abstract definitions of adequate cause and adequate idea. But they make sense with regard to mental acts of inference-drawing, and the relationship between the mind's endeavor to act in this sense, and the body's endeavor to survive as an organism is quite unclear. The human individual, we have learned prior to Part IV, is a physico-psychical organism whose identity, on the physical side, consists of a certain relatively constant ratio of motion and rest among the parts which make it up. The human body, so understood, is characterized by a homeostatic tendency to maintain its individual physical integrity though affected in various ways by things in the surrounding environment. This tendency toward self-maintenance Spinoza calls an individual's endeavor to persevere in its being, and this endeavor, we are told, is the very essence of the individual.

This same individual, when viewed not as a body under the attribute of extension, but as a mind under the attribute of thought, is a complex idea made up of the ideas of those many extended things which constitute the body, as well as the ideas of the ways in which the body is affected from without. The mind's power of understanding consists of its power to form what Spinoza calls 'adequate ideas', also called 'common notions', which are ideas of those things which all bodies have in common and which are equally in the whole and in the part of all extended things. The mind's endeavor to understand is an endeavor to form, and thereby become, such 'adequate ideas'.

Now I think that it will be agreed that it is not immediately apparent how the body's endeavor to maintain its homeostatic integrity as a complex physical organism can be equated with the mind's endeavor to form and become ideas of that which all extended things have in common. And yet, according to Spinoza, the body's endeavor to persevere and the mind's endeavor to understand are one and the same endeavor, viewed under the two attributes, extension and thought, respectively. Questions abound here. Why should my mind's attainment of the idea of that which all extended things have in common contribute to — indeed, *be* — an increase in my body's power for self-preservation? As my mind is attaining, and thus becoming adequate ideas, just what is going on in my body?

The real difficulties involved in Spinoza's view at this point sometimes escape us, for we assume a common-sensical point of view, and we know, as a matter of common sense, that knowledge and understanding are pragmatically valuable things, both for an individual and for us all, collectively. Fools, after all, often die young, and much of our evolutionarily adaptive advantage as a species is attributable to our powers of intelligence. Common sense is of course correct in telling us that knowledge is conducive to survival, but with regard to interpreting Spinoza on this point, common sense leads us astray, for this is not one of those points at which Spinoza is being common-sensical.

In order to see this, let us ignore, temporarily, the technicalities of Spinoza's system. We will have occasion, as we proceed, to reintroduce specific tenets of Spinozism, but for the moment, let us just think in general terms about various ways in which an increase in knowledge might yield an increase in one's ability to preserve oneself. I shall consider four suggestions — three quite briefly, and the fourth at some length. The four are arranged in order of decreasing common-sense familiarity and *in*creasing relevance to Spinoza's view.

(1) How can knowledge aid in the effort to survive? Faced with this question, the average child of the industrial revolution is likely to respond as a latter-day Baconian. Technology is the most dramatic way in which knowledge (especially scientific knowledge) can be put to use in the service of our will to preserve ourselves. One thinks of food production and storage, medicine, sanitation, construction, the whole long list of ways in which knowledge has been applied to give us more power over nature and render us less subject to her often life-threatening whims. It is interesting to note that Spinoza was aware of and sensitive to the promise and value of the limited technology of his time, but that he seems not to have thought it to be of central importance. In the treatise *On the Improvement of the Understanding* we read

... as health is no insignificant means for attaining our end, we must also include the whole science of Medicine, and, as many difficult things are by contrivance rendered easy, and we can in this way gain much time and convenience, the science of Mechanics must in no way be despised (Spinoza, 1955, p. 7; Gebhardt II, p. 9).

The emphasis in this passage is important. Neither gain in time and convenience, nor even health, is seen as an end in itself, but rather as conducive to the end which he seeks. That end, as we learn in the *Ethics,* is the self-preservation to be had in and through the understanding itself. For Spinoza, the positive power of knowledge lies not in the concrete environmental control and manipulation which it makes possible, but in the very act of knowing itself.

(2) The Socratic-Platonic tradition provides a second straightforward way in which increased understanding can contribute to one's success in the struggle to survive. On the whole, the more one understands — understands oneself and the world around — the more one will be able accurately to distinguish between that which is good for one and that which is not — between that which is truly good for one and that which only initially appears to be so. Now if, as Spinoza claims, self-preservation is our end, and it and all that conduces thereto are goods, increased knowledge can certainly be of service to us, for the more we know, the more able we will be to distinguish between those things and acts which conduce to our preservation and those which do not.

If one knows enough to be able to tell what is good for one, one is more likely, other things being equal, to do it or get it, and thereby to further one's self-preservation. I say "more likely" because Spinoza holds explicitly that a knowledge of good and evil does not, in itself, guarantee that one will choose or do the good. Even one who knows the good may be led down the road to perdition by destructive passions. Still, one increases the likelihood of one's doing that which is conducive to survival by coming to have knowledge of what is so conducive.

This second suggested way in which understanding can aid in the effort to preserve oneself squares both with common sense and with Spinoza's views. Understanding is valuable in that it makes possible our distinguishing the true from the merely apparent good. Unfortunately, this does not go very far in helping to explain Spinoza's view, for when we turn to the text to find out what it is that increased understanding reveals as truly good — truly conducive to self-preservation — the answer which we find, as mentioned earlier, is simply 'understanding'. Now if understanding is the true good, the understanding which reveals this fact to us is certainly of value, for realization of this fact will increase the likelihood that we will pursue understanding, and hence presumably increase the likelihood that we will achieve understanding. But the claim that understanding is good because it reveals to us that the good is understanding has a suspiciously circular ring to it. And even if it is not viciously circular, it is profoundly unhelpful if what one wants is to understand why the good is understanding.

(3) Knowledge can aid in the effort to preserve oneself through its application in technology and through revealing to us that understanding is that which is truly good. Neither of these benefits is to be sneezed

at, but neither gets at the root of Spinoza's claim that the effort to preserve oneself is identical with the effort to understand; and certainly neither satisfactorily explains the further claim that the activity of understanding nature and oneself as a part thereof *is* active self-preservation. Consider, then, a third possibility — less immediately obvious, and more in keeping with Spinoza's views. Among the greatest hindrances to one's self-preservation are certain emotional states which work at odds with one's essential conative endeavor to persevere in being — call them destructive passions. Spinoza shares the view of contemporary cognitive theorists of emotions according to which emotions involve, essentially, beliefs. If it should be the case that the beliefs presupposed by or involved in destructive passions are confused, misguided, false beliefs, then understanding and knowledge could be of real value. True understanding would reveal the confused and misguided character of those beliefs, and thereby undermine the cognitive support upon which the destructive passions rest. Thus the very act of understanding would involve a reduction of the power of those falsely-based passions which work counter to self-preservation, and hence would involve an increase in active power to persevere.

This suggestion is clearly in keeping with Spinoza's strategy for liberation as outlined in the first ten propositions of Part V of the *Ethics*. Destructive passions are to be dealt with by gaining a rational understanding of each. To gain a rational understanding of an emotion is for one's mine to attain to adequate ideas from which the idea of the occurrence of that emotion follows as a necessary consequence. To the extent that one succeeds in gaining such an understanding, one's own mind is the adequate cause of the idea of the emotion in question, and one is thereby active rather than passive *vis-à-vis* that emotion. Since adequate ideas, and any ideas which follow from such ideas, are necessarily true, the confused and fragmentary character of the inadequate idea which underlay the original destructive passion is cleared away, and one's active power of self-preservation is accordingly increased. Moreover, since the mind's attaining to adequate ideas from which the idea of the emotion in question follows is an increase in the mind's active power, the painful affective character of this emotion is countered by the pleasure of one's own increased power.

This account of rational therapy for destructive passions goes a long way toward explaining Spinoza's claim that the effort to understand is identical with the effort to persevere in one's being. But it, too, has its difficulties. As is too often the case, Spinoza relies heavily here upon the substantial identity of the mind and body which he has established earlier (in Part II). the entire account of our becoming active *vis-à-vis* previously passive emotions is couched in terms of the *mind's* attaining to adequate ideas from which the ideas of these emotions follow. The reader is left to figure out the bodily equivalent of the mind's becoming adequate ideas. If these adequate ideas are the common notions — the ideas of that which all things have in common — the substantial identity of mind and body would seem to require that as the mind becomes these ideas, these common notions, the body becomes that which all extended things have in common. But what could that mean? Furthermore, in recommending that we achieve an adequate understanding of passive, and especially painful emotions, Spinoza seems to be recommending that which is, on his own view, impossible. Pain is, by definition, a transition to a state of lesser perfection or power. If we could have an adequate knowledge of pain, it would have to be the case that the idea of a transition to a state of lesser power and perfection could follow from ideas adequate in our minds. Were this the case, our minds alone would be the adequate causes of this idea of transition to a state of lesser perfection. But this consequence is absurd, for nothing can follow from our minds alone except that which is conducive to the preservation of our being — and ideas of transitions to states of lesser perfection are not conducive to the preservation of our being. It seems, then, that in recommending that we gain an adequate understanding of destructive passions, Spinoza is recommending the impossible.

The obvious reply here is that this final point is not a damaging objection, but rather a confirmation of the efficacy of understanding in overcoming pain and pain-related emotions. Just as it follows that pain cannot be adequately understood, so too it follows that that which is adequately understood cannot be pain. Spinoza confirms this in the Scholium to Proposition LXXIII of Part IV in which he says ". . . the strong-minded man has this foremost in mind, that . . . whatever he thinks of as injurious or bad . . . arises from his conceiving things in a disturbed, fragmented and confused way".[3]

(4) ". . . [w]hatever he thinks of as injurious or bad . . . arises from his conceiving things in a disturbed, fragmented and confused way . . .". This passage suggests a fourth way in which we might interpret the claim that the effort to persevere in being is identical with the effort to understand — an interpretation not at all common-sensical, but I think

ultimately most consonant with Spinoza's views. Spinozistic understanding of nature, and most importantly of oneself as a part thereof, leads one to the realization that the concepts and categories in terms of which one has been thinking of oneself are inappropriate to the being one is. One is led to a conceptual perspective upon oneself from which point of view certain notions have no place, and certain questions do not and cannot arise. In language more familiar to us, it is as if one adopts a theoretical perspective upon oneself which is radically incommensurable with one's previous self-understanding. For example, one might come to think of oneself in a way which makes no mention of effort, conflict, struggle or perseverance. To the extent that understanding leads one to such a radically altered conception of oneself, it could be said to provide a radical solution to one's problems of self-preservation.

The sort of knowledge which I have in mind here is that which Spinoza calls knowledge of the essence of one's body under the aspect of eternity. I want to focus the remainder of my paper on this difficult aspect of Spinoza's positon. I do so because, as must be evident by now, I do not think that Spinoza's identification of knowledge with self-preservation is fully intelligible without reference to this view. Also, I think that this is a point at which recent ideas in the philosophy of science can be of assistance in understanding Spinoza, and perhaps a point at which Spinoza can assist us in our efforts to think creatively about the human significance of natural science and the "scientific image of man".

Proposition XXII of Part V of the *Ethics* states that "there is necessarily in God an idea which expresses the essence of this or that human body under a form of eternity". It is this very idea to which the individual's mind must attain if he/she would have knowledge of the essence of his/her own body. Such knowledge is the pinnacle of self-knowledge in Spinoza's view, and it involves a radically different way of thinking of oneself — a way which is without all those ideas of the imagination which characterize our normal perception of ourselves. This transformation in the way in which one knows oneself follows upon one's coming to know the essence of one's body and, I think, coming to know oneself to be that very essence.

The key concept here is the notion of the essence of an individual — one of the most important and least understood concepts in the entire system. A good grasp of just what the essence of an individual is would be of inestimable help, but unfortunately the textual data are scant.

Essence is power, we are told (*Ethics* I, Proposition XXXIV). The essence of an individual is an eternal truth (Spinoza, 1955, p. 37; Gebhardt II, p. 36). It is that which is described by an accurate definition (Spinoza, 1955, p. 35; Gebhardt II, pp. 34—35); it can be deduced from the "fixed and eternal things" (Spinoza, 1955, p. 37; Gebhardt II, p. 36); it is that, intuitive knowledge of which is eternal life (*Ethics* V, Propositions XXII and XXIX). Given the varied and cryptic character of Spinoza's remarks, any account of precisely what is meant by "the essence of a thing" is bound to be highly speculative. I hope that the following will be thought no more speculative than necessary.

As background for discussion of essences, I shall assume, without defense, an interpretation of substance and of divine causation which is similar to that of H. F. Hallett (1957).[4] On this view, God or substance is infinite structured power (*potentia*) which finds immediate, unhindered and complete expression as structured activity (*actualitas*), which activity *is* the world. There is no unactualized *potentia* in God — no power which is not expressed as activity, for as God's essence necessarily involves existence, so too His power, which Spinoza identifies with His essence, necessarily involves expression as activity. The necessity and completeness with which God's power is actively expressed leaves no room (so to speak) for introduction of distinctions among temporal stages of that expression. The necessity in question here is the sort which is characteristic of logical implication, and the atemporal, eternal character of logical relations is attributed by Spinoza to God's active self-expression as the world. Substance or Nature *is* God-in-act, where the divine activity is understood as timelessly complete, unhindered, necessary expression of infinite power.

This interpretation of substance suggests an account of modes which is nicely in keeping with the original Latin '*modus*'. A mode is a *way* in which the divine power is expressed as activity; a certain manner or way in which God acts. The essence of a mode is the power, the expression of which in a certain way is that particular mode.

In the treatise *On the Improvement of the Understanding*, Spinoza explains how we can best come to know the essence of individual things. First we are told that we must deduce all our ideas from "physical things — that is, from real entities . . . never passing to universals and abstractions" (Spinoza, 1955, p. 36; Gebhardt II, p. 35). This suggests that if we want to know the essence of a cup, we should begin with the ideas of the table, the saucer, the surrounding air, and other relevant physical things. But this suggestion is misleading, for Spinoza soon warns us that by "the series of causes and real entities, I do not mean the series of particular and mutable things, but only the series of fixed and eternal things" (Spinoza, 1955, p. 37; Gebhardt II, p. 36).The idea seems to be that knowledge of the fixed and eternal things will lead us to knowledge of the essence of things. But what are these fixed and eternal things?

This question is best approached, I think, by recalling that things, as ·modes, are ways in which the power of God expresses itself. A fixed, immutable and eternal physical thing is then a fixed and immutable way in which the divine power is eternally expressed under the attribute of extension. These fixed and eternal things are best understood, I think, as the general structural and dispositional features of God's activity which are always and everywhere the same — those characteristics of reality which we try to describe when we formulate 'natural laws'. It may seem strange to speak of general structural and dispositional features of reality as 'things', but Spinoza's view of what a thing is makes this less problematic. If a thing, as a mode, is a way in which God acts, then an unchanging way in which God everywhere and eternally acts is a fixed and eternal thing.[5]

I want to repeat this, for the rest of the paper depends on it. The fixed and eternal things are the structural and dispositional features of nature's activity which are always and everywhere the same. They are the regularities of nature's activity which we try to describe when we formulate natural laws. (I think, by the way, that these fixed and eternal things of the *Treatise* are the infinite and eternal modes of the *Ethics,* that of which the 'common notions' are ideas.[6])

Spinoza suggests that there are numerous fixed and eternal ways in which God acts — that is, numerous universally present structural characteristics of the divine activity expressed (in this case) through the attribute of extension. These ways, taken together, define configurations of motion and rest, some of which are composite configurations whose parts are so adapted as to maintain a certain inner ratio of motion and rest, though affected in diverse ways from without. Now the essence of an individual is that particular constellation of the fixed and eternal things which defines the characteristic structure and mode of activity which is that individual. The essence is the immanent cause of the individual in the sense that the individual just is a manifestation of the structurally and dispositionally regular ways in which God acts extendedly.

An example is no doubt needed here to draw together the above abstract claims and to conclude our discussion of the essences of individuals. The example which we shall consider is suggested, though never developed, by T. S. Gregory (1959) and Hans Jonas (1973). Let us consider, as a typical case of an extended thing, a candle flame. This may seem an odd choice among extended things, but given our understanding of a flame, it is quite appropriate as an example of Spinoza's conception of the finite modes of substances. A flame is essentially interactive, requiring other things for its existence. It is not likely to be thought of as inert. And it is blatantly evanescent in a way which nicely exemplifies Spinoza's conception of human existence ("Out, Out, Brief Candle"). We shall begin with a general scientific description of a flame and then relate this description analogically to the Spinozistic doctrines as interpreted above.

A flame, we know, is an oxidation reaction producing heat and light. For any given fuel source, there are certain percentage-parameters of fuel-to-air mixture within which combustion can occur. Moreover, a certain temperature must be reached in order to ignite this mixture. Once combustion begins, the reaction produces sufficient heat on its own to maintain the rate of reaction and continue the combustion process. The flame will continue so long as the conditions requisite for its existence are present.

Various features of this description deserve emphasis. The several elements involved in the reaction (oxygen, carbon, hydrogen) have definite delineable physical and chemical properties which are scientifically describable. Among the properties characteristic of these elements are those structural features which make possible their various combinations and recombinations. A full description of these elements, then, including dispositional properties, would entail that, given certain conditions, the elements combine in an oxidation process at a certain rate, releasing heat and light. One could also infer, from this full description, that the heat produced by the combustion process is sufficient to maintain the process indefinitely, so long as the other required conditions prevail.

The parallels between Spinoza's extended modes and the flame should be evident. First it should be noted that the flame is not defined by its spatial location, its size, or by identity of its constitutive parts. On the contrary, the flame is an interactive process involving continuous exchange of constituent parts. It is what it is by virtue of the character of that exchange and the rate at which it occurs.[7] So, too, in the case of Spinoza's extended modes. The parts change, and come and go, but the individual remains itself insofar as the interactions among the parts retain relative constancy of character and rate. Secondly, the structural and dispositional properties of oxygen, carbon, and hydrogen are such that taken together, under appropriate conditions, they produce a flame. Indeed, a flame just *is* the joint activity of these elements — these regular ways in which the power of nature is expressed physically.

Though this example is helpful, and though we shall have occasion to recur to it in the future, an emphatic historical *caveat* must be entered here. Spinoza knew nothing about molecular chemistry and nothing about the chemical features of the process of combustion. He did not conceive of the 'fixed and eternal things' in terms of elements, much less in terms of the molecular and atomic structure of these elements. These 'things' were, for him, simply unchanging ways in which God acts, describable by laws "according to which all particular things take place and are arranged". Indeed, he conceived these laws along the lines of Euclidean axioms of geometry. Nonetheless, that these 'laws' were the laws of nature, describing constant dispositional and structural features of God's activity is, I think, a plausible interpretation.

Spinoza holds that the fixed and eternal things are causally effica-cious in the production of finite modes of extension. I take it that this is how we are to understand the notion of God's 'immanent causation'. The fixed and eternal things are immanently present in, and immanently causative of, every finite mode in the sense that every finite mode is a manifestation of the active power of God as expressed in, as and through these ways in which God or nature eternally and everywhere acts. Summing up: the essence of an individual extended mode is a certain constellation of the most basic ways in which nature or God acts — a constellation immanently present in, as immanent cause of, that individual mode.

Now everything which has been said here about the structure of the expression of God's power under the attribute of extension can equally well be said of that expression under the attribute of thought. There are infinite and eternal modes under the attribute of thought — ways in which God eternally and (were spatial talk appropriate) everywhere thinks — and these ways are immanently present in, and causally efficacious in the production of every idea — that is, every finite mode of thought. Indeed, every finite mode of thinking just *is* an idea (a thinking) of a finite way in which God acts extendedly — an idea caused by, as manifestation of, the ways in which God eternally and everywhere thinks. The structure of God's thinking just *is* the structure of extension as expressed under the attribute of thought. Thus, when Spinoza tells us that we can attain to the common notions with which reasoning begins by noting that which all extended things have in common and which is equally in the part and in the whole of each extended thing, he is directing our attention to the fixed and eternal ways in which God thinks, which are present in, as immanent cause of, every idea. To repeat, every idea of an extended mode contains immanently, as its indwelling cause, those of God's ideas which are the mental expression of the ways in which God eternally and everywhere acts extendedly. This means that since the mind is the idea of the body, every human mind contains implicitly, as its immanent cause, ideas which are the mental expression of the fundamental laws of extended nature. It is for this reason that I suggested that all reasoning — all understanding for Spinoza — occurs by way of self-understanding. The mind has available the 'common notions', the ideas of that which all extended things have in common, because these ideas are immanently present in the mind as immanent cause of the idea which is the mind. Now the claim that every mind implicitly contains ideas of the fundamental laws of nature is an absurd claim if taken as a straightforward contemporary claim about the scientific sophistication of the man on the street. But it is not surprising that Spinoza held this, since he took geometry to be the science of extended nature, and geometry had long been held to be an *a priori* science.

The essence of the human body is a certain constellation of the most basic ways in which nature acts as extended — a constellation which is immanently present in, as indwelling cause of, that body. To come to know the essence of one's body is to come to know one's body as a manifestation of these ways in which Nature eternally acts. In achieving this knowledge, one's mind is becoming the ideas which are the mental expressions of these very same ways in which nature acts extendedly. One's mind is becoming a constellation of those ideas, immanently present in every idea, which are the ways in which God eternally thinks. To the extent that one knows in this way, one's mind is a finite eternal participant in, as manifestation of, God's infinite and eternal thinking. To know oneself in this way is for one's mind to be that idea in God which is the mental expression of the essence of one's body *sub specie aeternitatis*. As that idea in God, one's mind is without those first-level ideas which are ideas of oneself as imaginationally perceived *sub specie durationis*.

We began this digression on the subject of the essence of an individual in order to examine the change in one's way of thinking about oneself which follows upon one's coming to know the essence of one's body under the form of eternity. Having concluded that to know one's essence is to know oneself to be a finite eternal manifestation of and participant in the infinite and eternal activity of nature, I am more interested now in what one does *not* know about oneself. Insofar as one knows oneself in this way (as the essence of one's body, *sub specie aeternitatis*), one knows nothing of conflict — internal or external. The most fundamental ways in which nature acts do not conflict with each other, and to the extent that one knows oneself as a manifestation of these ways, one has no perception of conflict. One knows nothing of struggle or of effort, for there is no effort involved in God's self-expression as the world. Nature's power is fully and immediately expressed as activity, for there is neither external hindrance to be overcome, nor internal conflict to be worked out in the course of that expression. In the absence of such impediments, notions such as struggle and effort simply do not arise. It might also be mentioned that knowing oneself in this way, one knows oneself to be utterly without purpose. Nature does not act for a purpose, and to the extent that one knows oneself in terms of the ways in which nature acts, one knows nothing of purposes.

By way of analogy, we can profitably return to and fancifully extend our earlier example of the candle-flame. In the well-known *Letter LVIII* to Schuller (Gebhardt IV, pp. 265—268), Spinoza considers what a stone might think if it were conscious as it flies through the air. Let us follow Spinoza's lead, and imagine what our candle flame might think, and what it might say if it were not only mentally but also vocally endowed. Presumably it would tell us of the considerable effort involved in its continuous burning, and of its struggle to preserve itself in its conflict with hostile forces in an inhospitable environment. If we were well-trained physical chemists, we would see the flame as a manifestation of the structural properties of the elements involved in the combustion process — elements and properties with regard to which notions such as effort, struggle, conflict, purpose and other anthropomorphisms are just not appropriate. We would think the flame's claim to effort and conflict somewhat odd, attributable to the flame's apparent woeful lack of self-knowledge. If it only knew more about nature and about itself as a part thereof, it would know itself to be an effortlessly active participant in, as manifestation of, the fully active power of nature.

Now Spinoza is recommending that we come to understand ourselves in a way analogous to the way in which we understand the candle flame — a way of understanding which is without all those anthropomorphic misconceptions (confused and fragmented ideas) which have so long misled us about who and what we are. To the extent that we understand ourselves in this way, we know ourselves to be without conflict, effort, struggle, passivity, purpose, wishes, hopes, dreams and much else. We know ourselves to be effortlessly active participants in, as manifestations of, the fully active power of nature.

At this point one might well begin to suspect that that as which the individual knows himself *sub specie aeternitatis* is not really the individual at all. After all, what sense does it make to say that the eternal essence is the individual if that essence is without those things in terms of which the individual thinks of himself? To this objection Spinoza would reply that the essence *is* the individual, whereas that as which the individual has heretofore thought of himself is *not*. The individual eternally is a particular constellation of the ways in which God eternally and everywhere thinks and acts extendedly. If he has thought himself to be something else, this is a not-too-surprising consequence of the fact that ". . . the mind has not an adequate but only a confused knowledge of itself, its own body, and of external bodies, whenever it perceives things after the common order of nature" (*Ethics* II, Proposition XXIX, Scholium).

The fundamental suspicion which underlies this objection can be given voice in a different way as follows: When Spinoza recommends that one know oneself as the essence which one is, is he not really recommending that one become something different from that which one was? This question has no simple answer, for Spinoza has no simple account of the self. Indeed, he has no account of the self *per se* at all. The mind is not a thing, nor a faculty, nor a group of faculties. Rather, the mind is ideas and ideas thereof, where these are construed not as pictures but as acts of understanding. If, through attention to the common notions, which are the immanently present causes of all ideas, one's mind comes to be predominantly a constellation of these adequate ideas, shall we say that one has become something different from that which one was, or shall we say that one has recognized that which one is? If one holds that one's conscious self-conception is essential to or constitutive of one's identity, one would have to say that the individual has become something different when he achieves full self-knowledge. Yet Spinoza would be inclined, I think, to say that such an individual has become that which he is — or in less paradoxical language, he has become aware of himself as that which he is.

When one comes to know nature and oneself as a part of nature, known to oneself on its terms, one knows oneself to be a finite and eternal participant in, as manifestation of, the infinite and eternal activity which is *Deus sive Natura*. One's mind *is*, at this point, the constellation of God's adequate ideas which is the essence of one's mind, and one's body is that constellation of the ways in which God eternally and everywhere acts which is the essence of one's body. One knows oneself in terms of the concepts and categories appropriate to the most basic ways in which nature acts. One adopts a conceptual perspective upon oneself in which certain notions have no place, and certain problems do not (and cannot) arise. First among the problems which do not arise is the problem of self-preservation.

We began by asking how self-knowledge could contribute to one's effort for self-preservation. We must say, in conclusion, that on Spinoza's view, if one has achieved self-knowledge, this is no longer a meaningful question. A conscious participant in the eternity of God's active self-expression as thought and extension has no notions of effort or of duration. The extent to which one is concerned about preserving oneself is the extent to which one has, as yet, failed to know oneself. Of course if I think that I am struggling to persevere in my being, one of the best ways to help me to the self-knowledge which would dispel this confusion is to tell me that nothing is more conducive to self-preservation than knowledge, and especially self-knowledge. It is as if I desperately want to know the precise structure of the crystalline spheres in which the stars are mounted, and you tell me that nothing is more enlightening on precisely that subject than a serious study of Newton.

Throughout this paper I have been suggesting analogies between

Spinoza's view and that of a more contemporary natural science. This parallel can be overdone, and perhaps I have overdone it, for though Spinoza did indeed hold that all things were to be understood "through nature's universal laws", his geometrical conception of the structure of nature as extended is a far cry from the world-view of modern science. Nonetheless, I think that his suggestion that all things, including oneself, are to be understood in terms of the most basic ways in which nature acts is quite plausible, as is the view that such understanding will be couched in terms which make no mention of and have no conceptual place for notions such as effort, struggle, conflict, purpose, passivity, endurance or even temporal becoming.

When Sellars (1963) raises the specter of the "scientific image of man"; when Rorty (1970) tells us that the future's children might make non-inferential reports on brain processes, never experiencing that which we experience as pain; when Churchland (1979) argues for the "plasticity of mind", suggesting that we might adopt, and experience in terms of, a theory of ourselves which is very unlike our present everyday way of understanding ourselves — when we hear these things suggested, some of us feel threatened. We feel threatened by the prospect of an understanding of human beings and, more specifically, of ourselves, which makes no mention of our hopes and dreams, our struggles and conquests, our efforts, intentions, goals and purposes. At such times we do well to remember Spinoza — Spinoza, for whom effort, purpose and desire were sure signs of imperfection; Spinoza, whose view suggests that we should rejoice in having available a way of thinking about ourselves which is without these signs of imperfection. Spinoza would have us understand nature, and ourselves as part thereof, in order that we might consciously participate, in a finite way, in the effortless activity which is the expression of nature's infinite and divine power. Such self-knowledge brings a transcendence of those categories of self-understanding which misled us into thinking that we are engaged in a struggle to survive, an endeavor to persevere in being. Such self-knowledge brings — indeed, *is* — conscious participation in the life divine.

Rollins College, Winter Park, Fla.

NOTES

* The ideas in this paper grew and developed in conversation with John Lachs, Amélie

Rorty and the participants in Professor Rorty's NEH Summer Seminar in 1982. Special thanks, too, to Marjorie Grene, Debra Nails and the speakers at the Spinoza Symposium of the Boston Colloquium in the Philosophy of Science.

[1] Sellars' phrase — from Sellars (1963).

[2] Quoted by permission of the Houghton Library, Harvard University.

[3] The complete passage, as rendered by Shirley, reads: "Furthermore, as we have noted in IV, Pr. 50, Sch. and elsewhere, the strong-minded man has this foremost in mind, that everything follows from the necessity of the divine nature, and therefore whatever he thinks of as injurious or bad, and also whatever seems impious, horrible, unjust and base arises from his conceiving things in a disturbed, fragmented and confused way".

[4] See also Hallett (1930).

[5] My thinking has been influenced here by Curley (1969). His interpretation of the infinite and eternal modes as 'nomological facts' leads in the right direction, I think, but is insufficiently sensitive to the dynamic character of modes as ways in which God *acts*. See also Walther (1971, p. 65) for a similar view. The term 'constellation' is from Walther's '*Konstellation*'.

[6] In this I agree with Pollock (1880) and Curley (1969).

[7] See Jonas (1973).

REFERENCES

Bidney, D.: 1940, *The Psychology and Ethics of Spinoza,* Yale Press, New Haven.

Churchland, P.: 1979, *Scientific Realism and the Plasticity of Mind,* Cambridge University Press, Cambridge.

Curley, E. M.: 1969, *Spinoza's Metaphysics: An Essay in Interpretation,* Harvard University Press, Cambridge, Mass.

Gregory, T. S.: 1959, 'Introduction' to the Everyman's Library Edition of Spinoza's *Ethics,* Dutton, New York.

Hallett, H. F.: 1930, *Aeternitas: A Spinozistic Study,* Clarendon Press, Oxford.

Hallett, H. F.: 1957, *Benedictus de Spinoza: the Elements of his Philosophy,* Athlone Press, London.

Hampshire, S.: 1951, *Spinoza,* Penguin, Baltimore, Maryland.

Jonas, H.: 1973, 'Spinoza and the Theory of Organism', in M. Grene (ed.), *Spinoza: A Collection of Critical Essays,* Anchor, Garden City, N.Y.

Matson, W.: 1977, 'Steps Toward Spinozism', in *Revue Internationale de Philosophie* **31**, pp. 69—83.

Pollock, F.: 1880, *Spinoza, his Life and Philosophy,* C. K. Paul, London.

Rorty, R.: 1970, 'Mind-Body Identity, Privacy and Categories', in Borst (ed.), *The Mind/Brain Identity Theory,* MacMillan, London.

Sellars, W.: 1963, 'Philosophy and the Scientific Image of Man', in *Scinece, Perception and Reality,* Routledge and Kegan Paul, London.

Spinoza, B.d.: 1925, *Spinoza Opera,* ed. by Carl Gebhardt, 4 vols., Carl Winter, Heidelberg. (Cited in text as 'Gebhardt'.)

Spinoza, B.d.: 1951, 1955, *The Chief Works of Spinoza in Two Volumes,* transl. by R. H. M. Elwes, Dover Books, New York.

Spinoza, B.d.: 1966, *The Correspondence of Spinoza,* transl. and ed. by A. Wolf, Frank Cass, London.

Spinoza, B.d.: 1982, *The Ethics and Selected Letters,* transl. by S. Shirley, Hackett, Indianapolis.

Walther, Manfred: 1971, *Metaphysik als Anti-Theologie: Die Philosophie Spinozas im Zusammenhang der religions-philosophischen Problematik,* Felix Meiner Verlag, Hamburg.

GENEVIEVE LLOYD

SPINOZA'S VERSION OF THE ETERNITY OF
THE MIND*

One of the most elusive aspects of Spinoza's idea of science for contemporary readers is his assumption of a continuity between the individual mind and its contents, and the network of systematically interconnected ideas which makes up the totality of scientific knowledge. The individual mind is supposed to reach full self-knowledge through perceiving itself as inserted in the totality of thought; and this, we are told, is what it is for the mind to understand itself as 'eternal'. What is this perception? And how does it amount to the eternity of the mind?

The final sections of the *Ethics*, where Spinoza develops his doctrine of the eternity of the mind, is perhaps the most baffling part of a work renowned for its obscurity. It is not surprising that it has been treated as one of the incoherencies of his thought (Taylor, 1937, Part II), or seen as a lapse into mysticism (Wolfson, 1934, p. 350). Yet Spinoza's presentation of this material strongly suggests that he himself saw it as the climax of the work, thoroughly integrated with, and justified by, the preceding metaphysics. The eternity of the mind is supposed to be the ultimate conclusion from the central metaphysical thesis of the *Ethics* — the dependence of modes on one unique substance. It is in fully grasping the modal status of durational beings — and, in particular, its own status as a finite mode under the attribute of thought — that the individual mind attains the highest virtue and freedom, issuing in an understanding of itself as eternal.

The doctrine seems deeply paradoxical, even leaving aside the question of its relation to the rest of the system. It is presented in terms that suggest a continuity with traditional doctrines of immortality. "The human mind", Spinoza tells us, "cannot be absolutely destroyed with the body, but something of it remains which is eternal" (*Ethics* V, Proposition XXIII; Gebhardt II, p. 294). Yet what thus 'remains', he also insists, is not something that retains any continuity of consciousness with the mind's existence during life. "The mind can imagine nothing, nor can it recollect anything that is past, except while the body endures (*nisi durante Corpore*)" (*Ethics* V, Proposition XXI; Gebhardt

211

Marjorie Grene and Debra Nails (eds.), Spinoza and the Sciences, 211—233.
© 1986 *by D. Reidel Publishing Company.*

II, p. 294). The obvious response to this aspect of the doctrine was made by Leibniz. Without memory, he commented in the Paris notes, nothing that happens after death would pertain to *us* at all.[1] If Spinozistic "survival" involves the preservation only of "what is merely eternal in mind — the idea of body or its essence", Leibniz complained, the doctrine is not really one of the survival of the individual mind at all.[2] And in the *Discourse on Metaphysics* he pointed out that it is impossible to discriminate between a man's becoming King of China, without remembering anything of his former life, and his annihilation, succeeded by the existence of a King in China.[3]

On this point, Leibniz is surely right. But it does not in itself disqualify Spinoza's doctrine from being a respectable doctrine of immortality. Many mediaeval versions of immortality had stressed that only part of the soul survives bodily death.[4] And Aquinas insisted that the soul surviving bodily death, prior to the resurrection of the body, is not to be identified with the self. "My soul is not me."[5] However, Spinoza's eternal part of the mind differs from earlier doctrines of a surviving soul in a more radical way than that stressed by Leibniz. Not only does it lack all possibility of consciousness of its former existence; the concept of 'duration' is not applicable to it either. ". . . we cannot ascribe duration to the mind except while the body endures (*nisi durante Corpore*)" (*Ethics* V, Proposition XXIII, Schol.; Gebhardt II, p. 295). The 'common opinion of men' on immortality, we are told, is to be rejected not just because it clings to continued imagination and memory, but also because it confounds the mind's eternity with duration (*Ethics* V, Proposition XXXIV, Schol.; Gebhardt II, pp. 301— 302). Strictly, it cannot be said that after bodily death the eternal part of the mind 'continues'. Not only is this not a doctrine of the survival of the *self*; it seems to be not a doctrine of survival at all.

There are precedents for doctrines of intellectual immortality without commitment to individual survival. Aristotle suggested in the *De Anima* (Bk. III, Ch. 5, 430a20—25) that mind, as Active Intellect, was immortal (McKeon, 1941, p. 592). Such immortality did not involve the survival of individual minds. In Maimonides' version of it, individual minds participate in a transcendent Active Intellect which is the bearer of immortality.[6] Such an idea of the mind's eternity is not foreign to Spinoza. In the *Short Treatise*, what is eternal is a kind of cosmic intelligence into which individual minds merge through the right kind of knowledge. What is permanent and unchanging is a transcendent 'thinking thing'. Through contact with it during life, an individual mind can gain a temporary rest, a taste of eternity, without which it would live its whole life outside its element (*Short Treatise*, Pt. 2, Ch. 26; Gebhardt I, pp. 108—109). However, the mature doctrine of the *Ethics* is in sharp contrast with this. Here, what Leibniz dismissively described as merely eternal in mind — the idea of body or its essence — is much more intimately connected with the actual existence of the mind as an individual being.

What we have seems to be a strange hybrid of earlier doctrines of immortality. Individual minds are properly construed as eternal, and without surrendering their status as finite modes of Substance:

... our mind, in so far as it understands, is an eternal mode of thought, which is determined by another eternal mode of thought, and this again by another, and so on *ad infinitum*, so that all taken together form the eternal and infinite intellect of God. (*Ethics* V, Proposition XL, Schol.; Gebhardt II, p. 306)

The eternity of these individuals, however, seems not to involve survival. The Axiom of Part IV of the *Ethics* — that nothing indestructible exists in nature — applies to the human mind no less than to other finite modes. But, for all that, we "feel and know by experience that we are eternal" (*Ethics* V, Proposition XXIII, Schol.; Gebhardt II, p. 295). Spinoza's claim is not that the wise mind finds within itself some privileged invulnerability to change and decay. It feels and knows itself as eternal despite — indeed, because of — the dependent, modal status it shares with all other finite things.

In this paper I want to explore just what sense can be made of this unusual conjunction of stances. In doing so I will be leaving aside a major issue of interpretation. Attempts have been made to salvage Spinoza's doctrine as a version of individual survival of death.[7] It must be acknowledged that, despite his insistence that the mind can be said to endure only while the body does so, there are parts of the text where Spinoza does seem to maintain the continued existence of the mind after bodily death. We are told at the end of the *Ethics* that, in contrast to the ignorant man, who, as soon as he ceases to suffer, ceases also to be, the wise man "never ceases to be" (*Ethics* V, Proposition XLII, Schol.; Gebhardt II, p. 308). An interpretation of the eternity of the mind which stresses those parts of the *Ethics* which highlight the mind's destructibility must, to be fully persuasive, provide a plausible rendering of those apparently conflicting passages; just as the individual survival interpretation must provide a plausible rendering of those passages which stress the mind's destructibility.

On my interpretation, Spinoza's eternity of the mind is entirely a state which is to be attained during life — a state which in no way anticipates survival; but which is, for all that, supposed to reconcile the mind to its inevitable destruction. It is, moreover, a reconciliation which we are supposed to achieve precisely through fully understanding the very transience of the mind. But my aim in this paper is not so much to defend this against rival interpretations of the text, as to bring out what it is about the mind's understanding of itself as a finite mode that is supposed to pass naturally over into the perception — at first sight incompatible with finiteness — of itself as eternal. This will not of itself show that such self-knowledge during life was *all* Spinoza intended by the doctrine of the mind's eternity; but it will provide support for that stronger claim.

1. THE DEPENDENCE OF MODES ON SUBSTANCE

The idea that the understanding of *dependence* is the key to the mind's eternity is already present in the *Short Treatise*. There the point is that the mind draws its reality from that of its objects of knowledge or love. Echoing the principle on which Descartes relied in his Third Meditation proof of the existence of God, Spinoza argues that if the mind loves something weak and transitory it is to that extent itself transitory. But it can strengthen its hold on reality by attaching itself in love to stronger objects. The pursuit of transient objects of love is always fraught with misery; for love cannot withstand the confrontation with an object it sees as better than what it already has. We inevitably abandon what we come to see as inferior objects of love. But Reason shows us that all things depend on, and are hence inferior to, God; transient things cannot be understood without reference to the Being on which they directly depend. So Reason directs us to part from such transitory things and leads us to the love of God (*Short Treatise*, Gebhardt I, p. 63).

In the *Ethics* version, too, the knowledge of God is crucial. But the required perception involves an important difference in focus from the *Short Treatise* version. The mind does not shift its attention away from finite things to God; rather, in the light of the idea of God, it comes to a new perception of finite things themselves. These things are now understood under the form of eternity, and this is a way of understanding their actual existence.

> Things are conceived by us as actual in two ways; either in so far as we conceive them to exist with relation to a fixed time and place, or in so far as we conceive them to be contained in God, and to follow from the necessity of the divine nature. But those things which are conceived in this second way as true or real we conceive under the form of eternity, and their ideas involve the eternal and infinite essence of God . . . (*Ethics* V, Proposition XXIX, Schol.; Gebhardt II, pp. 298—299).

If we follow through the implications of this, we get a very different picture of the place of individuality in Spinoza's system from the one we have inherited from his many outraged commentators. A preoccupation with individuality is by no means what we are inclined to expect from Spinoza's system. He does, after all, say that individual things are "nothing but the affections or modes of God's attributes, expressing those attributes in a certain and determinate manner" (*Ethics* I, Proposition XXV, Corol.; Gebhardt II, p. 68). And this can be seen as downgrading the importance of individual existence — merging what we normally think of as separately existing things into a monistic totality of being. This was how Hegel saw Spinoza. Despite his admiration of the Spinozistic conception of Substance, he thought it could be embraced only at the cost of losing individuality. In his *Lectures on the History of Philosophy*, Hegel says that "thought must begin by placing itself at the standpoint of Spinozism; to be a follower of Spinoza is the essential commencement of all Philosophy" (Hegel, 1896, III, p. 257). But Spinoza's system, according to Hegel, was not up to the challenge of apprehending the unity of difference without letting difference slip. It tends to "divest all things of their determination and particularity and cast them back into the one absolute substance, wherein they are simply swallowed up, and all life in itself is utterly destroyed" (p. 288).

Perceiving the dependence of modes on Substance may well seem to demand that we divert attention away from individuals to the Substance on which they depend. It is then easy to gloss the later stages of the *Ethics* as being concerned with transcending particularity — shedding the illusions of individuality to focus on the generality of truth. Spinoza's path to freedom and virtue is readily seen as a journey of detachment from the objects of sensory awareness — the particular, the transient — to attach ourselves to the objects of Reason, which are 'common to all'. There is of course some truth in this picture. The cultivation of Reason brings freedom, and Reason does not focus on individuality. But it is not at all the whole picture. Reason passes over into Intuitive Knowledge which involves a return to the individual, now properly understood; and this understanding of individuality is supposed to yield the mind's grasp of its own eternity. Individuality is not an illusion to be shed by the mind bent on eternity; and the individuality of the mind itself, far from being transcended in the highest forms of knowledge, here reaches its ultimate expression. The required knowledge involves, not a withdrawal from individuals — merging the separate existence of things into a monistic totality of being — but an understanding of the nature and conditions of individuality. Understanding ourselves under the form of eternity is a way of understanding *our* actual existence. This is what is missed in Leibniz's dismissal of the merely eternal idea or essence of the body as having nothing to do with *us*.

If perception 'under the form of eternity' is taken as a diversion of attention away from the individual, all we are left with is a vague sense of the immensity of Substance — engulfing, as Hegel complained, all sense of individuality — or an equally vague sense of the wholeness of being.[8] Spinoza's sense of the eternal then comes through as something like the feeling which Freud, in *Civilisation and its Discontents*, labelled the "oceanic feeling" — the sense of eternity as of something limitless, unbounded — a "feeling of an indissoluble bond, of being one with the external world as a whole" (Freud, 1930, XXI, pp. 64—65). If we take seriously Spinoza's own insistence, in Part V of the *Ethics*, that the highest form of knowledge — unlike Reason — has access to the nature of individuals, and his conviction that the eternity of the mind follows from his earlier metaphysical themes, we can make of the doctrine something more concrete — and less Romantic — than this.

2. ESSENCE AND EXISTENCE; ETERNITY AND DURATION

The mind comes to understand itself as eternal despite its perishability. That this conjunction of mortality and eternity is not a contradiction is due to their being different — irreducible but not incompatible — ways of understanding the actual existence of an individual. And their compatability rests on two inter-related themes in Spinoza's metaphysics: the relations between essence and existence and the relations between eternity and duration. Spinoza, as Leibniz stresses, does describe the perception of things as eternal as an understanding of essences. But in Spinoza's system that is not divorced from considerations of existence in any way that would make the understanding of things under the form of eternity a turning away from their individual existence. It is true, too, that Spinoza defines eternity in terms of the self-caused existence of Substance; and duration in terms of the radically different, received and conditioned existence appropriate to modes. But this does not mean that a finite mode cannot rightly be perceived as eternal.

Duration is the "indefinite continuation of existence" (*Ethics* II, Definition 5; Gebhardt II, p. 85). Eternity is the "very essence of God, in so far as that essence involves necessary existence . . ." (*Ethics* V, Proposition XXX, Dem.; Gebhardt II, p. 299). It is because eternity and duration are thus superimposed on the existence of Substance and modes respectively that the one cannot be understood in terms of the other. Recognizing this has prompted some commentators to attribute to Spinoza a Platonic conception of eternity, according to which the eternal is altogether removed from the temporal — two realms excluding one another. But the novelty of the relations between Spinoza's Substance and modes makes this Platonic parallel a misleading guide in understanding his view of the relations between the durational and the eternal. Spinoza's Substance does not in that way transcend the existence of its modes; it is not removed from them in the way that Platonic Forms are removed from particulars. Spinoza is not saying that eternity is assigned to the conception of Substance alone, while duration is assigned only to the conception of modes. Indeed, the whole basis for the understanding of the mind as eternal is that a mode can be conceived as actual in both those ways.

However, as ways of conceiving the actuality of things, eternity and duration *are* radically distinct. We cannot understand something as eternal by construing it in durational terms. Here I take issue with what has come to be called, in contrast to the 'Platonic' interpretation of Spinoza, the 'Aristotelian' interpretation (Kneale, 1968—69 and Donagan, 1973). On the 'Aristotelian' reading, eternity is not the antithesis of time, but simply endless time; the eternal is what exists at all times, in contrast to the durational, which exists at some times and not at others. Whereas the 'Platonic' reading imposes too radical a separation between the *existence* of the eternal and that of the durational, the 'Aristotelian' reading fails to capture just how radical is the distinction between the two ways of *conceiving* things as actual. If the eternal is read as omnitemporally existent, Spinoza's second way of conceiving a thing as actual would involve conceiving it as existing at all times; whereas the first way would conceive it as existing at some times and not at others. But then the two ways could not both apply to the same thing. If a thing were truly conceived as eternal it would have to be a mistake to think that there are times at which it does not exist.

The natural move for the 'Aristotelian' interpretation to make in response is to suggest that, although it may appear that the same thing is perceived in the two ways, the two perceptions in fact have different objects. This kind of move is made by Alan Donagan (1973) in his ingenious use of the 'Aristotelian' interpretation as the basis for taking Spinoza's doctrine as an intelligible version of individual survival. The eternal 'essence', he suggests, is only part of an actually existing individual. So the thing, as a complex including that essence, can be said to perish. The essence itself, however, exists at all times. This interpretation would make Spinoza's doctrine akin to Aquinas's account of the immortality of the soul, though with the difference that the mind's essence exists before the body, as well as surviving its death. But this, I suggest, is not consistent with Spinoza's general treatment of the relations between essence and existence.

In conceiving a thing under the form of eternity we do not attend to a transcendent essence, abstracted from its immersion in actual existence and duration.

To conceive things . . . under the form of eternity, is to conceive them in so far as they are conceived through the essence of God as real entities (*entia realia*), or in so far as through the essence of God they involve existence. (*Ethics* V, Proposition XXX, Dem.; Gebhardt II, p. 299).

To understand things under the form of eternity is not to somehow tear aside the veil of existence to apprehend a pure eternal essence. It involves a drawing together of essence and existence; and this is precisely what we should expect, given Spinoza's insistence on their inseparability. For him essence is not only "that, without which the thing can neither be nor be conceived . . .". It is also that "which in its turn cannot be nor be conceived without the thing" (*Ethics* II, Definition 2; Gebhardt II, p. 84). Essence being given, the thing itself is also given.

Spinoza talks of non-existent essences, and the ideas of them, as lacking duration; they exist only as contained in the attributes (*Ethics* II, Proposition VIII; Gebhardt II, p. 91). This passage is notoriously obscure. But it is, I think, clear that he thinks ideas of non-existent essences cannot be differentiated out as separate ideas; any more than the non-existent essences themselves are separated out as identifiable individuals which happen not to exist. For Spinoza, as later for Peirce (1933, IV, par. 172, p. 147), it is only actual existence that yields discrete individuals. Existence is not incidental to Spinozistic essences; not because they are necessary existents — which is true of Substance alone — but because for him, unlike Leibniz, actually existing essences are not a sub-set of a wider realm of identifiable possible ones. We — and, for that matter, God — have access to Spinozistic essences only through actually existing things. We can think of an actually existing thing as not existing; finite modes, after all, have in the past not existed, and will cease to exist in the future. But we cannot contemplate an idea of a thing which in fact has never existed and raise the question of whether *it* might exist. There is no such idea to be contemplated.

3. THE MIND OF GOD

I have, so far, emphasised Spinoza's claim that the understanding of finite things under the form of eternity is a way of understanding their actuality. I want now to turn to its other strand — the relation of the essences of finite things to the idea of God. In the case of the mind's self-perception, this is supposed to be a matter of its perceiving itself as part of the mind of God. The mind, in so far as it understands, is an eternal mode of thought, determined by a multiplicity of other such eternal modes, making up in their totality the "eternal and infinite intellect of God" (*Ethics* V, Proposition XL, Schol.; Gebhardt II, p. 306). In understanding itself as eternal, a mind understands itself in relation to this totality of thought. What does this amount to?

It is at any rate clear that the 'mind' of which the human mind is to see itself as part is not Substance as *Natura Naturans* but as *Natura Naturata*. *Natura Naturans* has no parts; and Spinoza warns us that there is no further likeness between God as thinking thing and the human intellect than between the Dog which is a celestial constellation and the animal that barks (*Ethics* I, Proposition XVII, Schol.; Gebhardt II, pp. 62—3). In being part of the eternal and infinite intellect of God the human mind is not incorporated into a transcendent thinking thing, existing in a supernatural realm. The mind of God, of which the individual mind is part, is, rather, mind as *Natura Naturata* — the correlate in thought of the whole extended universe. But, given Spinoza's treatment of the relations between Substance, attributes and modes, this cannot be construed as just an extraneous product of Substance. It is to be understood, rather, as the full expression, under the attribute of thought, of the necessary being and power of Substance itself.

There is no transcendent realm in which Spinoza's God performs acts of thought. Wherever there are determinate acts of thought, we are in the realm of actually existing modes. God has no thoughts which do not belong to the totality of realised modes. His power is identical with his essence (*Ethics* I, Proposition XXXIV; Gebhardt II, p. 76) and there is nothing in his power which is not realised in the totality of modes. Whatever is conceived as being in his power must exist (*Ethics* I, Proposition XXXV; Gebhardt II, p. 77). To see the full implications of all this for the doctrine of the eternity of the mind, it is helpful to recall the consequences that Leibniz drew from Spinoza's version of God, and the very different picture of the mind of God which he opposes to it.

For Leibniz, it follows from Spinoza's position that any individual of which we can form an idea must exist; that all possible individuals must at some time and place come to be. He was appalled by what he saw as the consequences: King Arthur of Great Britain, Amadis of Gaul and the fabulous Dietrich Von Bern invented by the Germans all scandalously wander over some poetic regions in the infinite extent of space and time. God becomes a compulsive and indiscriminate creator, exercising no choice between the infinite number of possibles, lacking control over what should or should not have the privilege of actual existence. It is an opinion, Leibniz complains, which would "obliterate all the beauty of the universe".[9] But Leibniz is here foisting onto Spinoza his own models of individuality and creation. Since, for Spinoza, ideas of non-existent individuals are not themselves separately identifiable mental items, there is no question of his God having determinate ideas of individuals, concerning which a decision might be made. And his repudiation of the distinction between will and under- standing (*Ethics* II, Proposition XLIX, Corol.; Gebhardt II, p. 131) is inconsistent with the Leibnizian picture of an array of possible individuals, laid out in the mind of God, waiting to be made real. It is not — as Leibniz would have us think — that for Spinoza every possible individual must become real. Rather, the supposed array of possible individuals also disappears. All that remains of the idea of creation is the fact that actual individuals derive their existence from Substance.

In reaction against what he sees as the inevitable consequences of Spinoza's rejection of all distinction between will and understanding, Leibniz adopts a picture of God as having ideas of an infinity of non-existing individuals, from which he chooses to make actual the best. His 'possibles' are seen as individuals independently of whether or not they become real. The mind of God is the 'country' of such 'possibles'.[10] To avoid the Spinozistic doctrine of necessity which, he thinks, drives choice, reason and beauty from the universe, Leibniz makes actual individuals a sub-class of individuals. For Spinoza, in contrast, individuality and actuality coincide. And although his 'mind of God' does 'contain' ideas of all bodies, these are in no way 'prior' to the bodies themselves. (*Ethics* II, Proposition VI, Corol.; Gebhardt II, p. 89). Ideas in the mind of God are correlates of individual bodies; but they are not their exemplars. In fact, given mind's status as idea of body, bodies are, in a sense, 'prior' to minds in Spinoza's system; although the priority is not of course a causal one. The mind of God in which ideas and minds are contained is thus very far removed from the Leibnizian picture of the country of the possibles. To see what it does amount to we must turn, briefly, to what it is for the body to be contained in the universe as a whole.

The individuality of the mind, and the nature of its containment in the idea or mind of God, parallel the situation with bodies and the universe. We have seen that in the *Short Treatise* the mind is presented as drawing its reality from that of its objects of love or knowledge. Something of this conceptual priority remains in the mind-body relationship of the *Ethics*. The mind is the individual mind it is in being the idea of a particular body. But the body's being the individual body it is depends on its being enmeshed in ever more comprehensive individuals, reaching up to the universe as a whole. Spinoza's insertion of individual bodies into ever wider individuals seems bizarre if we think of it in terms appropriate to a multiplicity of distinct material susbtances. But that there is an individual body at all consists, for

Spinoza, in the continued existence of a certain proportion of motion and rest; and his point is that this depends in turn on the inclusion of that proportion of motion and rest in wider rhythms.

Spinoza's treatment of individuality here goes beyond the commonplace that individuals exist in wider environments and need constant sustenance from those environments for their continued existence and well-being. It could be granted that individuals are parts of wider systems without accepting Spinoza's theory of individuality. What is novel is the suggestion that these wider systems in which individual bodies occur are themselves individuals, in the same sense as the bodies themselves. To be a Spinozistic individual *is* to be part of such wider systems; individuality intrinsically involves being inserted in ever wider systematic interconnections. No individual — with the exception of the whole of nature — is complete in itself. But individuals are no less real, or really individual, for that.

4. THE REALITY OF INDIVIDUALS

Transposed to the attribute of thought, this stress on the systematic interconnections of modes in a totality means that the relationship between an individual mind and the totality of thought parallels the relationship between the 'fragmentation' of error and the 'wholeness' of truth. It is in some ways just a special instance of it. Spinoza's treatment of individuality maps, and can be seen as another version of, his treatment of truth and error. All ideas are in the mind of God. There is, after all, nowhere else for them to be. Yet all ideas, "in so far as they are contained in the mind of God", are supposed to be true. So there is, on the face of it, a problem in admitting the existence of false ideas; and it is a problem of the same kind as that of reconciling his emphasis on the 'whole' with the reality of individuals. If falsity consists in fragmentation, this may seem to suggest that the only really true idea is the idea of the universe in its completeness. Likewise, if nothing short of the totality of ideas suffices as an adequate idea of an individual, it may seem that God's adequate idea of each mode will be in all cases just the one totality of ideas. It may seem then that the appearance of separately existing individuals rests on inadequate knowledge.

There is a sense in which it is true that the reality of individuals, and of falsehood, depends on inadequate knowledge; but it is not a sense which makes the existence of individuals illusory. Spinoza's individuals are real; they are not illusions produced by the mind's incomplete grasp of the whole of reality. Individuals are real only as parts of wider wholes; but that is not to say that only the wholes are real. The ideas of individuals are inadequate ideas; but individuality is not thereby an illusion produced by inadequate knowledge. It is not as if the individual mind has a limited and distorted perspective which produces the illusion of individuality. For the individual mind is itself an inadequate idea. It is not a pure Cartesian ego, trapped in a body which inflicts on it a distorted perspective on the world. That our knowledge — and especially our self-knowledge — is inherently inadequate and perspectival arises, for Spinoza, from the nature of the mind, not from an extraneous intrusion of body. And this is crucial to understanding Spinoza's version of the eternity of the mind.

Individuals are not illusory; but nor are they there independently of the reality of sensation and imagination. In seeing just how this is so in the special case of self-knowledge we see the interconnections between the mind's mortality and its eternity. For Spinoza, the mind is not a real, self-contained 'whole', there to be grasped adequately by some means of knowledge it does not itself possess. Its reality *is* the reality of an inadequate idea. That is to say, it consists in an awareness of body from within the totality of modes. The individual mind is a direct, and hence inherently perspectival, awareness of bodily modification. In contrast to the metaphysical completeness which grounds the individuality of the mind for Descartes, the Spinozistic mind understands its nature by coming to understand its lack of insulation from ideas of the rest of nature. Self-awareness is mediated through a direct awareness of body; and, given the dependence of the body on wider material individuals, this means that self-awareness has as its base an inadequate awareness of what is happening in the universe as a whole. The unavoidably fragmented character of our awareness of body is paralleled by an unavoidably inadequate self-knowledge.

The individuality of the mind follows the same pattern of dependence on other finite modes as that of the body. Like all confused, inadequate ideas, the mind is isolated out from the totality of thought. It is itself an erroneous idea, in a way comparable to that in which its own subsidiary ideas can be seen as erroneous. This means that it can be included in wider, more adequate ideas, which embrace its non-existence. The boy imagining the winged horse avoids error by perceiving the inadequacy of the image; or by his image being accom-

panied by an idea which 'negates its existence.' (*Ethics* II, Proposition
XLIX, Schol.; Gebhardt II, p. 134). But the idea which is the boy's
mind itself stands in such subsidiary relations to other ideas, which
embrace it, and within which its inadequacy can be discerned. The
boy's mind cannot affirm the wider ideas which encompass it; for that
would involve his perceiving his body as excluded from existence by the
collective force of other finite modes. The mind cannot affirm its own
non-existence. But, for Spinoza, understanding this implies not a
Cartesian commitment to the independence and invulnerability of the
mind; but, on the contrary, a realisation of its interconnections with,
and dependence on, other finite modes. Like every inadequate idea, the
mind is inserted in a hierarchy of more adequate ideas, reaching up to
the mind of God in its entirety. In understanding this, it understands its
own inherent destructibility, but also the grounds of its 'eternity'.

The mind is the actually existing individual mind it is by being the
idea of an actually existing body. That being so, it endures only while
the body endures. Once that particular proportion of motion and rest
loses the struggle against the collectively greater power of rival
conatuses, which both sustain and threaten it, the mind also ceases to
exist. But, Spinoza insists, it is not *because* the body ceases to exist that
the mind does. And during life, too, the continued existence of mind is
by no means a merely passive reflection of the vicissitudes of the body.
The mind has its own *conatus*; and, as idea of the body, its actual
essence consists in the endeavor to affirm the continued existence and
well-being of the body. It is in this that its individuality resides; and it
depends on its insertion in the totality of modes of thought. The mind
endeavours to affirm and articulate the continued existence of an
individual body which is itself immersed in the flux of being — a body
which is what it is, and does what it does, only through a multitude of
forces which the mind can never comprehend with full clarity.

5. THE ETHICAL SIGNIFICANCE OF TRUTH

It comes easily to us to think of the fragmentary character of the human
mind, along with other inadequate ideas, as if it were just a matter of a
sub-set of ideas being considered in isolation from a surrounding
totality; as if a segment of a jig-saw puzzle were not being seen as part
of the surrounding picture by a mind with a perspective from outside.
This fosters the impression that Spinoza denies individuals in favour of

the picture of the whole to which they contribute. And it is not at all clear how, from such a picture, we could arrive at the mind's eternity. But we must keep in mind here the dynamic character of Spinozistic ideas. Inadequate ideas are not merely passive reflections of partial realities, which could be put together by a superior intelligence to add up to a total representation of the world. Each has its own *conatus*, which consists in a positive affirmation of the continued existence of some bodily mode. A body's interaction with its surrounding environment can either enhance or impede its own power of activity, the preservation of its own proper ratio of motion and rest. And the mind actively endeavours to affirm those things which help the body's power of activity. This yields the mind's own correlated transitions to greater states of activity, in which reside its own well-being, and ultimately its eternity.

It is easy to ridicule this aspect of Spinoza's thought as a vague programme for a 'healthy mind in a healthy body'. But there is something deeper in it than the suggestion of a correlation between mental alertness and physical fitness; or the apparently bizarre claim that reflective thinkers can be expected to have a better physique than those less committed to the cultivation of the higher forms of knowledge. Spinoza's discussion, in Part IV, of the well-being of the mind is obscure and difficult. But we can extract from it two general conditions which he sees as manifesting an unhealthy mind. Both conditions involve a false conception of individuality, and by transcending them the mind passes to the desired state of understanding its own eternity. On the one hand there is the mental analogue of a body cut off from the surrounding totality of modes which both sustains and threatens it. Spinoza, of course, sees such an isolation as in fact an impossibility — the body would cease to exist. The mind too, as idea of such a body, cannot in fact exist as an individual apart from its surrounding context of other modes of thought. But it can think of the body and hence of itself, falsely, as a self-contained individual; it can think of itself as a substance. The mind in such error remains in existence; for it remains the idea of an actually existing individual body, however wrongly conceived. But it is psychologically cut off from other modes of thought which are in fact crucial to its very existence as an idea.

The second condition which manifests an unhealthy mind may seem the opposite of this; but they are closely related. In this second state, the mind is altogether too subject to the intrusion of extraneous ideas.

In its state of ignorance it does not exert its own *conatus* against the onslaught of others. This is the state of passion, the state of bondage. Just as the body can be reduced to passivity by the impinging of forces which its own activity cannot withstand, so too the mind can be put into thrall by rival *conatuses* which threaten its own powers of active affirmation. This state of passion — contrasted with the transformed state of active, rational emotion — is also the state of obsession. The mind in the grip of passion, untransformed by Reason, loses its capacity to enjoy a wider range of activities. Its activity becomes circumscribed by the affirmation of something which it conceives, wrongly, as the sole cause of its pleasure or pain. Its self-affirmation becomes bound up, in the case of love, with the affirmation of a restricted segment of reality, which is given a distorted status. The mind takes its well-being to depend on a falsely construed individual. The status of the loved object as part of a whole is not grasped and, to that extent, the mind's own status as part of a whole is not grasped either.

In these states of error, based on a false conception of individuals, the mind remains 'part of a whole'; it remains in the 'mind of God'. But minds can — and for the most part do — live without realising their true status; unaware of their existence in the mind of God. Such minds live without transcending the limitations of inadequate forms of knowledge. The mind which manages to avoid boredom and obsession, understanding bodies and itself in relation to wider contexts, is the idea of a body engaged in a large variety of activities, without being thereby reduced to passivity. Such a mind is open to wider reality without its own activity's being subdued. The metaphysical error of seeing individuals only as 'wholes' — as if their interconnections with other things were merely incidental to their being individuals — is the illusion built into obsessive loves and hates. The mind is under constant threat of thus succumbing to a false individuality. With its range of affirming activity circumscribed through passion, it is incapable of seeing itself as part of a whole. Human life is a constant struggle between activity and passivity, autonomy and dependence, freedom and bondage; and true freedom of the mind is to be attained only by understanding individuals as interconnected modes.

The transition to more adequate forms of knowledge is at the same time the source of the mind's freedom from the destructive effects of passion. And, conversely, scientific knowledge remains grounded in the mind's awareness of agreement and disagreement between the body and impinging forces. This grounding of the higher forms of knowledge in directly experienced feelings of synchronisation and disharmony, joy and sorrow, sets the scene for Spinoza's version of the attaining of freedom through understanding. Spinoza's 'common notions' are firmly anchored to actual existence; they are grounded in the direct awareness of bodily modification. And the mind's grasp of these modifications depends on the actual existence of the body. It is this that makes the *Ethics*, despite its echoes of earlier doctrines of intellectual immortality, a treatise on how to live, centred on the understanding of transience.

6. REASON AND INTUITION

The process of increased freedom from the bondage of passion is, to a large extent, attainable through the exercise of Reason. And even without moving on to the third form of knowledge, this increased freedom has a bearing on the mind's attitude to death. The rational mind transcends fear and despondency through the very fact of its self-enhancing activity. The mind whose activity is enhanced by the pursuit of reason fears death less, precisely because it is active. Fear, despondency, the dread of death — all these Spinoza associates with passivity and bondage. They are passions which impede activity. For the wise man, the inevitable fact of death does not intrude on the active enjoyment of life.

There is another respect too in which the cultivation of even the second form of knowledge bears on attitudes towards death. To the extent that the mind cultivates Reason, it is concerned with common notions — with what is 'common to all'; and, to this extent, its affections are not bound up with what will inevitably perish. That which is common to all does not perish with the decay of particular objects; to the extent that the mind is concerned with such things, what it deeply cares about will not perish with it. There is here some remnant of the Maimonidean concept of intellectual immortality. But what survives is not a 'thinking thing' into which individual minds merge at death, but just the common notions which it loved during life and which are left intact by its death. What is 'common to all' survives because it is peculiar to none.

But none of this yet yields what we were promised — the eternity of the mind itself. The rest of the story comes with the contrast between Reason and Intuitive Knowledge. Reason grasps the general truth of the dependence of modes on Substance. But it does not — as Intuitive Knowledge is supposed to do — grasp this truth, directly, with respect to the particular individual which the mind itself is. Confined to a grasp of what is common to all, Reason does not comprehend the actual existence of a specific individual 'in relation to the idea of God'. It comprehends the general truth: "Every idea of any body or actually existing individual thing necessarily involves the eternal and infinite essence of God" (*Ethics* II, Proposition XLV; Gebhardt II, p. 127). But the mind's comprehension of its own actual existence in relation to the idea of God is left to Intuitive Knowledge. What, finally, does this amount to? And why is it presented as the mind's understanding of its own eternity?

The general truth on which the desired insight depends is the analogue for mind of the conditions of individuality of bodies. It is a matter of the mind's dependence on other modes of thought, which are, at the same time, the basis of its destructibility. In the case of mind, this, on the face of it, is something very strange indeed. The mind, Spinoza says, ceases to exist not *because* the body of which it is the idea ceases to exist; but because it is excluded from existence by another idea.

... it does not cease to affirm the existence of the body because the body ceases to be, but ... because of another idea excluding the present existence of our body, and consequently of our mind, and contrary, therefore, to the idea which forms the essence of our mind. (*Ethics* III, Proposition XI, Schol.; Gebhardt II, p. 149)

We can see, internally to his system, why Spinoza must say this. To say that minds cease to exist because bodies do so would be to suggest that causal relations hold across attributes. But what are we to make of this strange destructive efficacy of ideas? Part of the problem here is that it comes so naturally to us to think in terms of a Cartesian picture of a mind inspecting its ideas and deciding which it will affirm or negate. To see what Spinoza is saying we must, rather, think of ideas as themselves active; to be affirmations, they need no input from an external 'will'. And we must also think in terms of a hierarchy of encompassing ideas. The individual mind can be 'excluded from existence' by virtue of its inclusion in a wider network of thought; just as the image of the winged horse can be excluded from existence by virtue of its inclusion in the wider network which is equated with the boy's mind. The idea which is the boy's mind can be subsidiary in a wider network of ideas, in an analogous way to that in which the image of the winged horse is subsidiary to his mind.

It is impossible for the mind to understand itself in the totality of its relationship to other modes, although that framework is provided in the totality of scientific knowledge. The individual mind can, however, come to see itself as a mode — as a being conditioned, through the mediation of an infinity of other modes, by the necessary being of Substance. During life, there can be included in the mind affirmations of modifications of the body, of which it is the idea. But the very fact that an individual mind exists at all means, for Spinoza, that this holds also of the body as a whole. We need not think of this as involving the mind's being an idea in a super-intelligence; any more than we need think of the body's inclusion in larger individuals as involving its inclusion in some super-organism. It is simply a matter of the *idea* of the body as a whole having a place within wider ideas of other finite modes, which are bound up with its survival and the eventual source of its destruction. We do not have to think of the idea which *is* the mind as an act performed by a rational agent. That it does exist in this wider framework constitutes the mind as an individual. Analogously, the 'endeavour to persist in being' is not something an individual body does. Rather, that it occurs is what constitutes the existence of an individual body. For the mind to understand itself in relation to the idea of God is for it to perceive itself in this way — as included in a wider network of thought, on which it depends to be what it is, and which can encompass its having ceased to exist.

7. ETERNITY AND MORTALITY

But why does the mind perceiving itself thus perceive itself as eternal? Is it not the case that — whatever the individual mind achieves in the way of self-knowledge — all that 'remains' after death is the fact that it lived well — with knowledge 'of itself, of God and of things'? Is Spinoza guilty of misleading us here in presenting his doctrine in terms that suggest a continuity with doctrines of immortality? He has assured us that reflective self-knowledge is an enhancement of individual existence during life; and that what is common to all is incorruptible. But does the content of Spinozistic self-perception retain any continuity at all with the belief in individual immortality? Is there not an imaginative gap between the mind's perception, during life, of its dependence on Substance and its reconciliation to death? How is this perception of itself as 'part of a whole' supposed to reconcile it to death, in any way over and above what is accessible through the cultivation of mere

Reason? How is it supposed to reconcile it to the death of the all-important part which it is? Can we really see Spinoza's doctrine as salvaging something that answers to the traditional hope of immortality, despite his commitment to the mortality of all individuals?

I have claimed that it is not by shifting its attention to a dimly grasped 'whole' — wallowing in the 'oceanic' feeling — that the Spinozistic self is supposed to be reconciled to death. What then is the content of the required perception of self as part of a whole? The answer lies in Spinoza's distinctive conception of the totality of thought as a systematically interconnected, articulated 'idea'. The mind of God, of which the individual mind comes to see itself as a part, is the total articulation in thought of all that there at any time is. It is not a totality of omnitemporal ideas; each constituent idea ceases to exist, along with its correlated finite mode of extension. The eternal intellect of God is more appropriately seen as the totality of all that is ever true, than as a totality of omnitemporal truths. However, such a totality does, in a sense, transcend time.

The character of what is past is not altered by the passage of time. This is not to deny that there are important differences between past, present and future. But that an event is over, or that a thing has ceased to be, does not intrude on *what* it has been. The point does not involve the preservation of the past, as it were, in a cosmic memory. It is as *having been true*, not as *being now remembered*, that what is perceived 'under the form of eternity' can be said to escape the ravages of time. The point involves, in Dummett's sense (1968—9), a 'realism' about the past; rather than an 'anti-realist' reduction of the past to present memories. The perception of things 'under the form of eternity' would be impossible for an 'anti-realist', for whom there is no truth about the past, independently of the possibility of present recognition.

This aspect of the 'eternity' of finite modes derives from the relationship between truth and time;[11] and can be expressed without reference to truth as a unified totality. Even if finite modes were not parts of a unified whole the truths of their individual existence would be unaffected by the passage of time. Spinoza's point goes further than this. Each individual mode is what it is only because of its integration in the whole interconnected system. This gives the doctrine an additional content to what can be captured by a mere realism about the past. Each individual mode of thought — the human mind included — is what it is only as part of the totality of truth. Thus what it is — not just that it has been — transcends the limits of its own temporal span. For Spinoza there is no relevant distinction between what it 'essentially' is and mere facts about its historical relations to other things.

I have argued that the content of Spinoza's Intuitive Knowledge is not a perception of the 'whole'. The idea of a unified totality does nonetheless play, off-stage, a crucial role in the doctrine of the eternity of the mind. To understand itself as eternal the mind must see itself as a being whose very nature depends on the total expression of the one unique substance. Spinoza's monism remains in this way essential to his version of the eternity of the mind; although it is not a perception of the whole, but an understanding of individuality, that is supposed to yield the mind's eternity. The Spinozistic mind aspires to understand itself as an integral part in a total, unified articulation of the world, sustained by the necessary being of Substance. It is the mind's recognition that it has such a place in a systematically unified order of thought that sustains the full Spinozistic reconciliation to finitude.

For contemporary readers, imbued with the spirit of Hume, the idea of a necessarily connected unified whole must strike a discordant note. It is this aspect of Spinoza's rationalism that marks most significantly the gap between his metaphysics and contemporary thought. If the doctrine of the eternity of the mind amounts, as I have argued, to drawing out the full implications of the oneness of substance, it is not surprising that it should baffle contemporary readers. The very fact that the doctrine rests on a view of the universe as a necessarily inter-connected system — and on a view of knowledge as grasping those necessary connections — can make it seem a thoroughly alien doctrine. Our difficulties with it are compounded by the fact that the idea of the self as part of a unified whole has, despite Spinoza's rationalism, acquired retrospective associations with Romanticism. The combination may well make the system seem quite bizarre. Many a contemporary reader may find in the *Ethics*, especially in its final sections, nothing but the exotic charm of a thoroughly alien system of thought. But I hope I have shown that the eternity of the mind is, at least, a coherent development of Spinoza's metaphysics of Substance; and that we do not have to cast it as a doctrine of individual *immortality* to save Spinoza from Hegel's charge that he allows Substance to 'swallow up' individuals.

The Australian National University

NOTES

* Citations are to Gebhardt, Carl: 1925, *Spinoza Opera*, Carl Winters Universitaets-buchhandlung, Heidelberg. Quotations from the *Ethics* follow the translation of W. H. White and A. S. Stirling, 1899; reprinted in Wild, J. (ed.): 1958, *Spinoza Selections*, Scribner, New York. Where my translations depart from this version, the Latin is included in parentheses. Quotations from the *Short Treatise on God, Man and His Well-Being* follow the translation of A. Wolf: 1910, A. C. Black, London.

[1] Leibniz: April, 1676, *Paris Notes*, 12 *verso*, in Jagodinski (1913, p. 100); transl. as in Loemker (1969, p. 161).

[2] Leibniz: April 15, 1676, *Paris Notes, Recto*; in Jagodinski (1913, p. 110); transl. as in Loemker (1969, p. 162).

[3] Leibniz: 1686, *Discourse on Metaphysics*, Ch. 34; in Loemker (1969, pp. 325—326).

[4] For a discussion of the relationships between mediaeval treatments of the soul and Spinoza's, see Wolfson (1934), II, Ch. XIII, Section 3, and II, Ch. XX, Section 3.

[5] This aspect of Aquinas's doctrine is discussed in Geach (1969), Ch. 2, 'Immortality'. The quotation, cited in Geach, is from Aquinas's Commentary on I. *Corinthians*, 15.

[6] *Maimonides: The Guide for the Perplexed*, transl. Friedlander (1881). See esp. Part I, Ch. LXVIII, pp. 100—102; Part II, Ch. IV, pp. 156—158; Part III, Ch. XXVII, pp. 312—313. For discussions of the influence of Maimonides on Spinoza's treatment of the eternity of the mind, see Roth (1963, pp. 137—142); and Wolfson (1934, II, Ch. XX, Section III).

[7] See esp. Kneale (1968—69) and Donagan (1973).

[8] Errol Harris (1971) presents 'the whole' as the real object of the perception of things under the form of eternity; although he tries — I think unsuccessfully — to reconcile this with Spinoza's treatment of individuality.

[9] Leibniz: ca. 1679, 'On Fredom', in Fourcher de Careil (1857, pp. 178—9); transl. as in Loemker (1969, p. 263).

[10] Leibniz: July 14, 1686, Letter to Arnauld, Gerhardt (1875—1890), Vol. II, p. 55.

[11] This aspect of the eternity of the mind is expressed in Santayana's (1910, pp. xviii—xix) exposition of the doctrine: "A man who understands himself under the form of eternity knows the quality that eternally belongs to him, and knows that he cannot wholly die, even if he would; for when the movement of his life is over, the truth of his life remains"; and by Timothy Sprigge (1972): "Facts are not subject to change though not because they are prolonged through time. They cannot somehow be blotted out from reality and cease to be . . .".

REFERENCES

Donagan, A.: 1973, 'Spinoza's Proof of Immortality', in Grene, M. (ed.): 1973, *Spinoza: A Collection of Critical Essays*, Anchor Press/Doubleday, New York, pp. 241—258.

Dummett, M.: 1968—69, 'The Reality of the Past', *Proceedings of the Aristotelian Society*, 69, reprinted, 1978, in *Truth and Other Enigmas*, Duckworth, London, pp. 358—374.

Foucher de Careil, L. A. : 1857, *Nouvelles Lettres et Opuscules inédits de Leibniz*, Auguste Durand, Paris.

Freud, S.: 1966—74, *Standard Edition of the Complete Psychological Works*, Hogarth Press and Institute of Psycho-Analysis, London.

Geach, P. T.: 1969, *God And The Soul*, Routledge and Kegan Paul, London.

Gerhardt, C. I.: 1875—1890, 1960, *Die philosophischen Schriften von Gottfried Wilhelm Leibniz*, Georg Olms Hildesheim, Berlin.

Harris, Errol: 1971, 'Spinoza's Theory of Human Immortality', *The Monist* **55**, pp. 668—851; reprinted in M. Mandelbaum and E. Freeman (eds.): 1975, *Spinoza: Essays in Interpretation*, Open Court, La Salle, Illinois.

Hegel, G. W. F.: 1896, *Lectures on the History of Philosophy*, transl. by E. S. Haldane and F. H. Simson, Routledge and Kegan Paul, London; The Humanities Press, New York.

Jagodinski, I.: 1913, *Leibnitiana: Elementa Philosophiae Arcanae De Summa Rerum*, Kasan.

Kneale, M.: 1968—69, 'Eternity and Sempiternity', *Proceedings of the Aristotelian Society*, 69, 223—38; reprinted in M. Grene (ed.): 1973, *Spinoza: A Collection of Critical Essays*, Anchor Press/Doubleday, New York, pp. 227—40.

Loemker, L. E.: 1969, *Gottfried Wilhelm Leibniz: Philosophical Papers and Letters*, 2nd ed., Reidel, Dordrecht, Holland.

Maimonides, M.: *The Guide to the Perplexed*, transl. 1881, M. Friedlander, Dover, New York.

McKeon, R. (ed.): 1941, *The Basic Works of Aristotle*, Random House, New York.

Peirce, C. S.: 1933, *Collected Papers*, ed. by C. Hartshorne and P. Weiss, Harvard University Press, Cambridge, Mass.

Roth, L.: 1963, *Maimonides, Descartes and Spinoza*, Russell and Russell, New York.

Santayana, G.: 1910, *Introduction to the Ethics*, transl. by A. Boyle, Everyman, Dent, London.

Sprigge, T.: 1972, 'Ideal Immortality', *Southern Journal of Philosophy*, Summer, pp. 219—236.

Taylor, A. E.: 1937, 'Some Incoherencies in Spinozism', Part II, *Mind* **XLVI**, 281—301; reprinted in P. Kashap: 1972, *Studies in Spinoza*, University of California Press, Berkeley.

Wolfson, H. A.: 1934, rpt. 1958, *The Philosophy of Spinoza*, Meridian, New York.

PART V

SPINOZA AND TWENTIETH CENTURY SCIENCE

HANS JONAS

PARALLELISM AND COMPLEMENTARITY:
THE PSYCHO-PHYSICAL PROBLEM IN SPINOZA
AND IN THE SUCCESSION OF NIELS BOHR*

The psycho-physical problem named in the title of this paper was born together with modern science in the seventeenth century and is the twin brother of its guiding axiom that things corporeal must be explained by corporeal causes alone, or that the latter are sufficient to explain everything in the physical realm, neither requiring nor even admitting the cooperation of mental causes. Indeed, completeness of intraphysical determination excludes the introjection of any non-physical source of action. So put, the axiom — in a stunning break with pre-modern, 'Aristotelian' physics — amounts to the thesis of a causal redundancy of mind in nature. This redundancy found its first expression in the disavowal of 'final causes,' the concept of which is somehow borrowed from mentality. At the root of the radical turn was a reinterpretation of 'nature' itself in purely spatial or geometrical terms. This made measurement of magnitudes the main mode of scientific observation, and quantitative equation of cause and effect, i.e., of antecedents and consequents, the ultimate mode of explanation. This epistemic program of the new science received its metaphysical underpinning in Descartes' doctrine of two heterogeneous kinds of reality ('substances' in his language) — the *res extensa* and the *res cogitans* — each defined by its one essential attribute 'extension' and 'thought' respectively, and each having nothing in common with the other. The gain of this ontological dualism was the setting free of nature for the unrestricted reign of mathematical physics; the cost was that the relation of mind and body became an intractable riddle. Post-Cartesian continental philosophy is one persistent grappling with this riddle: whatever solutions were proposed, the most persuasive assertion of our untutored experience — *interaction* between body and mind — had to be discounted from the outset. This gave seventeentli century speculation the anti-common-sensical, intellectually violent flavor which characterizes it. The most familiar appearances had to be contradicted. It appears to be the case that an external object, through the intermediary of our bodily senses, 'produces' in our mind a perception of it — a mental presence. But this violates the principle that physical causes can have only physical effects.

237

Marjorie Grene and Debra Nails (eds.), Spinoza and the Sciences, 237—247.

Likewise, it *appears* to be the case that my will produces the raising of my arm, but this violates the even more vital principle that physical effects can have only physical causes and that, therefore, the actual rising of my arm must be completely determined and accountable for by its physical antecedents. Thus, the *apparent* interaction must be replaced by a different, non-interactionist theoretical model.

The philosophically most remarkable as well as least capricious such model was designed by Spinoza in his doctrine of one substance with different attributes. This model provided for the mind-body problem the solution known as 'psycho-physical parallelism'. Let us briefly recall the general principle of Spinoza's system. Its basis is the concept of one, absolute and infinite *substance* that transcends those specifications, viz., extension and thought, by which Descartes had distinguished between two different kinds of substances. The infinity of the one substance involves an infinite number of *attributes* that 'express' the essence of that substance — each expressing it truly though not completely because each does so under just this form or aspect. They are necessary aspects, not because our subjective cognition happens to be cast in that mold, but objectively inherent in substance itself as articulations of its plenitude. Thus there is no going behind them to a hidden ground: that ground itself exists in no other way than through its attributes, which are coeval with it and coequal with one another. None of them is more genuine than the other, and all of them are concurrently actual in each actuality of substance. The same can also be stated by saying that the attributes all together 'constitute' the essence of substance, not, however, additively but as abstract moments that are only abstractly separable (as shape and color are only abstractly separable from a visual object).

Now, of these infinitely many attributes of the one cause, we humans know only two, extension and thought. They are the two universal forms under which alone we can and must conceive all things. But these 'things' are not themselves substances; infinite substance leaves no room for a plurality of finite substances. Whatever is finite is not a substance but a modification or affection of infinite substance — a '*mode.*' With this term we have, after 'substance' and 'attributes,' the third key term in Spinoza's ontology, 'modes'. Individual items of reality, even what we are most want to regard as self-subsistent, like molecules and rocks, are states or occurrences rather than entities, variable local determinations of the self-subsistent One *in terms of* its invariable attributes — this particular body, this particular thought. And if in terms of any one, then in terms of all of them: From the very status of 'attributes' as merely different aspects of the same, it follows that each such finite affection of infinite substance as it occurs is exhibited, equally and equivalently, throughout all its attributes at once, and with a sequential necessity of succession *within* each attribute that reflects the eternal necessity of the divine nature. These necessities, then, run '*parallel*' in the diverse attributes, being in reality one and the same necessity. Its simultaneous, coequal manifestation in different essential forms means that it can be apprehended with equal truth under each such form by a finite mind that enjoys cognizance of some of them, and with more truth when this mind apprehends its sameness in several of them. Since, in the human case, this is limited to the two indicated, our world consists in fact of body and mind, and nothing else.

The point for our context is that what to Descartes were two separate and mutually independent substances are to Spinoza merely different aspects of one and the same reality, no more separable from one another than from their common cause. And he stressed that this common cause — infinite substance or God — *is* as truly extension as it is thought, or, as truly corporeal as mental; but there is as little a substance 'body' as there is a substance 'mind.' Now since both these attributes express in each individual instance an identical fact, the whole problem of interaction or of any extraneous interrelation vanishes. Each occurrence (mode) as viewed under the attribute of extension is at the same time, and equivalently, an occurrence viewed under the attribute of thought or consciousness, and vice versa. The two are strictly complementray aspects of one and the same reality which of necessity unfolds itself in all its attributes at once. It would even be too disjunctive to say that each material event has its 'counterpart' in a mental event, since what externally may be registered as a parallelism of two different series of events is in truth, that is, in the reality of God or nature, substantially the same. Thus the riddle created by Cartesian dualism — of how an act of will can move a limb, since the limb as part of the extended world can only be moved by another body's imparting its antecedent motion to it — this riddle disappears. The act of will and the movement of the body are one and the same event appearing under different aspects, each of which represents in its own terms a complete expression of the concatenation of things in God, in the one eternal cause.

If we had the time we should now go on to present the ingenious theory of organism with which Spinoza focused the general ontological scheme specifically on the biological sphere, where mentality is ordinarily seen to be conjoined to physical fact, and particularly on the case of man (see Jonas, 1974). It must be enough to say that Spinoza makes it beautifully intelligible from his general premises that the quality and power of a mind are proportionate to the complexity of the body to which it corresponds, so that the perfection of the human body as a piece of physical organization is a direct yardstick for the perfection of the human mind which, as it were, conformally (or: isomorphously) duplicates the body's physical performance on the plane of thought. Likewise, a horse's mind is the mind conformal to the horse's body, and so down the line. We need not follow that line in its dubious descent into inorganic matter. What Spinoza is really about, and we with him, is the human mind and the human body, and with these alone in view let us try a critique of this solution to the psycho-physical problem.

Surely, Spinoza's parallelism of attributes expressing differently but equivalently one and the same substance was a feat of genius and far superior to all other treatments of the problem at the time. Without interposing a synchronizing deity, as did others, it overcame Descartes' dualistic rift by a monistic reduction, yet retained the full severity of the disjunctions which that dualism had been designed to ensure. Both sides of the coin are evident in the following propositions from the *Ethics*.[1]

"The order and connection of ideas is the same as the order and connection of things." (*Ethics* II, Proposition VII)

"A body in motion or at rest must be determined to motion or rest by another body, which was also determined to motion or rest by another, and that in its turn by another, and so on *ad infinitum*." (*Ethics* II, Proposition XIII, Lemma 3)

"The body cannot determine the mind to thought, neither can the mind determine the body to motion nor rest, nor to anything else if there be anything else." (*Ethics* III, Proposition II)

In short, the parallels never cross, each continues as determined by itself.

My critique is two-fold: (1) the price for parallelism is a strictly necessitarian view of mind; (2) the alleged equal status of the attributes, the heart of the doctrine, fails to hold up in the execution of the system — notwithstanding all professions to the contrary, matter gains priority and mind in effect becomes an appendix or epiphenomenon of the body.

(1) The necessitarian point is obvious: if conformal parallelism is to hold, then the mental sequences must be as deterministic as the physical. Indeed, Spinoza is emphatic in declaring free will to be an illusion. I leave it to the judgment of the reader whether absolute psychological determinism, the denial of any spontaneity of mind, is not too high a price to pay for the solution of a theoretical problem, and an unwarranted price at that, since unlike physical determinism, intramental determinism is without evidence of its own and is merely postulated for the system's sake.

(2) The issue of determinism leads of itself to the second objection, viz., that the attribute of thought is, contrary to the claim of parity, in effect subordinated to that of extension. For where is 'determinism' itself a determinate concept and not merely a summary assertion? Surely in the realm of matter, i.e., of spatiality, and precisely by virtue of the attribute of extension, which alone offers the kind of manifold where single items can be distinguished by space-time location, and quantitative values can be assigned to them by measurement of defined magnitudes, and paths can be plotted in relation to coordinates, and their intersection predicted and its outcome as a causal event computed — and where all this can be verified or falsified by new measuring observation. There — in the *res extensa* — determinism is well-defined and testable, and even its limits, its margin of imprecision, can be precisely stated in quantum physics. The stream of consciousness, on the other hand, non-spatial and with only the dimension of time, has nothing to match these conditions for vector analysis and for quantifiability in general. Put plainly: neither Spinoza had, nor do we have, a predictive science of mind as we have a predictive science of body, and Spinoza's own heroic attempt at founding such a science by 'geometrizing' mind must be deemed a grandiose failure. Thus when I say that the present state of the mind determines its future state with necessity, and this in strict conformity to the parallel determinism of the states of the body, what I am really saying is that we must look to the causal history of the body, the only we have, to tell us by proxy about the otherwise unknown determinism of the mind. The fact that we cannot reverse the procedure, as the principle of parallelism would stipulate as equally eligible, puts the attribute of mind in the position of unilateral dependence on the other attribute of substance.

Add to this that the human mind, according to Spinoza, is already defined with reference to the body, namely, as having this body for the primary object of its thought, that its thinking is a thinking of the body, whereas the attribute of extension that defines the body has no intrinsic reference to mind, so that indeed we can consider body in its terms alone, as physics does, but mind only in conjunction with the body and as reflecting its condition in thought — and we see that matter is master in the match, and that the very rule of parallelism then reduces mind to one-sided dependence on it. At the end of this road stands the view of mind as a mere impotent epiphenomenon of matter, and matter as the one true substance — not at all what Spinoza had intended.

This critique is tinged with reverence for the might and originality of Spinoza's thought. I also confess to the philosophical hunch that, for all the inadequacies of the parallelistic formula, the monistic effort behind it, the vision of one reality in different manifestations, points in the direction where the truth may lie.

Now, with some trepidation, I pass to the question whether 'complementarity' in the technical sense evolved by quantum physics, or rather in analogy of it, offers perhaps a better model for handling the psychophysical problem. The question is prompted not so much by Niels Bohr's (1961, pp. 24 and 100ff) having at one time tentatively suggested it (never to take it up again) as by the tempting currency which the formal concept of complementarity has gained outside its native ground, and by its actual use, or abuse, in different spheres, such as the social sciences.

First a few words about the original, quantum mechanical meaning of 'complementarity' as coined by Bohr. It concerned the possibility of defining the 'state' of a system in terms of two mutually exclusive, conceptual representations. One may say that complementarity, in the sense of Bohr, means that we can ask one of two mutually exclusive sets of questions about a system, but not both. The answer to one question would describe the system as a particle; the answer to the other would describe it as a wave. But the two models are not just optional alternatives equivalent with one another. They stand for different observables; e.g., the position component x_1 is complementary to the impulse component v_1, thus the particle description answering to the measurement x_1 is complementary to, but not interchangeable with, the wave description answering to the measurement of v_1. The knowledge of one of them precludes the full determination of the other — the more I know about the one, the less I know about the other — yet both are required for a full account of the phenomenon, complementing one another in conveying its truth, i.e., the exhaustive knowledge of what is knowable about it. As quantum physicists are want to say, the entity 'is' a particle when I measure its position, and it 'is' a wave when I measure its momentum. Thus, whatever it is 'in itself,' only a dual account can do justice to the object (or, express the 'truth' about it), without therefore bespeaking a dual nature of things.

Now, it has proved tempting to think that something similar might also apply to the two-fold account of human action (and of conscious behavior in general), the 'outer' and the 'inner,' and provide a solution to the ancient problem of necessity and freedom. Descriptions of one and the same train of events in terms of physical necessity, on the one hand, and in terms of mental spontaneity, on the other, are 'complementary' in the sense of Bohr's principle: the either-or is one of representation, not of fact, and only both representations together *in* their difference convey the truth of the identical fact. As in the case of the particle and wave descriptions, both are equally genuine — and, we should add, equally symbolic. Somehow, the underlying reality thus doubly expressed is supposed to be one in itself, as was the case with Spinoza's 'attributes.' But the epistemic situation is quite different.

The suggestion is appealing, but I doubt that it works better than Spinoza's parallelism. First, we ask: is the transfer of the complementarity concept to this case *formally* faithful to the original, so that with all differences of content 'complementarity' retains the same logical structure? I think not. To 'complementarity' in its quantum physical conception by Bohr himself, it is essential that the two descriptions are clearly separate, each complete in itself and neither intruding into the other; the wave description is not to be contaminated by corpuscular terms, and vice versa. The two models are, in short, strictly alternative. But we cannot begin to describe anything 'mental' without referring to the 'physical,' the world of objects with which mind, sense, will, and action have actively and passively to do. That is to say, any speech about mind *must* also speak of body and matter. And when speaking of ourselves, we not only can and do but always *must* embrace with it our physical and mental being *at once*: precisely that *simultaneous* entertaining of *both* sides which quantum theory rules out for its complementary alternatives. From this original, *joint* givenness, after all, the psycho-physical problem arises in the first place. Here, the *isolation* of the two components is an artifact of abstraction, their interlocking co-presence being the primary datum. Even in abstraction the isolation does not really succeed, as the description of one side intrinsically refers to the other: the lines themselves do not run parallel, but cross. On this *transitive* relatedness alone, by which one description draws into itself elements of the other, the purported analogy with the quantum mechanical situation breaks down.

Even simpler, in this formalistic vein, is the objection that in the quantum mechanical case we begin with data of the same kind (space-time measurements) and end up with a duality of representations of our own devising to account for them, whereas in the psycho-physical case we begin with a duality of cardinally different data, not of our making at all, and try for a theoretical unification of them. If in such a unification they are found to be 'complementary' in some sense then 'complementarity' itself becomes the unitary representation for a dual phenomenon. Thus the direction of the logical operation in the two cases is opposite, the one yielding a divergent model, the other (hopefully) a convergent one. These purely formal objections, by the way, especially that of the 'crossing lines,' fall on *all* extramural uses of the complementarity principle I know of (e.g., in the social sciences): they are all forced to violate the (at least) *semantically exclusionary* character the duality has in the original model and come to grief already on this count alone.

More to the point of our discourse than this formal observation is the blunt reminder that what is substantively at issue in the psycho-physical problem is, *interaction* and, more particularly (surely so for Bohr's interest in 'freedom'), the question of an *intervention* of mind in the affairs of matter. This, if it takes place at all (which is just in question), is nothing like an invariable concomitant, an innocent complement of physical processes, but is a particular event affecting their course. Does this happen? Is it possible? How? Such a question is obviously meaningless in the case of the complementary wave and particle descriptions; i.e., to ask whether, to what extent, on what occasions the wave aspect of events leaves its mark on their particle aspect. But just those questions (with the appropriately substituted terms) are the most meaningful ones to ask in the psycho-physical setting. Complementarity, 'noninterventionist' by its formal nature, does not even allow them to be asked when seriously held to apply.

It is equally meaningful to ask *what* of our behavior, even of the mental state in back of it, is conditioned or circumscribed or prescribed by physical necessity, and what *we* truly initiate; i.e., to *apportion* the relative *shares* of the two sides in a given instance, which again is something wholly inapplicable to complementarity in the genuine sense.

In sum, I believe no faithful analogue of 'complementarity' as understood by Niels Bohr really applies to our problem, and philosophers should leave it where it belongs. The philosophical interest of its attempted enlistment lies in what it has in common with Spinoza's 'parallelism', viz., the *noninteractional* premise. Of this, it seems to offer a more sophisticated version, but it actually is, for that assigned purpose, inferior to the predecessor in one essential respect: Spinoza expressly acknowledges for the attribute of 'thought' an intrinsic reference to the attribute of 'extension' (but not vice vesa!), inasmuch as he *defines* the human mind from the first as being the 'idea' of an 'actually existing body': that very body, in all of its changing states, is the proximate 'ideatum' (= datum) in all of the individual mind's ideations. This permanent 'idea' of its own body, as complex as this body itself, is indeed 'the first thing' that '*constitutes* the actual being' of an individual mind and that mediates its ideas of all other things. Thus, far from interdicting physical terms in the description of a mental state (as complementarity must do), Spinoza's parallelism stipulates their very employment as integral to any description of mind: the 'crossing' of lines, which the evidence displays and complementarity forbids, is here part of the doctrine itself. This brings the doctrine into closer agreement with the facts, but also into some disagreement with itself. For by the general doctrine, all attributes have equal status and no interrelation other than through their common root, 'Substance' itself: no direct relationship of one attribute to another should have a place in their mere juxtaposition according to the system. Yet, 'thought' of all attributes does have just such a relationship — ideally to *all* the others (as asserted for God's thought) and actually to the attribute of extension in man's case. The asymmetry which this *uniqueness* of mind imports into the doctrine of attributes is one of the major tensions in Spinoza's system and an old crux for its interpreters. 'Thought' in God embraces all the other attributes and thereby excels them: it consists in reflecting them (and itself — there being also 'ideas of ideas') and thereby is subject to them. Likewise, 'thought' in man excels his one other attribute, 'extension', whose self-enclosure is deaf to the coexistence of its companion, and by the same token is subject to it because it must reflect it. Mind alone transcends itself towards all the other attributes, while these are simply and immanently themselves. From this, there results a double disparity of mind with the other attributes, a positive in scope and a negative in autonomy, the former a disparity at the conceptual origin, the latter one at the functional consequence. The necessitarian and ultimately epiphenomenalist consequence of the disparity has been noted before; its initial inconsistency with the terms of the system must now be added. The necessitarian consequence, which troubles us, did not trouble Spinoza, the great denier of free-will. The flaw of internal inconsistency (which would have troubled Spinoza indeed) must be seen by us as a great mind's tribute to the force of truth at the cost of systematic symmetry: by this very irritant Spinoza's parallelism is cognitively superior to a 'complementarity' model that *must*, under the uncompromising either-or of its representational scheme, rule out the physical reference from the mental description. That reference being, however, a stubborn datum of the evidence, the schema becomes simply inapplicable. What in Spinoza is a crack in an edifice which can still house many insights, would here annul the very foundation. But whatever the logical merits or demerits, these and all noninteractional models of the mind-body relation equally submerge the real sting of the psycho-physical problem: the question of a *power* of mind to intervene in the course of things. No speculative appeasement can ever disarm this sting.

The result of our critical journey, then, is chastening. The signs are that the psycho-physical problem, the mind-body riddle, is today, 300 years after Spinoza's death, still there to haunt us and still poses the same challenge it did to him. My intimation at the end is that the challenge is better met head-on by boldly trying once again the long tabooed 'interaction.' And for this, quantum mechanics may indeed offer an opening which classical mechanics denied — not by the principle of complementarity but by that of 'indeterminacy'. This, to be sure, would no longer be in the vicinity of Spinoza and will be discussed in another paper. Suffice it here to say that in the post-Newtonian age, aim can at least be taken again at a solution whose very concept was outlawed by deterministic natural science. After centuries of almost obligatory noninteractionism, of which Spinoza was the mightiest philosophical symbol, there may yet arise an interactional model of the mind-body relation that vindicates, insted of belying the irrepressible evidence of our thinking and acting experience.

New School for Social Research

NOTES

* Originally published as 'Parallelism and Complementarity: The Psycho-Physical Problem in the Succession of Niels Bohr', in Richard Kennington (ed.), *The Philosophy of Baruch Spinoza*, Studies in Philosophy and the History of Philosophy 7, Catholic University of America Press, Washington, D.C., 1980. Reprinted here by permission of the author and the publisher.

[1] Here quoted in the translation by William Hale White (Spinoza, 1930).

REFERENCES

Bohr, Niels: 1961, *Atomic Theory and the Description of Nature*, Cambridge.

Jonas, Hans: 1974, *Philosophical Essays. From Ancient Creed to Technological Man*, Chicago University Press, Chicago.

Spinoza, Benedicti de: 1930, *Ethica ordine geometrico demonstrata*, transl. by William Hale White, in John Wild (ed.), *Spinoza, Selections*, Charles Scribner's Sons, New York.

JOE D. VAN ZANDT

RES EXTENSA AND THE SPACE–TIME CONTINUUM

1.1. In the following pages I propose to suggest, rather than to demonstrate, the relevance of the philosophical perspective of Benedictus de Spinoza to a view of the current state and future progress of physical theory. To what extent these suggestions can be made more concrete or more firmly grounded is a task which cannot be fruitfully explored in a short exposition, but perhaps the desirability of further thought and exploration concerning these matters can be indicated.

1.2. The twentieth century has looked unfavorably on the intrusion of professional philosophers into the realm of what was once known as natural philosophy. Indeed, the movement of both physics and philosophy has been such as to preclude substantial interaction. But enterprises have become so technical and insular that representatives of either discipline seem curiously naive and superficial when intruding upon the other. Philosophers who pronounce on the freedom of will on the basis of the principle of uncertainty, for example, are usually thought not to understand its role in twentieth century physics, often with justification. On the other hand, physicists who give reign to their urge to speculate are thought to have retired from the profession of physics, but are rarely admitted into the company of philosophers as serious thinkers. Just as philosophers are ill at ease among the forest of elegant symbolism of modern physics, for which a few courses in elementary analysis cannot suffice, so are physicists ill-prepared for the sea of distinctions and problems that modern philosophy counts as its domain, which requires more than a keen intellect and wide experience for its practice. One reaction to this situation, common to both sides, is to discount the significance of one discipline for the other. There was a time in our recent past when even philosophers of science were willing to wear their ignorance of science as a badge, justifying this by claiming that issues of the logic or methodology of science transcend the particulars of research, the details of which might cloud the analytical process! On the other hand, the eminent physicist, Yuval Ne'eman (1981), has recently written the following:

Marjorie Grene and Debra Nails (eds.), Spinoza and the Sciences, 249–266.
© 1986 *by D. Reidel Publishing Company.*

In the case of philosophy, ... that discipline's specific historical role as an *ersatz* science suffered major collapse at his [Einstein's] touch (in the manner of the conception of time and of space-time) similar to that of the Church after its collision with Copernicus and Galileo. The Church — or religion — later wisely retreated into the domain of ethics, where it may be safe (for a while?) as a source of doctrine. ... Likewise, post-Einsteinian philosophy has abandoned metaphysics and the direct weighing of natural concepts, leaving them to the relevant sciences and entrenching itself in epistemology and the study of scientific method. In this field, it is doing well and has enriched our understanding. As to the rest of philosophy, logic has been reshaped after Cantor as a branch of mathematics. In the last century, politics has been replaced by the social sciences. Aesthetics and ethics are still there, in that region where scientific method has not yet found its foothold. (pp. xvii—xviii)

I cite this passage not in order to dispute the characterization it presents, but precisely because it would seem unexceptionable to the great majority of practising physicists. Whatever philosophy may offer, this passage asserts, it provides nothing for real physics. I dare say that even if the particulars of this characterization may be taken to task by philosophers, past and present, the general sentiment it expresses is not foreign to a great many philosophers of the twentieth century as well.

It seems to be the time to step into this breach, although, to be sure, the situation is not as bleak as it might have been a quarter of a century ago. It is apparent that many young philosophers are beginning to be better versed in the formalism of physics and its leading areas of research. There is also evidence that physicists are becoming more reflective concerning the hyper-theoretic, or meta-theoretic, aspects of their work.[1] Perhaps the time is now ripe to suggest that physics and philosophy may, indeed, be able to cooperate in the quest to understand nature. That is, in fact, the major intent of this paper, with one added element which makes it both more controversial and appropriate for this forum. I will try to show that much is to be gleaned for the presently spurned pre-Kantian metaphysical tradition of modern philosophy through its most remarkable representative, Benedictus de Spinoza.

1.3. It is natural to introduce Spinoza into the contemporary scene by way of juxtaposition with Albert Einstein. The similarities in the character, vision and even, to some extent, the circumstances of the two men are too obvious to miss and it would be excessive to enumerate them. While there is excellent reason to believe that Einstein was familiar with some of Spinoza's writings,[2] there is less reason to believe that any of Einstein's work was directly influenced by this.[3] Einstein was, as is well known, enormously original, innovative and independent in the development of his views. What I believe I can show is the confluence of their thought, not the influence of Spinoza. Neither does it appear fruitful to me to seek an anachronistic anticipation of twentieth century physics in the writings of Spinoza. Plainly, Spinoza could not have presciently grasped the complex dialectic of the development of physics which continues to the present and is likely to continue into the foreseeable future. What can be sought, however, is a philosophical view of sufficient richness to encompass the alterations in our concepts and intuitions that have undeniably attended the growth of our understanding of nature.

A brief digression is now unavoidable. If one accepts as representative the quotation from Ne'eman, and it seems most representative to me, one might well question the utility of the enterprise I propose. What need is there to concern oneself with metaphysics at all? Of course, a serious grappling with this question would require argumentation far more extensive than can be marshalled at present, but some preliminary effort is demanded if the reader is to be justified in continuing along this path.

The enormous tri-partite success of the mechanical world-view developed from Newton through Laplace, the development of a field-theoretic treatment of electromagnetism by Oersted, Faraday and, above all, Maxwell, and the statistical mechanics of Boltzmann and Gibbs can scarcely have prepared any physicist for the twentieth century, even those perceptive enough to discern some fundamental incongruence among these very grand theories. To be sure, the last quarter of the nineteenth century does exhibit, in retrospect, that phase of the science which might be described, following Kuhn,[4] as pre-revolutionary, namely, more and more puzzles began to accrue which were increasingly difficult to reconcile with the prevailing paradigms. With the turn of the century the flood-gates of change were opened. I need not recite here the remarkable achievements of Planck, Einstein and Bohr, nor the later achievements of Heisenberg, Schrödinger, Pauli and Dirac, to mention only a few of the major figures of this period. These have been exhaustively and careful recorded and analyzed in many excellent histories of this period. For now, the point is not the content of their achievement, but the incredible cascade of utterly new concepts, experiments and mathematical techniques, and the concomitant growth of technology and support for research from public and private sources which marked this time, from 1900 to the present. The active physicist has scarcely had time to catch his breath at any point in this century. The concrete difficulties of analyzing the newest data, developing new experimental techniques and learning to employ the growing number of powerful mathematical tools, not to say two World Wars, have mostly precluded reflection on the enterprise itself. This is not to say that there have not been Kuhnian paradigms at work, but these paradigms are largely pragmatically justified. The most common defense of quantum electrodynamics, if it needs a defense, is the remarkable precision with which its calculations predict experimental results; one no longer actively discusses Bohr's view of complementarity or Heisenberg's epistemological formulation of the uncertainty principle.

Now the present situation in physics is not precisely placid, but the very success of some of these theories provides at last not only the opportunity to reflect, but to some extent the demand that parts of them be carefully examined.[5] Such an examination cannot be done in an intellectual vacuum (though it seems, metaphysics aside, that such a vacuum can, in fact, exist). There must be some agreement, or at least reasoned discussion, of what is to be asked of physical theory, what principles underlie our investigation of nature.[6] This is, in the purest sense, metaphysics. It may or may not be what the term brings to mind to a generation weaned on the doctrine that metaphysics either is or ought to be dead, and it is also not the caricature which some analytic, linguistic, sceptical or pragmatic philosophers have presented. I also freely admit that it may not be the whole of what metaphysics means, for physics may not exhaust the concept of what there is, but I do not see how it can be seen as anything other than the metaphysical enterprise in nature.

Many physicists will have little interest in this pursuit, retaining a healthy dose of scepticism about its very possibility, and they will be joined in this by a substantial circle of philosophers. Such a bulwark against excess should be welcomed, but it should not deter the reflective process. Our best theoreticians and philosophers should not give serious thought to the current foundations of physics, where we already know some puzzles most certainly exist.

Sketchy as it is, this is my justification for urging a consideration of what follows in the next section and pursuing the project, whatever the outcome in this attempt.

2.1. To exploit Ne'eman's really rather innocent remarks further, let me note that it is in Einstein's analysis of simultaneity, that is, of "time and space-time" that Ne'eman locates the collapse of philosophy as *"ersatz* science". I believe that what he has in mind here is fairly clear. The mechanical world-view of the eighteenth and nineteenth centuries had found its most formidable philosophical defender in the person of Immanuel Kant. That Kant sought to establish apodictically the validity of Eucidean geometry as the only possible physical geometry, and the thorough-going deterministic order of phenomena is known by those who know nothing else of his work.[7] Kant's philosophy of nature was so dominant in the nineteenth century that the collapse of the mechanical world view, which Kant was most certainly not alone in affirming, has appeared to many as signalling the collapse of philosophy itself, at least as a discipline competent to examine and criticize the concepts of science. This is what I take the term *"ersatz* science" to mean. So, Ne'eman continues, philosophy has given up "the direct weighing of natural concepts", presumably having learned its lesson from Kantian mis-adventure. Again, I select these comments just because they strike me as concisely expressing the near-consensus view of physicists. The polemic, then, is aimed at the prevailing spirit and not against these offhand, undefended remarks which I am using as a vehicle. Even though it is arguable that Kant's work itself is rather more profound and far-sighted than some are wont to allow, it is not the aim of this paper to reconcile Kant's philosophy of nature with the vision of twentieth century physics. It will suffice to make a few elementary points using Kant as an example.

First, the view which Kant sought to establish was by no means idiosyncratic, but represented the best available scientific thinking of his day, in particular that of Isaac Newton. It may well be thought to be to his credit that Kant effectively eliminated the Cartesian and Leibnizian objections to Newton's theory through his critique of the traditional metaphysical doctrine which had been urged against the Newtonian system. Further, the view to which Kant gave voice, the mechanical world-view so-called, was virtually universally endorsed by the physicists of the nineteenth century, quite without Kantian coercion, even among those physicists, the results of whose work were fundamentally opposed in principle to such a view. The history of the luminiferous ether is just the attempt to mechanize the wave properties of light. Secondly, and more importantly, the Newtonian conception of space and time, which was considerably softened by Kant, by the way, is by no means the only, or even the most prominent philosophical view of space and time. The topic is, after all, an ancient one and views far from common sense have been suggested from time to time. This is perhaps most sharply brought home by taking as an example Newton's most illustrious and keenest competitor, Leibniz. Leibniz sharply objected to Newton's treatment of space and time as objective things or substances, advancing his own view that space and time are merely relational notions, indeed, internal relations of Leibniz's 'metaphysical points', the monads.[8] In point of fact, Kant attempted to synthesize these two contrasting points of view by stripping space and time of their quasi-substantial character found in Newton and transforming them into pure forms of intuition. The point I wish to draw out of this is that philosophy, for better or worse, exhibits far greater deviance from Kuhn's model of progress than the sciences; rarely does any 'paradigm' completely dominate the philosophy of any age. Lastly, it seems at least partly justified to describe philosophy as having retreated from metaphysics, particularly if one attends mostly to the philosophy of science, which has largely limited itself to issues of epistemology and methodology, or to the celebrated 'linguistic turn' in philosophy, which has centered largely on issues outside the philosophy of nature. It must be admitted that metaphysics in the grand style is not in vogue. But neither is it extinct. It may take rather unfamiliar forms, from Alfred North Whitehead to Gustav Bergmann, from Heidegger to Kripke, but the problems do not disappear by decree.[9]

What most certainly must be concluded, however, is that philosophy cannot proceed as though nothing had happened in the conception of nature during this century. A view which is utterly incompatible with our current understanding and holds no promise of reconciliation simply must be rejected. I therefore take seriously the challenge implicitly set forth by Professor Ne'eman: If recent philosophy has dropped out as *ersatz* science, is Spinoza's pre-Newtonian, pre-Kantian view of the nature of space and time compatible with our current views of space-time? This question is, finally, the central theme of this essay. If the answer is affirmative, then perhaps the story of the interaction of philosophy and physics is not yet concluded.

2.2. Spinoza's writings on physics are extremely limited. There is some ground to believe that he may have intended to compose a treatise on physics had he lived,[10] but as it stands we are left with his explication of Descartes's physics (Spinoza, 1961, pp. 57—106; Gebhardt I, pp. 181—225), a few short passages in the *Short Treatise* (Spinoza, 1963, esp. Chs. 4, 9; Gebhardt I, pp. 36—39, 48), a few scattered, but important, passages in his correspondence,[11] and a long scholium following Proposition XIII of Part II of the *Ethics*.[12] It has been widely assumed that Spinoza's physics was essentially the same as Descartes's, despite explicit rejection of at least one of Descartes's rules and the absence of explicit affirmation of the others,[13] and a sharp criticism of Descartes's definition of matter, and, indeed, Descartes's conception of extension itself. It seems quite arguable that Spinoza anticipated, to some extent, Leibniz's improvement on Descartes through the introduction of an active force in nature, notably absent in Descartes's system.[14] Lachterman (1978) has presented an excellent discussion of the physics of Spinoza's *Ethics* elsewhere, and while the topic may not be exhausted, it will not profit us here to pursue it.[15] Lachterman ends an 'Excursus' at the end of his paper with these words:

> However this issue [the gap between bodies, on the one hand, and motion and rest on the other] is to be resolved, it would be a deliciously seductive anachronism to see Spinoza moving toward the notion of a field of force, or, more venturesomely, toward a geometrodynamical theory in which material bodies (*Natura naturata*) are a dependent function of the geometry of space, while that geometry is, in turn, a function of the play of forces. (p. 103)

Since Lachterman has the explication of Spinoza as his goal in this essay, he wisely abstains from actually giving such an interpretation. Unlike Lachterman, I do not feel restrained by Spinoza's actual theory of physics, which must, of necessity, be placed in the context of the speculation and experimental evidence of his day. Rather, the question which I address is the suitability of the metaphysical view to a modern interpretation. I believe I may be deliciously seduced without being anachronistic. The discussion of space and time in Spinoza, then, will not be in the context of his brief physical remarks, but in the more general context of his system.

2.3. Let us start with time. Samuel Alexander put it well when he wrote "The trouble is that there is very little to say about Spinoza's conception of Time" (Alexander, 1972, p. 68). Spinoza does not often employ the term, to be sure, but fortunately we are not left wholly speechless.

Spinoza distinguishes eternal things and those involving (indefinite) duration. One should be clear from the outset that for Spinoza time has nothing whatsoever to do with eternity. Spinoza's notion of eternity is not that of what is called sempiternity,[16] that is, existence at all times. Perhaps Spinoza would have appreciated Plato's poetic expression of time as the moving image of eternity as an insight expressing darkly the distinction he intends to make in a clear fashion, but it is most unlikely that Spinoza's views here should be described as Platonic.[17] Time is connected only with duration; indeed, it is the measure of duration. Hence it is in the concept of duration that we must seek clarification of the temporal notion.

The definition of duration does not appear until Part II of the *Ethics*: "Duration is the indefinite continuance of existence." It is followed by the explanation:

I say 'indefinite' because it can in no wise be determined through the nature of the existing thing, nor again by the thing's efficient cause, which necessarily posits, but does not annul, the existence of the thing. (*Eth.* II, Def. 5 and Expl.)

Duration, then, is solely to be found in *Natura naturata*, the world considered as modified. It must be presumed that the reader is basically familiar with Spinoza's distinctions, but put briefly, Substance [*Deus sive Natura*] may be considered either from the point of view of eternity, in which case it is referred to as *Natura naturans* (mostly because no natural sounding English translation to the Latin is available); from this vantage point, the concepts of individuals, occurrences, times and places have no sense. One may, however, view Substance insofar as it is modified, that is, insofar as we consider an infinite nexus of things coming to be and passing away. This is referred to by the term *Natura naturata*, and only in Substance so considered do any of the above mentioned terms have any sense, and even here one must be most careful about what sense one attaches to some rather common terms. As opposed to Aristotle or the atomists, either the ancients or Spinoza's contemporaries, Spinoza holds that there are no natural individuals. All of these concepts are, for Spinoza, '*entia rationis*' — things of reason, aids to the imagination.[18] Spinoza seems absolutely clear on this point, but it is so far removed from the common view that few of his readers have ever been able to take this seriously.[19] Yet unless it is taken seriously, the unity of *Natura naturata* and *Natura naturans* is hopelessly obscure.

It is only from this second point of view that we may proceed. A modification of Substance — a thing — is always a product of a prior modification. Spinoza does not shrink from the view that the causal nexus is infinite, and not merely potentially so.[20] This prior thing is the efficient cause of the individual under consideration. Every existing individual is possessed of *conatus*, the striving to persevere in existence as that individual.[21] The efficient cause of that individual (speaking broadly here) is the ground of the nature of the individual so produced, but, as Spinoza's explanation makes clear, the efficient cause of a thing's coming to be cannot also be that which extinguishes a thing. Neither can this lie in the thing itself, for the *conatus* of a thing precludes this. Therefore neither from a consideration of the cause of a thing, nor from a consideration of the thing itself can the duration of an individual be determined. Hence Spinoza's definition is that of indefinite duration. How, then, do we get from the indefinite notion to a determinate one? Well, the power of *conatus* of a finite thing is finite; eventually the individual will be destroyed by interaction with a modification or individual with a greater power of *conatus*.[22] Thus, the duration of a thing is made determinate by the order of nature, which links all modifications in *Natura naturata*. Time, then, is a derivative notion: it is the measure of the duration of a thing, within the ordering of all things. Hardin (1978, p. 130) refers to this as a causal theory of time, a phrasing I find unfortunate, but the idea which he intends to express by it seems sound, namely that the conception of time is derivative from the order to nature.

I think several interesting things flow immediately from this. First, it becomes clear why Spinoza did not make time an attribute of God as Alexander (1972) attempts to work out as an hypothesis.[23] There is no need to do so, for duration and time flow naturally from the order of nature itself. Secondly, and most significant for this paper, every individual, as a member of the order of nature, is necessarily a space-time object. They are not to be separated, even in the immediate or mediate infinite modes of Substance.[24] Time is not something extractable from extension. Hence, extension becomes a space-time continuum in a most unexpected fashion. Lastly, but unfortunately outside the present discussion, the generality of this implies a mind-time continuum as well, since the order and connection of ideas is the same as the order and connection of things (*Eth.* II, Proposition VII). It is perfectly appropriate for this to be found in Spinoza but not therefore easier to interpret.

If this is correct, I think it must be clear that Spinoza's view of time is much nearer to the relational notion of time expressed by Leibniz than the abstracted notion of absolute time in Newton. Agassi (1969, p. 332) suggests that Leibniz took his lead from Spinoza here, and I think it entirely plausible that he is correct, though it is not easy to demonstrate. Are there similar reasons to expect a similar relational view of space? I think so. It does not have quite as catchy a name, but Spinoza refers to 'Time and Measure' in his *Letter XII* on the nature of the infinite (Spinoza, 1928, p. 118; Gebhardt IV, pp. 52—62). Time is related to duration, it would seem, as Measure is related to Quantity. Accordingly, space, as the place of modifications, would have a similar derived role. It should not be confused with the Attribute of Extension, which expresses infinite, eternal essence. It does not seem to me to present too great an obstacle to suggest that space and time are both 'things of reason' for Spinoza. What, in the order of nature, retains the character of the absolute is described by Spinoza with that wondrous phrase *"facies totius universi"* — the face of the universe, which, always changing, ever remains the same.[25] This is to say that the only true individual for Spinoza is the whole of nature so considered.

I believe I have given sufficient ground for at least entertaining the idea that Spinoza's *Natura naturata* is comprised of space-time objects. Are they of the right sort for our purpose, however? That is, if one considers the deeper insights of our century, can one find a home in such a view? It seems quite likely that the answer is "yes."

2.4. In order to support this affirmative judgment, however, one must distill from our current views of nature some of the essential insights. For the moment, I intend to focus on the theory of relativity in a rarified form. Both the special and general theories begin from certain postulates (which are not necessarily independent, it seems, but that is unimportant). For the special theory, the important postulate is the postulate of special relativity, which is that the laws of nature maintain their form when formulated in any inertial (i.e., non-accelerated) frame of reference.[26] [The selection of this nomenclature had unfortunate consequences for the understanding of this principle, as Einstein was acutely aware.[27]] The point has been often enough made that this is an extension of the Galilean principle of relativity, the applicability of which for mechanical phenomena was beyond serious question by Spinoza's time. The phenomena which required Galileo's principle to be extended were just beginning to be investigated at this time, however, and would not be completed in any important sense until the latter part of the nineteenth century. By this time, of course, the Newtonian-Laplacian-Kantian world-view was decidedly established. The special theory extends the principle to all laws of nature whatsoever. The general theory extends this principle even further, from inertial frames to non-inertial frames of reference, so that the laws of nature are required to maintain their form under covariant transformations of coordinates.[28] The central insight of the entire project is that it is not the ordinary concepts of space and time which are to be seen as fundamental to our concept of the world, but the thesis that physical law must be absolute (what those laws may be, I leave open) and invariant. A lot has to be added to this to get any equations, but that is not important here. To state this more Spinozistically, the laws of nature are to be understood as eternal truths. Another way of putting this which relates it more obviously to *Natura naturata* is that the laws of nature express the order of nature itself, or yet another way, they express the character of the *facies totius universi*. General covariance, then, seems to me to express the essential character of Substance insofar as it is considered as modified.

Now lest one object that I have not deduced the Lorentz group of transformations from the ratio of motion and rest, let me repeat that I do not seek to deduce physics from metaphysics, and still less twentieth century physics from seventeenth century concepts. What I have proposed is a way of viewing the metaphysics of Spinoza in light of, and in a way compatible with, the current view of nature. That the views of Spinoza have a deeper affinity with those of Einstein than I have shown I think is reasonably obvious[29] and deserves to be more fully drawn out, but it seems sufficient to indicate the general compatibility of these points of view to vindicate the efforts of this paper. Metaphysics and physics are not by their nature exclusive, but complementary.

Let me conclude this section by pointing out that I have thus far considered only what Spinoza calls *Natura naturata*. There seems to me to be a perfectly natural place for the consideration of *Natura naturans* as well. General relativity, too, admits of two points of view. One is the way just considered, namely how the laws of nature are preserved in the unfolding of space-time. The other view is, of course, that of the space-time manifold considered as a whole. From such a viewpoint, the world exists eternally, without becoming or passing away, for these dynamic notions are locked inside space-time itself.[30] No one can fail to see how closely this parallels the view of Spinoza. Whether Einstein had this partly in mind when he declared his God to be that of Spinoza I simply do not know, but I feel certain that the connection would not have displeased him.

3.0. Clearly, what I have done to this point is schematic at best and I will not try to flesh it out at this time. It does seem incumbent upon me, however, at least to mention some problems, of which there are many. First, Spinoza's view was absolutely and completely deterministic and rationalistic. Now this may indeed suit Einstein too, as is well known, but probably not too many other modern or contemporary physicists would approve. I simply state (and hope at some time to defend) my view that the Spinoza—Einstein view is fundamentally correct, the current state of theory notwithstanding. The force of my general argument, however, need not rest on this. It is just as important for the quantum field theories to find a suitable framework for this view of the world. Secondly, my attention here has focused on the most general part of the general theory of relativity, and not on its role as a theory of gravitation. We know that our understanding of nature is going to have to encompass much more than this. The ultimate aim of all investigators should be a truly unified theory of nature and of the fundamental forces therein. Neither philosophy nor physics should be content until this unification be completed.

Thirdly, the implications of real singular solutions in the equations of nature provide the hardest of philosophical as well as physical puzzles. Of course, no physicist is happy with them, and it is a truism to say "something has gone wrong" with physics at or near such singularities. Philosophically, the question may be even more fundamental, for it brings clearly into focus the tension produced by the demands of rationality in the world and is, to some degree, a measure of the truth of the ancient Parmenidean dictum: To think and to be are the same.[31]

It is not without considerable trepidation that I submit these reflections on Spinoza and modern science. No two figures produce greater awe for me than Einstein and Spinoza; I necessarily fear that I have done neither justice. My best hope is that someone better qualified than I can proceed from these reflections toward something worthy of these two great philosophers and their vision.

University of Kansas

NOTES

[1] This activity should be sharply distinguished from the popularization of modern physics which has also shown a new spurt in recent years. That, too, has its place, but this introspection must be on a higher level. Perhaps Davies (1983) may be a step in the intended direction. I feel that Penrose's essay (1979) also represents the proper spirit of this enterprise.

[2] Maurice Solovine mentions the *Ethics* as among the texts read and discussed in the days of the 'Olympia Academy' in Bern. References by Einstein to Spinoza are dealt with by M. Paty in the essay that follows.

[3] John A. Wheeler (1980, p. 354) describes Spinoza as "hero and role-creator to Einstein in youth as well as later life". He credits Hans Küng for suggesting that Spinoza may also have led Einstein astray, in adopting the "cosmological constant" and the assumption of determinism. I frankly doubt that Spinoza can be so credited with either, leaving open the question whether they are equally errors.

[4] I am making rather free use of Thomas Kuhn's ideas and terms developed in Kuhn (1962). I do not wholly endorse his model and the use of his terminology is not as careful as I am certain Kuhn would wish. I trust the meaning is, however, clear.

[5] A couple of examples might suffice for the point. Relatively reasonable assumptions in stellar evolution lead to solutions involving real singularities (i.e., not dependent upon coordinate choices) in space-time. Roger Penrose has expounded the widely recognized 'cosmic censorship principle', asserting that nature is such that no such 'naked' singularity can ever be observed. Granting that this is a reasonable presumption, though not yet fully established, it seems precisely the sort of principle in need of philosophical examination. Sheldon Glashow, speaking informally at the Eidgenössische Technische Hochscule in Janaury 1984, indicated that, despite the profound success of particle physics (ignoring quantum gravity for the moment), the sheer number of seemingly arbitrary constants which must be supplied seemed unsatisfactory for the final form of a theory. He did not invite philosophical participation, but it seems to me that this is another of the sorts of questions to which philosophers of nature should address themselves.

[6] In a certain sense this is what Kant attempted for Newton's *Philosophiae naturalis principia naturalis*, most generally in the *Kritik der reinen Vernunft* and more particularly in the *Metaphysische Anfangsgründe der Naturwissenschaft* (see Volumes 3 and 4 of Kant, 1902—56). Perhaps it is the power of these works in the light of post-Newtonian physics that makes one doubt the utility of any similar attempt.

[7] The middle and later parts of the eighteenth century produced many attempts to demonstrate the necessity of the parallel postulate. Kant was certainly aware of this work and was himself interested in it (see Vol. 14 of Kant, 1902—56, especially the notes by Erich Adickes, pp. 1—61), and it is plain from his discussion of the differences between mathematical and philosophical knowledge in Kant (1929, B 741/766, as well as B 202/207) that he saw no logical problem in denying the Euclidean parallel postulate, but one of the construction of an appropriate object in pure intuition. This is a far more subtle approach to the issue, and, after all, practically no one outside of Riemann himself thought there could be any physical consequence of his geometry until Einstein actually employed it. Physical space, which was the real object of Kant's investigation, was most certainly thought to be Euclidean. Whether Kant's notion of construction leads in-

exorably to Euclidean concepts is still open for debate. It is instructive for following Kant's exposition of space and time to see his correspondence with Johann Lambert, in particular a long letter from Lambert to Kant dated October 13, 1770, in which many of the Kantian themes are easily recognizable (Kant, 1902—56, Vol. 10, pp. 103—111). Lastly, I would like to point out that there are now and then attempts to make Kant and Einstein compatible. See Strohmeyer (1980) for one such effort.

[8] See Leibniz's reply to Samuel Clarke's letter, for example, in Leibniz (1951, pp. 222—224).

[9] It is clear that not all of these figures would think of themselves as doing the same thing, and even less so if philosophers like David Lewis or Donald Davidson were added to the list, but it is fairly clear that these are not just doing logic, scientific methodology, ethics or aesthetics, or at least not these merely.

[10] To be slightly more honest, I cannot find much reason to suppose that Spinoza had any such treatise in mind except that he exhibited such obvious interest in the physical controversies of his day and clearly had views distinct from the Cartesians, so that one might hope that he intended to systematize his observations and theoretical considerations at a later time. I do not agree with the opinion expressed by Hubbeling (1978, pp. 65—70), which seems to characterize Spinoza as having taken over the Cartesian physics wholesale, more or less. There are too many critical references in the correspondence (see *Letters XXX* (Spinoza, 1928, pp. 205—206; Gebhardt IV, p. 166), *XXXII* (1928, pp. 209—214; IV, pp. 169—176), *LXXXI* (1928, pp. 362—363; IV, p. 332), and *LXXXIII* (1928, p. 365; IV, pp. 334—335 — misnumbered as '*Letter LXXIII*') for examples) to make this seem plausible.

[11] See especially *Letters XII* (Spinoza, 1928, pp. 115—121; Gebhardt IV, pp. 52—62), *XXVI* (1928, pp. 198—199; IV, p. 159), *XXX, XXXII, LXXXI*, and *LXXXIII* (all cited in Note 10).

[12] I have used Shirley's translation (Spinoza, 1982) in preference to Elwes's (or White's) on aesthetic rather than scholarly grounds.

[13] In *Letter XXXII* (cf. Note 11) Spinoza affirms his belief that Descartes's sixth rule is defective, but that Huyghens's principle, too, is incorrect. He cautiously corrects Oldenburg's belief that he had denied Descartes's rules, stating that Huyghens had done so. On the other hand, he does not explicitly affirm any of the rules either. An enlightening discussion of Descartes's rules can be found in Desmond Clarke (1977). Given the relatively frequent criticism of Cartesian science, already referred to, I think we should assume that Spinoza had reservations about the Cartesian system itself, which is, after all, quite different from Spinoza's own views. The use of *The Principles of Descartes Philosophy* and the *Cogitata Metaphysica* is probably unavoidable in this discussion, but seems to me to require great circumspection.

[14] Spinoza's criticism of Descartes's notion of matter and extension in *Letters LXXXI* and *LXXXIII* (cf. Note 10) to Tscirnhaus make it very likely that Spinoza conceived 'motion' and 'rest' actively, but there is no evidence that he supported Leibniz's notion of *vis viva* as the conserved quantity in nature. Indeed, Spinoza's conservation of the proportion of motion and rest is not equivalent either to Descartes's or to Leibniz's formulation. For the opposing view, see Hubbeling (1978), who seems to follow P. van der Hoeven in this.

[15] A particularly important point in this essay is Lachterman's discussion of *conatus* in Spinoza's physics. I agree with Lachterman, against Gueroult, that the concept must be

read into Spinoza's 'Physical Digression' despite the conspicuous absence of that important term. Much of what I write depends upon this interpretation.

[16] See Kneale (1973) for an excellent discussion of the history of these notions, although I do not subscribe to her conclusions regarding Spinoza's notions of eternity and duration.

[17] I assert this because it does not seem to me that ultimately the Platonic and Spinozistic views agree in general, although if pressed to decide whether Spinoza's views had a greater affinity for the Platonic or the Aristotelian conception of time and eternity, I would naturally choose the former. As usual, Wolfson (1934, Vol. I, pp. 331—369) has an exhaustive discussion of the topic.

[18] See *Letter XII* (cf. Note 11) p. 118, for a relatively clear discussion of this.

[19] A failure to take Spinoza seriously in what he says is what leads most initial, and some advanced, readers to believe his views must be coherent. For an example of advanced misunderstanding, see Taylor (1972) who, it must be said, is rather benign in his objections.

[20] This is made quite clear in the next to the last paragraph of *Letter XII* (Spinoza, 1928, p. 122; Gebhardt IV, pp. 52—62) where Spinoza explicates his understanding of the ancients (apparently as he interpreted them through Crecas) which does not depend upon the impossibility of and infinite regress of causes, but upon the necessity of a self-subsisting cause. Here, he seems to agree with Kant that the arguments for God's existence depend upon the ontological argument, although not quite in the way Kant (1929, B 658) explicates it.

[21] The definition of *conatus* does not appear until Proposition VII of Part III of the *Ethics*, but, as I have mentioned in an earlier note, seems to be implicitly used in his 'Physical Digression', and, indeed, it seems to me to be implicitly present in his explication of the concept of duration in Part II.

[22] See the single Axiom of *Ethics*, Part IV: "There is in nature no individual thing that is not surpassed in strength and power by some other thing. Whatever thing there is, there is another more powerful by which the said thing can be destroyed" (Spinoza, 1951, p. 156).

[23] The issue of the infinite Attributes remains a fascinating area for consideration in Spinoza scholarship, but I do not think Alexander's hypothesis is necessary or workable.

[24] The immediate infinite mode under the Attribute of Thought is "absolutely infinite understanding" and under Extension "motion and rest"; the mediate infinite mode is the "face of the whole Universe." See *Letter LXIV* (Spinoza, 1928, pp. 306—308; Gebhardt IV, pp. 277—278) and the Scholium to Proposition XIII of Part II of the *Ethics*.

[25] See Note 24 above and the Scholium to Lemma 7 of the Scholium to Proposition XIII, Part II of the *Ethics*.

[26] See Einstein (1923b, p. 41): "The laws by which the states of physical systems undergo change are not affected, whether these changes of state be referred to the one or to the other of two systems of co-ordinates in uniform translatory motion".

[27] For some time after its christening as the 'Theory of Relativity', Einstein continued to refer to it as the 'so-called' theory of relativity. Gerald Holton (1980) quotes a letter from Einstein to E. Zschimmer, dated September 30, 1921:

Now to the name relativity theory, I admit that it is unfortunate and has given occasion to philosophical misunderstanding. The name 'Invarianz-Theorie' would describe the research *method* of the theory but unfortunately not its material content (constancy of light-velocity, essential equivalence of inertria and gravity). Nevertheless the description you proposed would perhaps be better; but I believe it would cause confusion to change the generally accepted name after all this time. (p. 63, Note 21)

The philosophical misunderstandings have only slightly subsided.

[28] See Einstein (1923a, p. 113): "The laws of physics must be of such a nature that they apply to systems of reference in any kind of motion. Along this road we arrive at an extension of the postulate of relativity". Einstein (1949, p. 69) also writes:

The general theory of relativity, accordingly, proceeds from the following principle: Natural laws are to be expressed by equations which are covariant under the group of continuous co-ordinate transformations.

Einstein goes on to add that this does not suffice to derive the basic concepts of physics, but that is consistent with the contentions of this paper, too.

[29] The obvious case is the demand for a deterministic theory, which Einstein never abandoned even after conceding the rightful place of the quantum theory. Wheeler (1980, p. 354), after quoting Proposition XXIX of Part I, writes "Einstein accepted determinism in his mind, his heart, his very bones. What other explanation is there for his later-life position against quantum indeterminacy than this 'set' he had received from Spinoza?" As I mentioned in Note 3 above, this seems far too strong a claim for influence, but I do not doubt that Einstein perceived the world in much the same way as Spinoza. The desire to extend the General Theory to a unified theory of fields is another way in which Spinoza and Einstein are connected, the insistence that our theory of nature be as complete as possible. Sachs (1976) gives an interesting, if slightly obscure, discussion of the latter connection.

[30] The point of view that this expresses is the most abstract, geometricized form of the theory, of course. It is hardly possible from such a point of view to see how a theory of gravity is obtained from it. It is also important here to emphasize that it is clear that space-time, as we have been considering it, is going to have a far more complex and rich structure than a pure theory of gravity in order to incorporate the other 'fundamental' physical forces. Extension is going to have to be enriched to include at least a high-dimensionality Hilbert space and who knows what else. Although I have not shown that this can be done in the Spinozistic framework, I feel certain it can, and in any case this does not affect the current discussion, for general covariance is going to be a formal constraint on any such theory, one would suppose.

[31] The existence of singular solutions to Einstein's field equations, both as an initial cosmological singularity and local singularities of collapsed objects are sources of philosophical as well as physical puzzlement. Einstein himself did not believe that the final form of the theory could contain singularities and that was a part of the project of a unified theory of fields. For a fascinating discussion of his views, see his correspondence with Elie Cartan (Debever, 1979). Some have suggested that the problems of singularities will not be resolved until a unification of quantum mechanics (broadly construed) and gravitation is achieved. Even the standard practice of renormalization of

infinities in quantum electrodynamics is not without its puzzles, though here success in practice has made the puzzles seem unimportant. In general connection to these points, I should like to commend Penrose's essay (1979) as the kind of synthesis of physics, speculation and reflection which would have pleased Einstein or Spinoza, even if his conclusion would not.

REFERENCES

Agassi, Joseph: 1969, 'Leibniz's Place in the History of Physics', *Journal of the History of Ideas* **30**, 331—344.

Alexander, Samuel: 1972, 'Spinoza and Time', in S. Paul Kashap (ed.), *Studies in Spinoza: Critical and Interpretive Essays*, University of California Press, Berkeley, pp. 68—85.

Clarke, Desmond: 1977, 'The Impact Rules of Descartes' Physics', *Isis* **68**, No. 241.

Davies, P. C. W.: 1983, *God and the New Physics*, Simon and Schuster, New York.

Debever, R. (ed.): 1979, *Elie Cartan-Albert Einstein: Letters on Absolute Parallelism, 1929—1932*, Princeton University Press, Princeton.

Einstein, Albert: 1923*a*, 'The Foundations of the General Theory of Relativity', in *The Principle of Relativity*, by H. A. Lorentz, A. Einstein, H. Minkowski and H. Weyl, transl. by W. Perrett and G. B. Jeffery, Dover, New York.

Einstein, Albert: 1923*b*, 'On the Electrodynamics of Moving Bodies', in *The Principle of Relativity*, by H. A. Lorentz, A. Einstein, H. Minkowski, and H. Weyl, transl. by W. Perrett and G. B. Jeffery, Dover, New York.

Hardin, C. L.: 1978, 'Spinoza on Immortality and Time', in R. Shahan and J. Biro (eds.), *Spinoza: New Perspectives*, University of Oklahoma Press, Norman, pp. 129—138.

Holton, Gerald: 1980, 'Einstein's Scientific Problem: The Formative Years', in Harry Woolf (ed.), *Some Strangeness in the Proportion*, Addison-Wesley, Reading, Mass.

Hubbeling, H. G.: 1978, *Spinoza*, Verlag Karl Alber, Freiburg.

Kant, Immanuel: 1902—56, *Kants gesammelte Schriften*, 23 vols., de Gruyter, Berlin.

Kant, Immanuel: 1929, *Critique of Pure Reason*, transl. by Norman Kemp Smith (ed.), St. Martin's, New York.

Kneale, Martha: 1973, 'Eternity and Sempiternity', in Marjorie Grene (ed.), *Spinoza, A Collection of Critical Essays*, Doubleday, Garden City, pp. 227—240.

Kuhn, Thomas: 1962, *The Structure of Scientific Revolutions*, University of Chicago Press, Chicago.

Lachterman, David: 1978, 'The Physics of Spinoza's *Ethics*', in R. Shahan and J. Biro (eds.), *Spinoza: New Perspectives*, University of Oklahoma Press, Norman, pp. 71—112.

Leibniz, Gottfried Wilhelm: 1951, *Leibniz Selections*, Philip P. Wiener (ed.), Charles Scribner's Sons, New York.

Ne'eman, Yuval (ed.): 1981, *To Fulfill A Vision: Jerusalem Einstein Centennial Symposium on Guage Theories and Unification of Physical Forces*, Addison-Wesley, Reading, Mass.

Penrose, Roger: 1979, 'Singularities and Time-asymmetry', in S. W. Hawking and W. Israel (eds.), *General Relativity: An Einstein Centenary Survey*. Cambridge University Press, Cambridge.

Sachs, Mendel: 1976, 'Maimonides, Spinoza and the Field Concept in Physics', *Journal of the History of Ideas* **37**, 125—131.

Spinoza, Baruch: 1925, *Spinoza Opera*, ed. by Carl Gebhardt, 4 vols., Carl Winter, Heidelberg. (Cited as 'Gebhardt' in the text.)

Spinoza, Baruch: 1928, *The Correspondence of Spinoza*, trans. by Abraham Wolf, George Allen and Unwin, London.

Spinoza, Baruch: 1961, *Principles of Descartes' Philosophy*, transl. by Halbert Britan, Open Court Publishing Company, LaSalle, Illinois.

Spinoza, Baruch: 1963, *Short Treatise on God, Man, and His Well-Being*, transl. by Abraham Wolf, Russell and Russell, New York.

Spinoza, Baruch: 1982, *The Ethics and Selected Letters*, transl. by Samuel Shirley, Seymour Feldman (ed.), Hackett Publ. Co. Indianapolis.

Strohmeyer, Ingebor: 1980, *Transzendental-philosophische und physikalische Raum-Zeit-Lehre*, in Peter Mittelstaedt (ed.), *Grundlagen der exakten Naturwissenschaften*, Bibliographisches Institut-Wissenschaftsverlag, Mannheim.

Taylor, A. E.: 1972, 'Some Incoherencies in Spinozism, I & II', in S. Paul Kashap (ed.), *Studies in Spinoza: Critical and Interpretive Essays*, University of California Press, Berkeley, pp. 189—211, 289—309.

Wheeler, John A.: 1980, 'Beyond the Black Hole', in Harry Woolf (ed.), *Some Strangeness in the Proportion*, Addison-Wesley, Reading, Mass.

Wolfson, H. A.: 1934, *The Philosophy of Spinoza*, Harvard University Press, Cambridge.

MICHEL PATY

EINSTEIN AND SPINOZA*

Einstein was often asked to write on Spinoza, or to provide prefaces for books on him,[1] but he constantly refused, claiming to be largely ignorant of the work of the man for whom he had a deep veneration, speaking of him as "our master Spinoza" (*unseres Meisters Spinoza*, letter to Willy Aron, October 17, 1946), who "was the first".[2] "Unfortunately", he wrote, "to like Spinoza is not sufficient to be allowed to write on him: one must leave it to those who have gone further into the historical background" (letter to S. Hessing, September 8, 1932),[3] i.e. into a precise knowledge of his life and work. Probably I should have followed such an illustrious example of modesty and deep respect, leaving to those better acquainted with Spinoza's thought — who are numerous — the task of speaking competently of him. However, after much hesitation, I finally made up my mind and stopped hesitating to speak, as I had been asked, on the topic 'Einstein and Spinoza', because I have become more and more convinced that these two thinkers have indeed a deep relation to one another. Knowing Einstein's thought somewhat, I have been, so to speak, naturally led to go further in my knowledge of Spinoza's; and I finally came to believe that I could perhaps help, though in a modest way, to make inteligible the nature of this relation, which in fact illuminates at the same time the depth of Einstein's philosophical thought and the exceedingly contemporary importance of Spinoza.[4]

1. ON MISINTERPRETATIONS

In such a matter, one should be careful about misinterpretations. I shall stress two of these here. The first would be to think it possible, useful or legitimate to find in a given field of investigation of contemporary thought — and notably in science — some kinds of illustration or justification — as well as, on the contrary, of refutations — of Spinoza's thought considered as a system capable of application.[5] Some have seen, for example, in the space-time of special and general relativity a modern equivalent of Spinozistic substance. Indeed, Hermann Min-

Marjorie Grene and Debra Nails (eds.), Spinoza and the Sciences, 267—302.
© *1986 by D. Reidel Publishing Company.*

kowski, who expressed four dimensional space-time mathematically, himself used the words "world's substance", structured by world-lines, insisting on the fact that, in such a formalization, "space by itself, and time by itself, are doomed to fade away into mere shadows" (Minkowski, 1923, p. 75). The physicist Boris Kouznetsov, who is right on the one hand in saying that modern physics did in fact eliminate that aspect of classical physics which stood against Spinoza's ideas — and that is nothing else than mechanism — declares, on the other: "Obviously, the unified field, self-interacting and responsible for the existence and the substantial predicates of particles, is the analogue of Spinoza's substance, *causa sui* and *natura naturans*" (Kouznetsov, 1967, p. 38). Bernard d'Espagnat succumbs to the same temptation, seeing in the attribute of substance that is extension, according to Spinoza, a pre-figuration of the realist conception of Einstein's space-time; he further considers that quantum non-locality demands, "on this precise point, a revision of their common doctrine" (d'Espagnat, 1979, p. 96).[6] Thus reincorporating in Spinoza's ideas all the modifications which these new facts would require, he destroys the whole logic and coherence of these conceptions: because according to him one would then have to consider "thought and extension not as having each an existence in itself, but as reciprocally engendering one another within Being" and this is expressly contrary to the most fundamental statements of the *Ethics* (Part I, Proposition X and Scholium 1). Actually, one cannot make a projection of such and such a Spinozistic concept independently of the system of thought in which it acquires its meaning; we shall come back to that later.[7]

But the misinterpretation is still more flagrant when Kouznetsov refers *natura naturata* to the classical physics of determined facts and *natura naturans* to the universe of ultra-relativistic quantum phenomena, which are governed by quantum indeterminacy;[8] he reads Spinoza and the *Ethics* as if it were a work on the physics of the period, praising its penetration for having foreseen the transcendence of classical physics, and apparently without understading what 'substance' and 'attributes' really mean; he reads Spinoza as he would read a scholastic writer. But his own illustration is itself of a scholastic character, and he even sees in Bohrian complementarity a modern version of the irreducibility of *natura naturans* and *natura naturata* (Kouznetsov, 1967, p. 51);[9] a concept such as complementarity would at most be related to the modes of only one attribute, the attribute thought, according to

Spinoza's categories. Such exercises are totally artificial because, however deep a philosophy might be, it obviously cannot substitute for particular phenomena, results of experiments and scientific procedure — that is, the characterization of *modes*, in Spinoza's terminology. We could show, on the other hand, *more geometrico* in Spinoza's own way — simply by referring to some demonstrations of the *Ethics* — that, for example, physical space, even relativistic space, is not to be identified with *extension* considered as an attribute of substance;[10] it would merely be a *mode*, like those objects which are contained in it and which, according to general relativity, determine its structure. Such scholastic games of identifying contemporary scientific concepts with Spinoza's categories have been tentatively justified by invoking the deductive character of the demonstrations performed by the author of the *Ethics*, and by noticing the absence, in that work, as well as in the *Tractatus de Intellectus Emendatione,* of any considerations of physics and of any reference to experiment. But it is obviously a misunderstading of both the very nature of Spinoza's reasoning and the form he gave the *Ethics,* and of the situation and the role, according to his conceptions, of the particualr modes, and of the experience which makes them known.

The second type of misinterpretation would be to evaluate Spinoza's thought in terms of contemporary conceptions, be they in philosophy, in natural science, or in politics,[11] thus trying to illuminate it *a posteriori.* But this would lead to a mere distortion, as if Spinoza's thought could receive its true significance from our point of view. Such a bad reading, or the absence of reading, leads to the misidentifications described above; whereas the only path to follow, when considering any kind of thought, undoubtedly — but especially in the case of Spinoza's thought — is first to consider this thought as it stands, and not to superimpose on it external elements of meaning. An example of that kind of inadequacy is provided by the misunderstanding of Spinoza as it has been expressed by one of the most important contemporary physicists, Richard Feynman (1979): he finds the philosopher's reasonings absolutely childish, but imbedded in such a mixture of attributes, substances and other sorts of nonsense that he just laughs at it. Spinoza had no excuse, he explains, at a time when Newton, Harvey and others were making scientific progress. Take any one of Spinoza's statements, he adds, transform it into its opposite and look around: I challenge you to decide which one is correct. The conclusion, for Feynman, is that it was possibly somewhat courageous to deal with important questions; but, he asks, is it any use to be courageous if it comes to nothing?

Trying to superimpose Einstein's and Spinoza's thought one on the other would result in a similar misunderstanding. Each is relative to its own distinct context, to its own perspective and requires, for its elaboration, the proper use of its own concepts. What is needed, therefore, is the consideration of the peculiar logical structure of these two kinds of thought, respectively; only in this way can we find similarities — or differences — that are really significant, and only in this way can we understand how and why the deep logic of Einstein's thought meets the deep logic of Spinoza's. There is no interest in asking whether Einstein is reasoning in terms of attributes, essences and modes; and it would be meaningless, for these are not his philosophical categories. What matters is to know whether the conceptual ensemble that Spinoza developed in order to manifest the ground, range, and truth value of a philosophy of the world would not find some kind of resonance in a thinker like Einstein — I say, indeed, a thinker, and not only a scientist. ... Now, it is indeed highly instructive to note that, notwithstanding the great difference in their respective historical and philosophical environments, as well as in the problematics of the philosophy and of the science of their times, Einstein and Spinoza join together deeply in the following concerns: the significance they both attribute to their investigations, the relation they assign between thought and nature, the possibility of attaining a knowledge of nature and of acting on it beyond the possible content, quite different indeed, of these two styles of knowledge each in its own right.

Far more than the Spinozistic sources of Einstein's thought, what concerns us here is a convergence of themes and manner of approach: both are assimilated to Einstein's personality as well as to Spinoza's and these are not to be identified with a tradition or with influences, on which they are, however, obviously dependent. But both set, rebuild and reinvest the elements of their quest, by themselves and in an original way. If we want to give meaning to such a question as *to what extent is Einstein's thought Spinozistic?* we must understand 'Spinozistic' not as a model, a system, or even a tradition, but as a way of being, as a thinker, in the world.

Those who have had this experience of finding themselves Spinozistic without having intended to be, can say with Romain Rolland that what they find when reading the text is nothing but themselves:

in the inscription written in the opening passages of the *Ethics,* in these definitions written in blazing letters, I deciphered not what he said, but what I meant to say, those words which my own childish thought tried to spell out from its inarticulate tongue. (Rolland, 1980, p. 285)

If it has meaning to say that somebody nowadays finds himself close to Spinoza, it is probably in such a sense; I mean when someone has reflected starting from intellectual practices which can be very different — physicists, Maxists, modern lens-polishers. . . . In this respect it is highly remarkable that problems set out by contemporary scientific knowledge often bring back on the scene Spinoza's conceptions, as we shall see, notably from the point of view of methodology, and without any artificial projection or reduction. Such is precisely the "actuality of Spinoza's thought", beyond any erudition, which will emerge from a confrontation with Einstein's conceptions — Einstein who is not, I insist, contrary to what has often been said, the last nineteenth century scientist (though indeed his personality and his type of quest are at variance with his time), but whom I consider as one of the beacons of contemporary scientific thought.[12]

2. EINSTEIN'S EXPLICIT REFERENCES TO SPINOZA

I shall now consider for a moment the explicit statements of Einstein about Spinoza, although we know that Einstein's Spinozism is not that of a scholar and that it is a question not so much of influence as of resonances or affinities of his thought with Spinoza's.[13] These statements are numerous and short, scattered in various essays, in interviews with the press, and, mostly, in his letters to his friends and to the innumerable correspondents who wrote to him from all over the world to get his opinion, not about scientific questions, but on daily problems concerning existence, life in society, peace, God. . . . These letters, gathered at the Einstein Archives, some of which already have been published by Banesh Hoffman and Helen Dukas (1979), touch on this aspect, revealing the deep humaneness of that man who was something of a myth for some people. Spinoza is often invoked in these letters, as if Einstein had a daily familiarity with him. The peom entitled 'for Spinoza's *Ethics*', written in 1920, begins, "How much do I love that noble man/more than I could tell with words. . ." (Einstein Archives, unpublished). In 1921 he happened to go to Vienna, and on this occasion went to visit the philosopher Josef Popper-Lynkeus, a socialist of Jewish origin, who was a friend of Mach and whom Freud also admired, speaking of him as "one who came as near to being a man 'wholly without evil and falseness and devoid of all repressions' as he had ever heard of" (quoted by Feuer, 1974, p. 57).[14] Popper-Lynkeus was at the time an old man of eighty. Einstein took note of this meeting, and its sounds like a *cri du coeur*.

So much goodness and mildness! When he entered, I thought at once: it is Spinoza! Such a physiognomy is only to be found in Jewish people, indeed among Jews we find the most extreme contrasts. At most among Italians would we find such a face. I mean among Italian saints: Francis of Assisi, for example. (Einstein Archives, unpublished)

To a Brooklyn Rabbi who asks him about Maimonides's philosophy and relativity, he answers first that he has never read Maimonides and that the theory of relativity has nothing to do with this kind of philosophical discussion, and then writes, "Answering your questions would fill up many books [. . .]. I can only say in a few words that I share exactly Spinoza's opinion and that, as a convinced determinist, I have no sympathy at all for the monotheist conceptions" (letter to Rabbi A. Geller, September 4, 1930, Einstein Archives, unpublished). To another correspondent, in 1932, "all that I think of that extra-ordinary man, I can express as follows: Spinoza was the first to apply with true consistency to human thought, feeling and action, the idea of the deterministic constraint of all that occurs" (letter to D. Runes, September 6, 1932).[15] On another occasion he notices that a limited causality "is no longer a causality, as our wonderful Spinoza was the first to recognize with all precision" (letter to E. B. Gutkind, January 3, 1954, Einstein Archives, unpublished). Questioned about God, he answers, "I believe in Spinoza's God, who reveals Himself in the orderly harmony of what exists, not in a God who concerns Himself with fates and actions of Human beings" (*The New York Times,* April 25, 1929, p. 60, col. 4, as quoted in Schilpp, 1951, pp. 659—660).

In 1948, to Michele Besso, who talked about the love which should be given to one's enemies, he wrote, "For me, however, the intellectual basis is the belief in an unlimited causality. << I cannot hate him, because he *must* do what he does. >> Consequently I am nearer to Spinoza than to the prophets. This is the reason why, for me, there is no sin" (Einstein, 1972: letter of January 6, 1948). And his friend, shortly before Einstein's death, answers him like an echo, "You profess to admitting Spinoza's God; this impels me to take *The Ethics* once again into my hands. . ." (letter of January 29, 1955).

With Max Born, another privileged correspondent, there is practically no evocation of the Dutch philosopher; the reason is that the two have quite different philosophies of nature and that, evidently, Born's Spinoza, if Born should refer to him, would be quite different from Einstein's. And Niels Bohr relates how Einstein conjectured with humor, during one of their endless talks, Spinoza's refereeing between them — supposing he had lived long enough to know the contemporary developments of physics (Schilpp, 1951, pp. 236—237).[16]

So we see it is Spinoza Einstein likes to invoke; he is the master of wisdom and truth, without any possible rival among other important philosophers he had been acquainted with since his youth: Hume and Mach, whom he read following Besso's advice, or Kant, with whose philosophy he had been familiar since he was thirteen (Hoffman, 1972, p. 24) and to whom he constantly referred in his philosophical and historical considerations. With Solovine and Habicht, and the small circle of the Olympia Academy, he studied these philosophers (Einstein, 1972, letter to M. Besso, January 6, 1948), read (and probably reread) Spinoza. Later on, he would nourish himself again with them, reading Kant's *Prolegomena* and Spinoza's *Correspondence*.[17] From Hume he learned to mistrust sense impressions and deceitful empirical 'evidence', as well as the criticism of causality. But instead of denying causality, as Hume himself had done, he retains from Hume's critique the provision that causality must be submitted to the critique of conditions of observation. He also accepts Hume's criticism of induction. From Kant he learns that besides the results of experience, the statements of science involve an element that originates in reason. But, contrary to Kant's view, for Einstein this element is not *a priori*.[18] If he accepts Mach's criticism of the absolute conception of Newtonian mechanics, he does not adopt his positivist epistemology for which sensations are the only real elements and observation is the ultimate reference of scientific discourse. These great authors whom he read early helped him in clarifying his own ideas, and in developing progressively both in his scientific work and in his epistemological conceptions, an epistemology and a philosophy of knowledge centered in realism and in rationality, from which he would never depart.[19]

And from Spinoza, we can now ask, what did he learn? The collection of quotations helps us to understand, but only from afar, as general bits of information which do indeed show a deep understanding of the Spinozistic way, but which are not yet enough to enlighten us on the true nature of this closeness.

In order to understand it, we must admit that Einstein's thought is not restricted to that of the material space-time of relativity or to the criticism of quantum physics, even if, as has often been emphasized, those major contributions to contemporary science have deep philosophical implications. Einstein's thought is itself philosophical thought, explicitly, and not just at an implicit and subterranean level, even though it is at work in the accomplishments of physical theory and in the development of his own research. True, the way it is expressed is different from most philosophers' treatises. First, his scientific contributions themselves often have, not only a philosophical content, but a philosophical tone, by the very nature of the fundamental questions dealt with.[20] Second, his works on general subjects[21] show a deep coherence of his quest. Third, his explicit epistemological writings, although they are not numerous, are of prime importance for philosophers of science. Such are 'Geometry and Experience' (Einstein, 1947, Ch. 8), 'Physics and Reality' (Einstein, 1950, pp.59—97), 'The Fundaments of Theoretical Physics' (Einstein, 1950, pp. 98—110)[22] not to mention his 'Autobiographical Notes' (Schilpp, 1951, pp. 2—95), and 'Reply to Criticisms' (Schilpp, 1951, pp. 663—688).

But if we want to understand Einstein's thought in the movement that governs it, we should recall some biographical elements by which we may understand a spiritual kinship between Spinoza and Einstein which will possibly make more palpable and vivid the comparison between their ways of thinking; for, in such men, biography and thought are one and the same.

3. A BIOGRAPHICAL PARALLEL

This biographical parallel is characterized, first by an early experience of inner freedom in an environment, a society, a religious community where the rule is to obey. Such was Spinoza's experience; he made a theory of it, and the purpose of his political treatises, in that state of things, was to seek a society permitting freedom of thought. Or, in Einstein's case, within a constraining milieu; he could never tolerate the military spirit of the German gymnasium; he was a pacificist in the Germany of the first world war, and he would always refuse to submit to public opinion, always making in science, as well as in politics, a lucid and personal judgement. " I . . . have never belonged to my country, my home, my friends, or even my immediate family, with my whole heart", Einstein wrote in 1931. He adds,

... in the face of all these ties, I have never lost a sense of distance and a need for solitude — feelings which increase with the years. One becomes sharply aware, but without regret, of the limits of mutual understanding and consonance with other people. No doubt, such a person loses some of his innocence and unconcern; on the other hand, he is largely independent of the opinions, habits, and judgments of his fellows and avoids the temptation to build his inner equilibrium upon such insecure foundations. (Einstein, 1954, p. 9)

However, Einstein appreciated the presence of a few friends and liked to try out his ideas on them. He also felt passionately concerned with social justice notwithstanding his "pronounced lack of need for direct contact with other human beings and human communities".[23]

This kind of distance from others, which determines Einstein as well as Spinoza to flee the crowd and common opinion, shows itself also in an ambiguous, or multiple, cultural adherence. The Jewish community of Amsterdam in the seventeenth century included ancient Maranos originating in Spain and Portugal; their culture was not only a Judaic one, and Spinoza himself, impregnated with biblical culture, thinks in reference to other traditions and to other criteria, even before his excommunicationin 1656. The parents of Einstein, not practicing Jews, sent their son to a Catholic school; he even embraced that faith in his twelfth year. Then he was, in the gymnasium, instructed in Jewish religion and learned the Old Testament. Then he abandoned all religious preoccupation, seeing in it only superstition, and refusing the idea of submission to fear, the argument for the very idea of which Spinoza had dismantled in his *Tractatus Theologico-Politicus.* The situation of Einstein with respect to German culture also bears the mark of separation. He speaks and writes in that language, is impregnated with classical German culture and would always be sensitive to that tradition when, later, unforgiving of Germany for having given itself without reserve to Nazism, he would break his ties definitively.

As to Spinoza, he gets himself expelled from the Synagogue and frequents free-thinkers, liberal Christians, and is close to the Republican party of Jan de Witt. His excommunication forces him to abandon business; he becomes an artisan, cuts and polishes lenses. His friends unite in a circle around him to hear his philosophy. Einstein has, in his youth, the greatest difficulties in obtaining a position because of nonconformity. He is offered the opportunity to go to Bern where he finds an unassuming place in the patent office. It is there that he composes his first articles, which revolutionize physics. It is only a few years after this *annus mirabilis,* 1905, that he will obtain a university position.[24]

Einstein was never an artisan as Spinoza was; but he was in the patent office in Bern. And later, at the time of McCarthyism, he declared that if he were again young and deciding the path to follow, perhaps rather than choosing "to become a scientist or scholar or teacher", he would "choose to be a plumber or a peddler in the hope to find that modest degree of independence still available in present circumstances" (Nathan and Norden, 1960, p. 613).[25] Let us not strain the analogies; the artisan lens-maker in the time of Spinoza had the benefit of high intellectual standing. At the same time, Spinoza, like Einstein, although one can speak in both cases of a certain 'solitude', did not live apart from the world. Spinoza belonged to the intellectual circles close to the ruling class; Einstein, Nobel prize laureate, benefited from universal consideration. The essence, here, is what the state of the one or the other expresses about their independence. We know this about Spinoza. We also know what independence of spirit Einstein manifested, at the moment of the witch hunt, calling on American scientists to refuse to reply to the commission of inquiry of Senator McCarthy.

He wrote to a student:

... the practices of those ignoramuses who use their public positions of power to tyrannize over professional intellectuals must not be accepted without a struggle. Spinoza followed this rule when he turned down a professorship at Heidelberg and (unlike Hegel) decided to earn his living in a way that would not mortgage his freedom. The only defense a minority has is passive resistance. (Letter to Arthur Taub: Cranberg, 1979, p. 9)[26]

For those Jews who broke with tradition, atheists in the sense of an anthropomorphic God, such as Spinoza and Einstein, *Jewishness* is nevertheless an important element of their personalities. Spinoza said that it is non-Jews who are responsible for the permanence of the Jewish people because, without their attitude, the Jews would assimilate.[27] Einstein notes that Spinoza's conception of the world was penetrated by the thought and sensitivity characteristic of Jewish intelligence in its very life, "I feel", he wrote, "that I could not be so near to Spinoza if I were not myself a Jew" (letter to Willy Aron, January 14, 1943: Einstein Archives, unpublished).[28] Although he was obliged, by force of circumstance, to be in favor of a Jewish state, Einstein always drew attention to the danger of a religious state, and the priestly tutelage (*ibid.*). Judaism, for him, was no longer a creed either; to base a moral law on fear is a "regrettable and discreditable attempt" (Einstein, 1954, p. 186) — one might be listening to Spinoza. For him, the Jewish God signifies the refusal of superstition and of the images, the acceptance of natural law, the will of justice and reason, an essentially moral attitude of personal independence (pp. 173f, 176f, 185f).[29] That is why Spinoza seemed to him "one of the deepest and purest souls our Jewish people has produced" (letter, 1946, quoted in Hoffman, 1972, p. 95).

One ought — but room is lacking here — to terminate this biographical parallel by speaking of the respective commitments of Spinoza and Einstein in their times, both choosing justice, the reign of reason and of liberty. It is not a matter of indifference that this common preoccupation was part of their philosophy and their vision of the world.

4. THE EXPERIENCE OF RUPTURE

I do not know if, at the moment of writing his 'Autobiographical Notes', which are not a biography in the usual sense, but a meditation on the road he had travelled intellectually, Einstein had in mind the admirable opening of the *Tractatus de Intellectus Emendatione* of Spinoza, where the philosopher expresses himself as follows:

After experience had taught me that all the usual surroundings of social life are vain and futile; seeing that none of the objects of my fears contained in themselves anything either good or bad, except in so far as the mind is affected by them, I finally resolved to inquire whether there might be some real good having power to communicate itself, which would affect the mind singly, to the exclusion of all else: whether, in fact, there might be anything of which the discovery and attainment would enable me to enjoy continuous, supreme, and unending happiness. (Spinoza, 1955, p. 3; Gebhardt II, p. 5)

Spinoza then describes his search for the sovereign good that is

. . . the union existing between the mind and the whole of nature. This, then, is the end for which I strive, to attain to such a character myself, and to endeavour that many should attain to it with me. . . . In order to bring this about, it is necessary to understand as much of nature as will enable us to attain to the aforesaid character. . . .[30]

That is a project for which, above all, he will seek "a means . . . for improving the understanding and purifying it" to "direct all sciences to one end and aim, so that we may attain to the supreme human perfection which we have named" (pp. 6—7; Gebhardt II, pp. 8—9).

It matters little whether Einstein remembered Spinoza's terms or even the existence of this passage. What is interesting and significant for the reader of today is to meet under his pen a text so overwhelming and of so great a kinship of intellectual and spiritual experience. "Even when I was a fairly precocious young man", he writes,[31]

the nothingness of the hopes and strivings which chase most men restlessly through life came to my consciousness with considerable vitality. Moreover, I soon discovered the cruelty of that chase, which in those years was much more carefully covered up by hypocrisy and glittering words than is the case today. (Schilpp, 1951, p. 3)

Einstein then evokes, after the religious experience of his youth, his attempts to free himself "from the chains of the 'merely-personal', from an existence which is dominated by wishes, hopes and primitive feelings".[32] He continues,

Out yonder there was this huge world, which exists independently of us human beings and which stands before us like a great, eternal riddle, at least partially accessible to our inspection and thinking (*unserem Schauen und Denken*). The contemplation of this world beckoned like a liberation, and I soon noticed that many a man whom I had learned to esteem and to admire had found inner freedom and security in devoted occupation with it. The mental grasp of this extra-personal world within the frame of the given possibilities swam as highest aim half consciously and half unconsciously, before my mind's eye. Similarly motivated men of the present and of the past, as well as the insights which they had achieved, were the friends which could not be lost. The road to this paradise was not as comfortable and alluring as the road to the religious paradise; but it has proved itself as trustworthy, and I have never regretted having chosen it. (p. 5)

This path, common to Einstein and Spinoza, is, when one has disengaged oneself "from the momentary and the merely personal", that of "the striving for a mental grasp of things" (p. 7).

The goal, the sovereign good towards which Spinoza aspires, and to which he decides to consecrate himself, is then this superior human nature which is "knowledge of the union existing between the mind and the whole of nature" (Spinoza, 1955, p. 6; Gebhardt II, p. 8). And for Einstein, "Joy in looking and comprehending is nature's most beautiful gift" (Einstein, 1954, p. 28). The intellectual love of God, which sums up their conducts, is a spiritual adventure, to be sure; but it is, out of their contingent being, which is founded in the universal nature of things, a search for the knowledge of Being. This is the highest form of human perfection, disengaged from illusory value.[33] Spinoza, in the *Tractatus de Intellectus Emendatione*, expresses how, in view of that goal, it is necessary to provide the means to have a knowledge of Nature of that order. The necessary condition is to cure this intellect in order to arrive as exactly as possible at things, to avoid the error and illusion that hinder and forbid that access. Afterwards, and this is also part of his program, it is a question of finding a way to establish a society which permits the access of the greatest number to the same goal.

From this end, and the means taken to attain it, one can derive some propositions, valuable as much for Spinoza as for Einstein, and fundamentally linked to one another. The experience of truth in freedom takes with it the definition of an ethic which is neither a morality nor a system, but the pursuit and the bringing to light of a project; this ethics is a manner of living and of practicing the search for knowledge. The knowledge of the truth is the highest joy and does not close back on itself, but opens upon the world. At the same time, truth cannot be disassociated from an essential solitude. The authenticity of the quest for truth is in fact guaranteed by this essential solitude, which is entirely the opposite of a closing off, because this truth is found nowhere else than in ourselves; it is not opinion which makes it or — to speak in the language of the physicists of the Copenhagen school whom Einstein opposed — intersubjectivity, which would reduce it to a purely psychological aspect. This solitude is not to be confused either with individualism or with asceticism. As to the second, the quest for truth is for Spinoza a spiritual adventure, but not desiccated or ascetic; how admirable is this "it is necessary that . . . we should carry on our life" in the *Tractatus de Intellectus Emendatione*, while we work to keep our intellect on the right path! As to the first, the search for truth is inseparable in Spinoza from communication, and this is precisely the interest of the idea of 'common notion' that appears in the *Ethics* to manifest the practical nature of ideas, conceived as instruments of communication.[34]

These propositions seem to me to define well enough the spiritual kinship between Spinoza and Einstein, over three centuries of distance and through the difference of their research. What establishes this kinship is the Spinozistic rupture — which made Einstein say, "he was the first" — this rupture which the affirmation of the oneness of Being and of his absolute necessity determines, with a force and a coherence unequalled before.

To free the understanding — that first task for Spinoza — this is to give it totally to Reason, to avoid aporia and contradiction, to think through the intelligibility of the definitions we start from, to draw all the consequences, without restriction. God, absolutely infinite Being, leads us by this path to nature, equally infinite: *Deus sive Natura*, God, that is to say, Nature. The inevitable logic of the *Ethics*, which carries through the program of the *Tractatus de Intellectus Emendatione*, sketches exactly the arguments of the rupture. Only one substance with an infinity of attributes, things and creatures being modes of these attributes; that is a doctrine of Totality which effects a radical overthrow of received ideas. The doctrine of immanence breaks with the philosophy of Descartes, from that point it founds certainty not like the latter on the existence of God, but on the idea of God (*The Cartesian Principles*: Spinoza, 1963; Gebhardt I). The idea of God, to which the self can reach, makes knowledge possible. Substance is not reduced to extension; it has as its attributes extension and thought; thus thought participates in Nature, and the latter, which produces an infinity of bodies, produces equally the infinity of the ideas of these bodies. The ontological point of view, in which Spinoza is anchored, starting with some definitions of infinite substance as cause of itself — *causa sui* — of its attributes and its modes; this is the self-sufficiency of Nature, of which we will speak again *a propos* of necessity and of determinism (*Ethics* I, Proposition XXXIII, Scholium 2).

One can read Spinoza as atheist, materialist or pantheist. He is atheistic, certainly, toward the anthropomorphic sense of God (*Short Treatise on God, Man, and his Well-Being*: Spinoza, 1910, pp. 53—55; Gebhardt I, pp. 9 [from summary], 45—47). The *Short Treatise* already shows that a bad logic accompanies false attributions of God or of Being, and that constitutes an implicit justification in advance of the geometrical method adopted for the *Ethics*.

Causa sui, God cannot be omniscient, merciful, wise; the supposed will of God is nothing but the refuge of human ignorance, and Spinoza rejects teleology as anthropocentric prejudice (Appendix to *Ethics* I). This is a complete decentration from the point of view of man, whose existence is contingent; it could equally well have happened that such and such a man exist as that he not have existed (*Ethics* II, Axiom 1); that is to say that he is a particular being, a mode of attributes of substance, a link in the infinite chain of causes and effects. By the same motion, all the anthropocentric values are dethroned, and we have one of the first really rigorous critiques of ideology; there is neither good nor evil (ontologically), but only 'well' or 'badly' (in practical life). Good and evil "are in our understanding and not in Nature", they are entities of reason. They are neither "things nor actions which exist in Nature" (Spinoza, 1910, pp. 59—60; Gebhardt I, p. 49; cf. *Ethics* III). Morality is wiped out, and obedience too, for the sake of knowledge and freedom, since knowledge is what allows us to know what is 'well' and what 'badly', insofar as they are simply modes. Einstein here follows without restriction the teaching of Spinoza. He protests — we have seen — against the notion of an anthropomorphic God who awards or punishes, who inspires fear. In a text of 1930, on 'Science and Religion', he writes for example the following, which has a perfect Spinozistic resonance:

The more a man is imbued with the ordered regularity of all events the firmer becomes his conviction that there is no room left by the side of this ordered regularity for causes of a different nature. For him neither the rule of human nor the rule of divine will exists as an independent cause of natural events. (Einstein, 1954, p. 48)

He speaks, concerning this joy of adherence to Nature and to the order of things, of cosmic religiosity which knows neither dogmas, nor God, nor church. He sees its adepts among the heretics of all time, atheists or saints; among them, Saint Francis of Assisi and Spinoza. It does not depend on any theory, it does not lead to any formal idea of God, and its only means of communication among men are art and science, whose chief function is to keep it alive for those who work at it (*ibid.*).

If the *Ethics* of Spinoza marks in a decisive manner the theoretical end of anthropocentrism, the sciences of our times — physics, biology, psychology, history — may well recognize in its author a contemporary. One knows how the question has been debated in physics, where it has been declared that the observer has come back to the front of the stage;[35] but, in the face of the subjectivist or positivistic interpretation, heirs to the second part of the nineteenth century, revived *a propos* of quantum physics, Einstein does not cease to affirm the absolute necessity of that decentration and demotion of the observer. This leads to the everburning question of determinism and, obviously, to the doctrine of determination in Spinoza.

5. EINSTEIN AND DETERMINISM

To affirm determinism, for Einstein, is to derive the necessary con-

sequence of the self-sufficiency of Nature. In his eyes, this is also
Spinoza's position. However, the word 'determinism', in connection
with Einstein and obviously also with Spinoza who did not use it, is not
to be taken without precaution. It would be a misinterpretation to
identify the idea of determinism in Einstein with the definition of
determinism in the sense used by Laplace who, if he did not employ the
word, described the thing to which the word generally refers. The
determinism of Laplace is that, in the mechanistic sense, of a causal
chain which goes from past to present to future, but which is, so to
speak, imposed on things, imposed on a nature which a Sovereign
Being might consider from the outside: the mechanistic metaphysics
belonging to that determinism is, in the last analysis, that of a great
watchmaker. There is nothing of that in Einstein, for whom it is in the
interior of itself that nature finds its determination. Einstein has not
forgotten the lessons of Hume's critique of causality, and if the events
of space-time are intertwined by a rigorous causality, the causality
which is in question is precisely circumscribed, at the heart of a
conceptual theoretical structure, that of relativity. It is precisely by
avoiding the imposition of a general causality, unrelated to the concepts
among which it functions and to the conditions of its use, that in 1905
he performed a critique of simultaneity. True, causality is one of the
forms of realization of determinism, but it is not to be confused with it,
as was the case in Laplace. This holds even though causality among
spatiotemporal sequences seems to be privileged in physics. But this is
only one aspect of his position with respect to the apparent defeat of
the causal ideal in quantum phenomena and quantum theory. This
defeat is often translated in terms of a role given to chance in nature by
the fundamental role that probability is playing in that theory. It is true
that, to the degree that the laws of probability in quantum physics are
reduced to the expression of an ignorance that is, so to speak, ontological, unsurpassable, denying determination and necessity, Einstein
could only be fiercely opposed to it — although he is one of the first to
have imported probabilistic laws into the quantum domain (Paty, 1981;
Paty, in press b). If the adherents of the Copenhagen school, who adopt
a positivist view, claim to superimpose on things existing in nature and
on nature itself — the pertinence of the concept which they contest —
'interpretations' which are philosophical propositions external to that
consideration alone (it is thus, in particular, that the principle of
complementarity must be viewed), such a claim can only be profoundly

false in Einstein's eyes. If that is the sense of the theory, it is because it is incomplete. He shows it with precise technical arguments — or, rather, that is what he believes he shows; I am alluding to the Einstein—Podolsky—Rosen argument which contributed to the exhibition of quantum 'inseparability' (Paty, 1982a). But the refutation of his demonstration by Bohr cannot be satisfactory to him with regard to the demands that he considers primary — the affirmation of a physical reality independent of observation and of the means of knowing it — since that refutation is referred, precisely, to the epistemological principles which initially deny the well-foundedness of those demands.

Einstein wrote to Max Born in 1944, "You believe in the God who plays dice, and I in complete law and order in a world which objectively exists, and which I, in a wildly speculative way, am trying to capture" (September 7, 1944: Born and Einstein, 1971, p. 149).

Such is, in fact, what is essentially at stake, in the struggle of Einstein against the dominant ideas proposed by the scientists of his time. This is not a senile refusal of new theories and new ideas, as one has often pretended to believe. It would be inappropriate to try to sketch here what could be today Einstein's position in the face of the recent developments of these questions. What is certain is that he would not refuse a further theory — indeed, that is what he wanted — on grounds that its formulation would be too abstract and apparently too distant from the intuitive description of the phenomena, provided that it respected the demand of objectivity and of determinism — since he distinctly thought that that should be its character.

This abandonment of determinism in Einstein's sense, that is to say of necessity and of non-contingency in nature, clearly leads us to Spinoza. But not only for reasons linked to physical theory, since the determinist position of Einstein, and the indeterminist position of Bohr, Born and the adherents of the Copenhagen school, is affirmed as much in relation to the physical phenomena as in relation to human freedom.

"I cannot understand", writes Born to Einstein,

how you can combine an entirely mechanistic universe with the freedom of the ethical individual. [. . .] To me a deterministic world is quite abhorrent — this is a primary feeling. Maybe you are right, and it is as you say. But at the moment it does not really look like it in physics — and even less so in the rest of the world. [. . .] Your philosophy somehow manages to harmonise the automata of lifeless objects with the existence of responsibility and conscience, something which I am unable to achieve. (October 10, 1944: Born and Einstein, 1971, pp. 155—156)

On this subject, Einstein's position is entirely Spinozistic: to Maurice Siguretx who asks him about the conflict between the purely causal point of view of Spinoza and his action for social justice, he replies that there is no conflict, since our psychic tensions, and not only the passions but our desire to arrive at a more just society, are elements which participate, with all the others, in causality (letter of July 14, 1935: Einstein Archives, unpublished).

6. NECESSITY AND DETERMINATION IN SPINOZA

Spinoza does not use the term 'deterministic',[36] but words like 'determination', 'determined', or the verb 'to determine' itself occur frequently in the *Ethics*,[37] as do the terms 'necessity' and 'causality'. Rather than the use of terms, what is important is the sense given to the thing. And if one agrees — with Negri (1981) and others — that Spinoza is the first philosopher to have gone all the way with the idea of the absolute necessity of nature which carries all determinations, some considerations on the place of that idea and on its consequences in Spinoza's thought will not be superfluous.

Let us recall the distinction given in the Scholium of Proposition XXIX of Book I of the *Ethics* between *natura naturans* and *natura naturata* (cf. Spinoza, 1910, p. 56; Gebhardt I, p. 47). The former is nature insofar as it is itself its own cause. The second, which follows from the necessity of Nature, is the totality of things (modes of attributes) insofar as they follow from Nature and cannot be conceived without it.[38] "Nothing in the universe is contingent", asserts Proposition XXIX, "but all things are conditioned to exist and operate in a particular manner by the necessity of the divine nature". This proposition is further extended — or, rather, applied — to bodies in the first definition of the second part of the *Ethics*, according to which "By *body* I mean a mode which expresses in a certain determinate manner the essence of God, in so far as he is considered as an extended thing". "[A]ll things which come to pass ..." the *Tractatus de Intellectus Emendatione* had already announced, "come to pass according to the eternal order and fixed laws of nature" (Spinoza, 1955, p. 6; Gebhardt II, p. 8).[39] What is essential in Spinoza's doctrine of determination is there, with Propositions XXVII and XXVIII of *Ethics* I defining for their part the infinite series of determinations of particular things.

It is a question here of determination by necessity, that is to say determination in the things themselves. It is imposed by the self-sufficiency and the intelligibility of nature, given that the only individual thing that is not determined by any other thing, "that of which a conception can be formed independently of any other conception" (*Ethics* I, Definition 3), is substance.[40] In this sense, determination is, from the point of view of knowledge, what results from necessity, and this is precisely the knowledge of necessity.

If we now consider, after the order of things, the order of knowledge, that is to say the order of ideas of things, it follows from Proposition XXXIII of *Ethics* I, Scholia 1 and 2, that there exists nothing in things that can make us say contingent; the indetermined and the contingent — that is to say, that of which one says only that it is possible, not that it is necessary or impossible — resulting only from a lack of knowledge on our part. To believe in such contingency in nature would result in attributing to God (to Nature) a liberty different from that defined by Spinoza, of being *causa sui*. For this would be "to set up something beyond God, which does not depend on God, but which God in acting looks to as an exemplar, or which he aims at as a definite goal" (*Ethics* I, Proposition XXXIII, Scholium 2), which would be absurd. Such an attribution, Spinoza stresses, is a great impediment to science (*Ethics, ibid.*; cf. *Short Treatise*, Spinoza, 1910, pp. 48—49; Gebhardt I, pp. 40—41). Thus determination as Spinoza understands it is not sensibly different from determinism in the sense of Poincaré or of Einstein, who make it the synonym of science, or a condition for science (Poincaré, 1963, p. 244; Poincaré, 1920, p. 50): not finished science, closed in categories and concepts of our present knowledge, as mechanism is, but the science that we have in view, that of our project, or of our program.[41] The polemic against the idea of chance in nature is taken up several times by Spinoza, notably in his letters to Hugo Boxel.[42] Spinoza specifies, on the other hand, that "It is not in the nature of reason to regard things as contingent, but as necessary" (*Ethics* II, Proposition XLIV), and that the idea that things are contingent results only from imagination (Corollary 1 to Proposition XLIV).[43]

Finally, we should speak of the doctrine of determination as applied to the behavior of individual entities: "By the right and ordinance of nature, I merely mean those natural laws wherewith we conceive every individual to be conditioned by nature, so as to live and act in a given way" (Spinoza, 1951, p. 200; Gebhardt III, p. 189).[44] The consequence that Spinoza demonstrates is in the last analysis the demand for freedom of thought and, as a political system, the superiority of democracy.[45]

7. SPINOZA'S METHOD AND THE DOCTRINE OF PARALLELISM

Although elaborated in a given time, confronted by a certain state of knowledge — of which Bacon, Descartes and Hobbes are significant representatives — Spinoza's thought transcends this contingent condition; it proposes to us, under a form marked, it is true, by this condition, a set of considerations which are attached for many reasons to the most fundamental questions now debated on the subject of knowledge. And this, probably because it was directed straight, immediately and accurately to what appears to be most essential. One cannot deny that this corresponds to a remarkable penetration, unless we dismiss, as the logical positivists do, these questions, and with them Spinoza, but also Einstein, into the outer darkness of metaphysics. The force of Spinoza's thought resides in the demands that it poses to the use of reason — by the reform of the intellect and by the lack of concessions once the intellectual way has been defined. By these exigencies, reason finally informs the scientific procedure — speaking here of this specific field of application — when the latter is concerned with the definition or the reflection of its conditions. Hence the link between this thought and what I should call, after Einstein, the idea of an 'epistemological program'.[46]

What is at stake is to pose the demand of reason, ". . . reason is the light of the mind, and without her all things are dreams and phantoms" (Spinoza, 1951, pp. 194—195; Gebhardt III, p. 184; cf. *Ethics* V, Preface), but also in associating it with the demand that orders reason; for it is not only reason which is at stake.[47] The origin of knowledge is Nature itself. It is the primacy of Nature, or of the Real, or of Being, that reveals itself, determining when we consider Spinoza's method, as well as in Einstein's epistemological program.

We may consider that the *Ethics* is at the same time an ontology and that it is not one. It is an ontology insofar as it is a theory of infinite Being identified with Nature, a theory of Nature as infinite Being which contains all that exists, and from which all that exists proceeds. It is not an ontology insofar as the whole work is only axiomatic (in the sense of the books of Euclid), deducing logically all the implications of the definitions and axioms posed at the start, posed by construction, and which, with regard to the definition of Being, on which everything rests, are at the same time so fundamental. These implications determine the character of all that exists. Insofar as it is axiomatic, this would be a transparency, which is totally opposed to the fullness that one usually claims for an ontology. What justifies it is the adequacy of the proposition from which we start to the coherence of what results as far as our experience of it goes; this is, in summary, an *a posteriori* justification of the fact that it is, without seeming to be so, an ontology. (An experimental ontology?)

On the other hand, one can consider that the *Ethics* is, and at one and the same time is not, a theory of knowledge.[48] It is not one to the extent that it does not propose the specific analysis of modes, insofar as it does not constitute a cosmology; true, it puts forward, notably in Part II, a characterization of individuality conceived as persistence in being, which evades all demarcation between physics and biology. In this respect it is situated in a lateral course with respect to the science of this time, which permits it to avoid the dualistic antinomy of physicalism and organicism. If it is not a cosmology strictly speaking, it is a theory of what one wants to speak about when one initiates a theory of knowledge. It is a theory of the necessity — the demand — of the intellect, of the logical coherence of the intellect's statements about modes (things, ideas), essences, whatever they be according to a description. Logically, after the *Ethics*, one would expect a cosmology, a politics. Now, the latter does exist: it is the two treatises, *Tractatus Theologico-Politicus* and *Tractatus Politicus*.

It would be inexact to say that the method followed by Spinoza, *more geometrico*, in the *Ethics*, gives no place to sense data and to experience. It is not because Spinoza places himself at once at a central point of view, in order to apply the deductive method defining relations and conditions on possible propositions, that the knowledge of particular objects, of the modes of the attributes of substance, could dispense with experience. The sense of the deductive method employed in the *Ethics* marks it out radically from scholastic demonstrations or even from those of Descartes. His project is not to deduce particular statements on the properties of modes; it is to situate these modes with respect to the attributes and to substance, that is to say, to know what knowledge — what science — is with respect to nature from the point of view of adequacy and truth.[49] That is precisely what science cannot provide by itself and of which it has need in the moment of reflection[50] in which one proposes to situate it, to appropriate it, to give oneself the means for a criterion of its foundations; in this way the method and the ontology are minimal.[51] In contrast to the method and to the ontology of Descartes which sinned by imposing *a prioris* and, above all, by dualism, that of Spinoza assures the autonomy of knowledge with respect to all preconceived ideas; such is the reversal effected by the ontology of absolute realism. "[I]t is evident", Spinoza writes in the *Tractatus de Intellectus Emendatione*,

that, in order to reproduce in every respect the faithful image of nature, our mind must deduce all its ideas from the idea which represents the origin and source of the whole of nature, so that it may itself become the source of other ideas. (Spinoza, 1955, pp. 15—16; Gebhardt II, p. 17)

And as Definition 6 of *Ethics* II tells us, "*Reality* and *perfection* I use as synonymous terms". Let us notice in this respect that Proposition XXXI of *Ethics* I and its Scholium prevent us from misinterpreting the geometrical method and the first definitions in the sense of a contemplation of substance; it speaks in fact of the *acting intellect* which is related to *natura naturata*; it has as its object the attributes and the modes of Nature. By (non-contradictory) thought one can know about these what results from self-sufficiency of Nature.[52]

In order that the method followed by the intellect may be perfect, it is thus necessary that we have "the idea of the absolutely perfect Being", Spinoza indicates in the *Tractatus de Intellectus Emendatione* (Spinoza, 1955, p. 18; Gebhardt II, p. 19),[53] and this will be indeed the program of the first part of the *Ethics*. To know if such an idea agrees with its object that follows, Spinoza points out, "truth . . . makes itself manifest" (Spinoza, 1955, p. 16; Gebhardt II, p. 17). May we interpret: this is the only reasonable postulate?

I come now to another particularly interesting aspect of the deployment of Spinoza's method. That is the so-called doctrine of parallelism, which we will consider with respect to what it indicates about the relations between the idea or the concept and the thing or the object.[54] Proposition V of *Ethics* II stipulates that "The actual being of ideas" (that is to say, the idea insofar as it is a real mode of the attribute of Thought) recognizes as its cause only the attribute of Thought. In other words, the idea is not caused by the object of which it is the idea, but by the attribute Thought only. This "absolute autonomy of the attribute of Thought in the production of its modes, that is, of its ideas", to repeat Gueroult's expression, implies that no idea is to be explained by its object. Gueroult says that this is a refutation of realism; it would be more exact to say that this is a refutation of empiricism, or of induction. (Inversely, no object can be explained by its idea; this refutation of idealism is the object of Proposition VI of *Ethics* II.) It follows that things and ideas are produced in nature starting with their own respective attributes and in particular that "Things represented in ideas follow, and are derived from their particular attribute, in the same manner, and with the same necessity as ideas follow (according to what we have shown) from the attribute of thought" (*Ethics* II, Proposition VI, Corollary).

The next proposition is that of parallelism: "The order and connection of ideas is the same as the order and connection of things" (*Ethics* II, Proposition VII).[55] It constitutes the foundation of truth, truth being defined as the agreement of the idea and the thing. We thus dispose of the indispensable complement, for knowing things, of that remark of the *Tractatus de Intellectus Emendatione* that "A true idea . . . is something different from its correlate (*ideatum*); thus a circle is different from the idea of a circle" (Spinoza, 1955, p. 12; Gebhardt II, p. 14).

Further, parallelism applied to ideas themselves — or intra-cogitative according to Gueroult, in opposition to the preceding, extra-cogitative, parallelism — has the consequence that, to arrive at the truth, the method is to deduce, by reflection, all our ideas from the given true idea. This parallelism is for us the foundation of our possible knowledge of truth: which once more justifies the deductive form of the *Ethics*.

8. EINSTEIN'S EPISTEMOLOGICAL PROGRAM

We already see, without forcing our conceptions, how Einstein — and probably without having considered it in detail — in the gap established by Hume with his critique of induction, and in the direction taken by the conventionalism of Duhem and Poincaré,[56] how Einstein, then, opens the path for Spinoza to sweep in. Although the propositions, the concepts and the theories are chosen in an arbitrary fashion, truth is possible, the conformity of the idea to the object is guarantor of truth, despite the absence of a common measure between the two chains — that of causal connection of ideas and that of causal connection of objects. And this results, in the final analysis, from the absolute primacy given to the proposition about Being — that is, to the postulate of realism. Einstein believes precisely, because he founds himself on this postulate and on the intelligibility of the real, that theory attains to a depth that anchors it in physical reality, quite differently from the way it would be if it were a question of a simple convention. Further, the form itself of physical theory — deductive, starting from principles small in number and mutually coherent — guarantees its depth with respect to the representation of the structure of reality.

Einstein was perfectly well aware of the properly philosophical dimension of his research, and this consciousness marks even the style of his scientific work. It is not too much to say that in elaborating his scientific conceptions, he in fact composed a vast philosophical work. He did not think separately about the elaboration of physical theory and the problem of its foundations. Epistemology and science cannot be separated, he explained; epistemology without science is an empty schema, but science without epistemology is primitive and . . . . Nevertheless, scientific activity adapts itself only to an open epistemology, not a systematic one, in the sense of an epistemology that must respect a certain number of conditions which rigid systems usually mutually exclude (Schilpp, 1951, pp. 683—684). These conditions define his own epistemological demands. They permit us to understand how Einstein's scientific work is, in a certain fashion, a philosophical practice. Of this philosophy, which subtended his research, he would make elements explicit little by little, going directly to the essentials.

The first reference is the Real. It is accessible to Reason. "In a certain sense", he wrote in 1933, "I hold it to be true that pure thought is competent to comprehend the real, as the ancients dreamed" (Einstein, 1933, pp. 12—13). And, in the 'Autobiographical Notes': "Physics is an attempt conceptually to grasp reality as it is thought independently of its being observed. In this sense one speaks of 'physical reality' " (Schilpp, 1951, p. 81). Let us note well that there is on the one hand physical reality and, on the other, its conceptual grasp; the concept of the object is not the object.

Theory is construction and, like its concepts and its principles, it is not contained in the empirical facts; it is a rational construction, speculative, abstract, mathematized. Einstein speaks, in that sense, of "the essentially constructive and speculative nature of thought and more especially of scientific thought" (Schilpp, 1951, p. 21). Thus, the elements of reality and those which, in the theory or in our concepts, designate them, have no logical link, despite their narrow correspond-ence.[57] (It makes us think of the theory of parallelism of Spinoza; in fact, it is from Hume and perhaps from Poincaré that Einstein draws this conception, which he develops in a realist direction; it is this realist direction that leads us to Spinoza; and, from there, we are reminded of the 'parallelism'. It is not that we are here imposing Spinoza on Einstein; let us say that there is a coherence of the one which at a deep level rejoins that of the other.)

Theory, as a rational dscription of physical reality, must be simple, based on a small number of independent principles; it must express a logical coherence, which is realized in its mathematical formalization. From his first work on quanta, taking account of the difficulties of classical theories,[58] which could not "claim exact validity", Einstein went on "to the conviction that only the discovery of a universal formal principle could lead us to assured results" (Schilpp, 1951, p. 53). Logical perfection is one of the criteria that permits us to choose the theory (pp. 21—23);[59] it holds to the logical simplicity of the premises and of the structure, to its fruitfulness as regards the propositions that can be deduced from it, to the absence of any arbitrary character — the theory, like its object, physical reality, must be sufficient in itself without external additions . . . . Another criterion is evidently that it must not be put into contradiction with the empirical facts;[60] but this is difficult to apply, in virtue of the possibility of invoking supplementary hypotheses, arbitrary parameters.

A word here about mathematics, an indispensable tool for theories becoming increasingly abstract. For Einstein, the relation to mathe-matics is an indirect conceptual process, whose adequacy to the empirical data effects only the last stage. There is no innate mathe-matics, like Euclidean geometry, for example. Indispensable to theory, mathematics does not suffice to guarantee its validity, although one owes it, if there are any, elements of certainty; for this, a trial of the facts is required — and it is there that we find the necessity of experi-mental activity. But the facts are reliable to the degree of the radical coherence they display between different experiments (the principle of relativity and the principle of equivalence of inertial mass and gravita-tional mass are 'facts' of that kind, according to him). Repeatedly, Einstein raised himself against the empiricist illusion, prejudice of numerous scientists, which consists "in the faith that facts by themselves can and should yield scientific knowledge without free conceptual construction". The concepts chosen are, from a logical point of view, arbitrary to the facts, and it is "through verification and long usage" that they appear "immediately connected with the empirical material" (Schilpp, 1951, p. 49).

In his 'Reply to Criticisms', written in 1949, Einstein returned to the question of the difference between concepts and sense impressions and, in particular, to the distinction between the latter and what one can call .the real, insisting on the character of non-evidence — in the sense of perceptions — of that distinction, but on its necessity if one wants to avoid solipsism (Schilpp, 1951, p. 673).[61] Further it is required as a presupposition by the physical thought in whatever domain it be. Far from incurring the reproach of "the metaphysical 'original sin' ", such a distinction is a category that conditions the possibility of understanding anything in the empirical world of immediate sensations (p. 673),[62] in order to make it intelligible. It is thus that one must admit a distinction between what is subjective and what is objective, which corresponds to a "programmatically fixed sphere of thought" (p. 675);[63] "the 'real' in physics is to be taken as a type of program, to which we are, however, not forced to cling a priori" (p. 674).[64] Thus Einstein expresses, in an admirably concise manner, the impossibility of grasping the nature of the object itself, in its relation of belonging to the Real, other than by definition — but a definition which founds the very possibility of science and which, in other words, determines an epistemological program, centered on realism and critical rationality.

9. A KIND OF INTUITION OF CERTAINTY

Einstein's overall attitude is summed up in this confession of his 'Autobiographical Notes', about his research in physics: "I soon learned to scent out that which was able to lead to fundamentals and to turn aside from everything else, from the multitude of things which clutter up the mind and divert it from the essential" (Schilpp, 1951, p. 17). This is an indication, largely confirmed elsewhere, of Einstein's research tending toward something that would approach certainty.

Certainty is a term that recurs often under his pen. For example, with respect to the relation between sensible experience and the conceptual universe, a relation which must be intuitive since it cannot be logical, and which results, as he says elsewhere, from a sort of 'suggestion' coming from experience. Einstein writes, "The degree of certainty with which this connection, viz. intuitive combination, can be undertaken, and nothing else, differentiates empty phantasy from scientific 'truth' ". Further on, "Although the conceptual systems are logically entirely arbitrary, they are bound by the aim to permit the most nearly possible certain (intuitive) and complete co-ordination with the totality of sense-experiences" (Schilpp, 1951, p. 13). There is a sure path that leads to the knowledge of physical reality, and we can arrive at it "in all certainty"; the guarantee of that certainty is mathematics, and the fact that "in Nature is actualized the ideal of mathematical simplicity" (Einstein, 1933, p. 12). For it is the nature of mathematical knowledge to procure evidence. Recalling a memory of his youth, his first encounter with mathematics as a system of demonstration based on axioms, Einstein speaks of the evidence that he grasped, for example about the determination of the ratios between the sides of a right triangle and one of its acute angles (Schilpp, 1951, p. 11), in terms which are consonant with the example considered by Spinoza to illustrate knowledge of the third kind.

If the model, and the manner, of certainty, is mathematics, what leads to it is a mental work, the description of which he does not insist very long on, and which, indeed, does not need words.[65] It is work on concepts, but the latter are not necessarily expressible in words. A concept is a tool, an element that organizes mental images which it links one to the others, and which takes shape from them: "the transition from free association or 'dreaming' to thinking is characterized by the more or less dominating role which the 'concept' plays in it". Einstein adds, "For me it is not dubious that our thinking goes on for the most part without use of signs (words) and beyond that to a considerable degree unconsciously". Without that, would we have immediate reactions, in the face of some experience, like astonishment? Indeed one can say that thought takes its flight starting from astonishment (Schilpp, 1951, pp. 8—9).[66]

It is clearly tempting to drive this conception close to Spinoza's knowledge of the third kind, privileged as to certainty, as to its access to truth.[67] Knowledge of the second kind, relative to (deductive) reasoning[68] or to experience, does not guarante the absolute truth of the generalization, and one can consider it as a negation of induction, "[F]or how" he says in the *Short Treatise*, "can he possibly be sure that his experience of a few particulars can serve him as a rule for all?" (Spinoza, 1910, p. 68; Gebhardt I, pp. 54—55). The third kind, that of 'clear intuition', is, as one sees by the example of immediate evidence of the fourth number of a proportion, like the perception of a condensation of the properties of the things. This is not an intuition in the psychological sense, but an intuition of the intellect, a kind of very high proximity and familiarity of the things we are considering, ". . . there is the perception arising arising when a thing is perceived solely through its essence, or through the knowledge of its proximate cause" (Spinoza, 1955, p. 8; Gebhardt II, p. 10).[69] In *Letter LXXVI* to Burgh, Spinoza writes,

> I do not presume that I have found the best Philosophy, but I know that I think the true one. If you ask me how I know this, I shall answer, in the same way that you know that the three angles of a triangle are equal to two right angles. That this is enough no one will deny whose brain is sound, and who does not dream of unclean spirits who inspire us with false ideas which are like true ones: for the truth reveals itself and the false. (Spinoza, 1928, p. 352; Gebhardt IV, p. 320)

This affirmation of the true philosophy sounds to us like an echo, through several centuries, of the sentence quoted earlier from Einstein, deciding to bring his research definitely to the essentials.

What shall we conclude? Nothing but the following: the rapprochement between Spinoza and Einstein, occurs without externally superimposing fragments of their thoughts on one another, but by trying to situate them in relation to the inner sense of their project; this rapprochement suggests a sort of common perspective on the most fundamental subjects of knowledge and of being: affirmation of the real, relation of thought to the real, unity of the world and of the approach to it, profound logic of reason and of things, and consequences that flow from this to the attitude toward life and to the organization of human society. This rapprochement suggests a community of inspiration — not an imitation or an application of Spinoza by Einstein, but an intense intellectual sympathy. This results from an intimacy and a maturation of problems that transcend a given ethic, a given place, a given knowledge, a given way of being face to face with the world and the power of thought — which send us to their personalities that are so far from common and yet so much concerned by this world.

It would be senseless to speak, in this respect, of Spinoza as a possible precursor of Einstein and modern thought — at least in some of its orientations . . . . It is not a question, here, of occasional ideas, of ideas in the air which would be repeated, which one would find again in a more or less analogical manner. One might say, rather, that the rupture inaugurated by Spinoza in the intellectual universe by the radical manner in which he conceives the unity of being, the self-sufficiency of nature and of its possible understanding, and its coherence centered on that decision, oriented to that program, is today, perhaps more than ever, productive of effects, beyond the determinations of his time. It is not a question therefore of adapting the thought of Spinoza, but of settling with him in that rupture, in that essential point of departure that changes all our perspectives on knowledge and on the appropriation of knowledge, on action, on our adherence — wise, joyful, active — to this world, to this nature of which we are modes, according to the indivisibility of substance, that is, being. In this sense, Einstein's thought, beyond its own interest, which is in itself of a considerable importance, could be for us, in some way, a 'Spinoza: directions for use'.

Centre National de la Recherche Scientifique, Paris

NOTES

* Presented at the *Association des amis de Spinoza*, Université de Paris-Sorbonne, March 19, 1983. I gratefully acknowledge the contributions of Marjorie Grene to the translation of this paper.
[1] Some of the quotations from Einstein on Spinoza mentioned here are unpublished and are located in the Einstein Archives. Their reproduction is with permission of the Hebrew University of Jerusalem. I thank John Stachel, as well as Marie Farge who devotedly transcribed the documents.
[2] Letter to Dagobert Runes, September 6, 1932; letter to Siegfried Hessing, September 8, 1932. True, Einstein finally agreed to provide a preface for a dictionary on Spinoza prepared by Runes (1951), but it is short and remains at a very general level.
[3] Cf. letter to Willy Aron, March 7, 1949.
[4] I owe valuable comments and several items of information to Etienne Balibar, Marie Farge, Martine Escoubès-Westphal, and John Stachel. I thank them, and also Renée Bouveresse: it is on her friendly insistence that I have undertaken this work.
[5] For example, in physics, in biology, in ethology . . . .
[6] See also d'Espagnat (1982).
[7] However, d'Espagnat (1979, p. 96) notes correctly that Spinoza's substance, being that which exists in itself, could be "neither a collection of particles nor a collection of observables". But he adds, thus destroying his prudence and rejoining Kouznetsov, that it resembles the "universal reality" to which certain symbols of the quantum field theory refer, for example $| 0 >$, representing the state of vacuum, "full of things all located half way between the virtual and the actual". Nevertheless, further on (p. 97), he sees clearly that, in Spinoza, "independent reality" is different "from purely phenomenal thought",

and he stresses the fact that Spinoza's language, in differentiating *natura naturans* and *natura naturata* — against the single use of the word 'nature' in effect today — "is better adapted to the truth than that of modern authors". (This last appreciation is probably because of d'Espagnat's own philosophical view of the veiled real as distinct from the empirical. But this is a particular interpretation of Spinoza, different from the analysis sketched in the present work.)

[8] (And, because this does not make much sense, he goes on to say the opposite a few pages later.) One hesitates to report so naïve a view: in the domain of energies of 400 to 1000 billions of electron-volts — the horizon of physics in 1967, accessible to investigation today — "the modal categories lose their meaning and [. . .] one is thrown directly upon the attributes of substance, upon *natura naturans*". If *natura naturans* is of an indeterminate character, Kouznetsov (1967, p. 46) draws the inference: "perhaps, unlike Einstein's God, Spinoza's does nevertheless play at dice". "In the quarrel that pitted Einstein against Bohr, Spinoza would, to a certain degree, have approved the opposing versions, at the same time that he would have recognized their incompatibility."

[9] Kouznetsov seems to me to conjoin two orthodoxies: that of the Copenhagen interpretation of quantum mechanics, which speaks of indeterminism, and that of dialectical materialism in the dogmatic version; the latter delighted to propose illustrations, through such and such scientific concepts, of such and such a philosophical category, in the scholastic fashion.

[10] Precisely to the extent to which Spinoza breaks with Cartesian thought, whose dualism would make possible such rapprochements.

[11] See on this subject the remarks of Macherey (1979) and Balibar (1982).

[12] Of course, some people relegate him to the attic, arguing that the sciences today follow other methods and other paths. But those who despise Einstein (beyond all necessary and legitimate critical debate, like that with Bohr) seem to me, as to method and to basic ideas, to propose only some wooly and opportunistic ideas, or even terrorizing ones as far as the demand for philosophical rigor is concerned.

[13] This is what Tonnelat (1979, p. 315) also emphasizes in her article.

[14] Cf. Freud (1942, pp. 88ff), and Freud's correspondence.

[15] Letter quoted in part by Hoffman (1972, p. 95); original in Einstein Archive. Expressing himself in similar terms in a letter to C. Van Slotten of Heelsum, The Netherlands, September 23, 1953 (Einstein Archives, unpublished), he adds, "He achieved an independence that few men attain".

[16] Bohr explains that he was "strongly reminded of the importance of utmost caution in all questions of terminology and dialectics" (Schilpp, 1951, p. 237).

[17] See Einstein's letter to Leo Szilard, September 1, 1928 (Einstein Archives, unpublished): "I read with great pleasure letters of Spinoza. He knew the freedom of evening solitude".

[18] "I did not grow up in the Kantian tradition", writes Einstein in 1949 [that is, although he had read Kant when very young, he had kept his distance from him] "but", he continues, "came to understand the truly valuable which is to be found in his doctrine, alongside of errors which today are quite obvious, only quite late" (Schilpp, 1951, p. 680).

[19] See especially, on these points, Einstein's 'Autobiographical Notes' (Schilpp, 1951,

pp. 2—96). On the relation between the thought of Mach and that of Einstein, see Holton (1970). But Holton gives an excessive interpretation to a declaration by Einstein in his autobiography, on his youthful adherence to the conceptions of Mach (in fact, he never adopted Mach's epistemology, but only his critique of the absolute and the *a prioris* of the classical conception), in the sense of a radical philosophical revulsion. See Paty (1979) and Paty (in press *a*).

[20] Einstein also wrote on various occasions that he was rather a philosopher, even a metaphysician, than a physicist.

[21] For example, Einstein (1921) and Einstein and Infeld (1938).

[22] See the collections, Einstein (1934), Einstein (1954), and Einstein (1950).

[23] See Einstein (1954, p. 9); originally published in *Forum and Century* 84 (1930) 193—194; original German in *Mein Weltbild*, Querido, Amsterdam (1934).

[24] With regard to his non-conformism and his independence *vis-à-vis* authority, which were legendary, let us mention one anecdote: at the gymnasium, when he was fifteen, his principal teacher, the teacher of Greek, who predicted that he would never amount to anything, complained of him that "your mere presence spoils the respect of the class for me" (Hoffman, 1972, p. 25).

[25] See also Cranberg (1979, pp. 9—11).

[26] Another quotation on the same subject, from a letter of Einstein to Robert Moss, of Brooklyn, in December, 1954: "Spinoza was wholly dominated by the striving for truth. But he rejected a professorship in the University of Heidelberg and earned his living in an independent profession in order to preserve his independence. The present situation is indeed comparable with that |of the seventeen century|" (Einstein Archives, unpublished).

[27] And Einstein: "When I came to Germany fifteen years ago I discovered for the first time that I was a Jew, and I owe this discovery more to Gentiles than Jews" (response to an article by Professor Hellpach which appeared in the *Vossische Zeitung*, 1929; see Einstein, 1954, p. 171).

[28] But preceding this statement: "it seems to me above all that you are right, that the abyss between Jewish theology and Spinoza cannot be bridged" (Einstein Archives, unpublished).

[29] "In modern times this tradition has produced Spinoza and Karl Marx" (conference in London, 1930: Einstein, 1954, p. 174). See also letter of M. Besso to Einstein, January 29, 1955 in Einstein (1972).

[30] Nor let us omit the continuation of the sentence, which indicates the meaning of his political preoccupations and announces his great treatises, the *Tractatus Theologico-Politicus* and the *Tractatus Politicus*: "and also to form a social order such as is most conducive to the attainment of this character by the greatest number with the least difficulty and danger" (Spinoza, 1955, p. 7; Gebhardt II, p. 9).

[31] Einstein was sixty-seven when he wrote his autobiographical notes in 1946.

[32] Einstein in fact says "*Nur-Persönlichen*".

[33] Two testimonies to this spiritual experience: "I feel such a sense of solidarity with all living things that it does not matter to me where the individual begins and ends", Einstein declared to Hilde Born; she recalled this sentence to him in her letter of October 9, 1944 (Born and Einstein, 1971, p. 152). And the other, from Spinoza, in the last Scholium of the *Ethics*: "Whereas the wise man, in so far as he is regarded as

such, is scarcely at all disturbed in spirit, but, being conscious of himself, and of God, and of things, by a certain eternal necessity, never ceases to be, but always possesses true acquiescence of his spirit".

[34] There is an objectivity of the idea in Spinoza; 'intersubjectivity' on the contrary would imply that one first posit the subject. On the importance of the '*multitudo*', in opposition to the subject, and on the importance and at the same time the ambiguity of Spinoza's position, see Balibar (1982).

[35] Let alone, according to certain interpretations — Wigner's for instance — objectivity or consciousness as psychic act.

[36] On the use of this term, see for example Balibar and Macherey (1984). 'Determinism' is a term that belongs to the vocabulary of moral theology. Leibniz uses it in the sense of moral philosophy — and it is determinism in the Leibnizian sense (tied to the principle of sufficient reason to the idea of a preformed development) that Kant criticizes. The term does not appear in French until 1836 (it does occur, according to Foulquié (1962) in the 1835 edition of the *Dictionnaire de l'Académie Française*). It appeared much sooner in German, in the above sense. See also Lalande (1980).

[37] 'Determination' (*determinatio*) appears seven times in the *Ethics* (in Part III, Proposition XLVII, Scholium, four times; Proposition II, Scholium, twice; in the introduction to Part V). 'Determined' appears sixty-seven times, and the verb 'to determine' one hundred and two times. See Guéret *et al.* (1977). See also Giancotti-Boscherini (1970). In reality, one ought also to report on the term 'necessity' as well as 'causality' and on their derivatives. For all of them, the meaning is clearly dependent on context.

[38] This proposition is a copy of Spinoza's text, replacing God by Nature.

[39] See also the *Short Treatise* (Spinoza, 1910, pp. 43, 46; Gebhardt I, pp. 4, 7 [from summary], 37, 39).

[40] "Things could not have been brought into being by God in any manner or in any order different from that which has in fact obtained" (*Ethics* I, Proposition XXXIII).

[41] Einstein uses the term 'program' (see 'Reply to Criticisms' in Schilpp, 1951). See Paty (1982*b*) and Paty (in press *b*).

[42] *Letter LIV*: ". . . who asserts that the world is the necessary effect of the divine Nature also absolutely denies that the world was made by chance" (Spinoza, 1928, p. 277; Gebhardt IV, p. 251).

[43] See also *Ethics* III, first paragraph before the definitions. *Letter LVI* to Boxel recalls that it is reason that wants the necessary and the free not to be opposed. See also, along the same lines, the refutation of miracles in the *Tractatus Theologico-Politicus* (Spinoza, 1951, pp. 81—97; Gebhardt III, pp. 81—96) and of ghosts in *Letters LI—LVI* (Spinoza, 1928, pp. 270—290; Gebhardt IV, pp. 241—262).

[44] On man, see *Ethics* IV, preface.

[45] In addition to the preceding, see the *Tractatus Theologico-Politicus*, Ch. 16. See also the *Tractatus Politicus*. Freedom is defined in *Ethics* I, Proposition VII. See also the *Short Treatise*, (Spinoza, 1910, pp. 44—46; Gebhardt I, pp. 9—10 [from summary], 38—39). Einstein takes an analogous position on democracy; see his lecture 'The Goal', given at a summer conference at Princeton Theological Seminary in 1939 (Franck, 1947).

[46] See Note 41 above.

[47] On the one hand, reason results ultimately from the action of bodies (*Ethics* IV and

V); on the other, Nature overtakes it (*Tractatus Theologico-Politicus*, Spinoza, 1951, pp. 201—202; Gebhardt III, pp. 190—191).

[48] Hamelin (1982) said that Spinoza did not produce a theory of knowledge, and reproaches the author of the *Ethics* for speaking of certainty as by dogma, by definition; for not introducing it with a critique of knowledge. But Hamelin's critique is rather an idealist critique of realism, which he calls dogmatic. However, if we read Spinoza well, we see that he does not leap from particular knowledge to ontology, and that there is a place for mediations. Epistemology in Einstein's sense might be one of the possible mediations.

[49] "... the true method teaches us the order in which we should seek for truth itself, or the subjective essences of things, or ideas, for all these expressions are synonymous ... method is not identical with reasoning in the search for causes, ... it is the discernment of a true idea, by distinguishing it from other perceptions, and by investigating its nature, in order that we may thus know our power of understanding, and may so train our mind that it may, by a given standard, comprehend whatsoever is intelligible, ..." (*Tractatus Intellectus Emendatione*: Spinoza, 1955, p. 14; Gebhardt II, pp. 15—16). Deleuze (1983, pp. 23—24) speaks of the geometrical method as of a vital and optical rectification, of an "optical geometry".

[50] "... method is nothing else than reflective knowledge, or the idea of an idea ... that will be a good method which shows us how the mind should be directed, according to the standard of the given true idea" (*Tractatus Intellectus Emendatione*: Spinoza, 1955, p. 14; Gebhardt II, p. 16).

[51] A defective logic accompanies the false attributions to God (or Being): *Short Treatise*, Spinoza, 1910, p. 54 (Gebhardt I, p. 46); *Tractatus Intellectus Emendatione*, Spinoza, 1955, p. 16 (Gebhardt II, p. 17). This is an implicit justification of proof *more geometrico*.

[52] See also the *Ethics* I, Proposition XXXIII, Scholium 2: if things were other than they are, it would be necessary "to set up something beyond God, which does not depend on God, but which God in acting looks to as an exemplar, or which he aims at as a definite goal".

[53] Against the skeptic who would express some doubts on the subject "of our primary truth, and of all deductions we make, taking such truth as our standard", Spinoza (1955, p. 17; Gebhardt II, p. 18) replies in terms that could be addressed to the neo-positivists.

[54] *Ethics* II, Propositions III—VIII. Cf. Gueroult (1974, Chapter 4).

[55] Things, that is, modes in the different attributes. This parallelism is called by Gueroult (1974) 'extra-cogitative parallelism'.

[56] Or even in Mach's fashion. See Mach (1976).

[57] This is constantly reaffirmed all through the epistemological writings of Einstein.

[58] Actually it was about mechanics and thermodynamics on the one hand (kinetic theory) and of electromagnetism on the other.

[59] This is found also in numerous other texts.

[60] Refutation or falsification in Popper's sense is, in Einstein, only *one of the criteria* that characterize a scientific theory; it is never *the* criterion of demarcation.

[61] This is evidently directed against the conceptions of Mach and, following him, the logical empiricists on the one hand and on the other the Copenhagen school.

[62] It is Einstein who uses the term 'category' (Schilpp, 1951, p. 673).

[63] "So long as we move within the thus programmatically fixed sphere of thought we are thinking physically" (Schilpp, 1951, p. 674).

[64] Einstein adds that he sees nothing in the quantum phenomena that would oblige him, on the contrary, to choose a thesis according to which the description of nature must be of a statistical kind. The adherents of the Copenhagen interpretation are doing nothing else than founding themselves on another program. He expresses this in a letter to Born of September 15, 1950 (Born and Einstein, 1971, pp. 188—189), ". . . you are . . . convinced that no (complete) laws exist for a complete description, according to the positivist maxim *esse est percipi*. Well, this is a programmatic attitude, not knowledge. This is where our attitudes really differ . . .".

[65] The role of the word in relation to the concept is to make it communicable. Einstein gives a brief description of the mental process which seems to him to operate in his own case in a reply to Jacques Hadamard, reproduced in Einstein (1954, pp. 25—26). Cf. Hadamard (1945).

[66] Einstein evokes one of his first wonders in front of physical phenomena; in his prime youth, he observed the behavior of a compass needle. The linguist Roman Jakobson has emphasized the interest of Einstein's conception described above concerning the formation of concepts and the role of language (Jakobson, 1982).

[67] *Short Treatise*: Spinoza, 1910, pp. 67—68 (Gebhardt I, pp. 54—55); *Tractatus Intellectus Emendatione*: Spinoza, 1955, pp. 8—10 (Gebhardt II, pp. 10—12); and *Ethics* II, Proposition XL, Scholium 2.

[68] Deductive reasoning consists in explaining things by their properties.

[69] ". . . certainty is identical with such subjective essence" (*Tractatus Intellectus Emendatione*: Spinoza, 1955, pp. 13—14; Gebhardt II, p. 16). See *Ethics* II, Proposition XLIII, Scholium: one knows that one has an idea that agres with its object: "truth is its own standard". It is true that, in the *Ethics*, knowledge of the second kind is validated in relation to the earlier considerations of the *Short Treatise* and the *Tractatus de Intellectus Emendatione*. But the knowledge of the third kind ('clear intuition') is of a different nature.

REFERENCES

Balibar, Etienne: 1982, 'Spinoza: la crainte des masses', Séminaire d'Histoire du matérialisme, Université de Paris I, December 11, 1982.

Balibar, Etienne and P. Macherey: 1984, 'Determinism', *Encyclopedia Universalis*, new edition, vol. 6, Universalis, Paris.

Born, Max and Albert Einstein: 1971, *The Born—Einstein Letters*, transl. by Irene Born, Walker, New York.

Cranberg, Lawrence: 1979, 'Einstein: Amateur-Scientist', *Physics Today* **32**:12 (December), 9—11.

Deleuze, Gilles: 1983, *Spinoza, philosophie pratique*, Minuit, Paris.

Einstein, Albert: unpublished, Einstein-archives, Spinoza folder, reel 33, No. 7.

Einstein, Albert: 1921, *The Meaning of Relativity*, Princeton University Press, Princeton, New Jersey.

Einstein, Albert: 1933, *On the Method of Theoretical Physics*, The Herbert Spencer Lecture, Oxford, June 10, 1933, Clarendon Press, Oxford.

Einstein, Albert: 1934, *The World as I See It*, Covici-Friede, New York.
Einstein, Albert: 1947, *Methods of the Sciences*, transl. by G. B. Jefferey and W. Perrett, Chicago University Press, Chicago.
Einstein, Albert: 1950, *Out of My Later Years*, Philosophical Library, New York.
Einstein, Albert: 1954, *Ideas and Opinions*, transl. by Sonja Bargmann, Crown, New York.
Einstein, Albert: 1972, *Correspondance avec Michele Besso 1903—1955*, in French and German, ed. and transl. by Pierre Speziali, Hermann, Paris.
Einstein, Albert and L. Infeld: 1938, *The Evolution of Physics: the Growth of Ideas from Early Concepts to Relativity and Quanta*, Simon and Schuster, New York.
Espagnat, Bernard d': 1979, *À la recherche du réel*, Gauthier-Villars, Paris.
Espagnat, Bernard d': 1982 (to be published), 'Spinoza et la physique contemporaine', Colloque Spinoza, Cerisy-la-Salle, September, 1982.
Feuer, Lewis Samuel: 1974, *Einstein and the Generations of Science*, Basic Books, New York.
Feynman, Richard: 1979, 'Interview with Monte Davis', *Omni* 1, 96—98, 113—114, 138.
Foulquié, Paul: 1962, *Dictionnaire de la langue philosophique*, Presses Universitaires de France, Paris.
Franck, Philipp: 1947, *Einstein: His Life and Times*, transl. by George Rosen, A. A. Knopf, New York.
Freud, Sigmund: 1942, 'My Contacts with Josef Popper-Lynkeus', *International Journal of Psychoanalysis* **23**, 85ff.
Giancotti-Boscherini, E.: 1970, *Lexicon Spinozarum*, 2 vols., Martinus-Nijhoff, The Hague.
Guéret, M., A. Robinet, and P. Tombeur: 1977, *Spinoza, Ethica: Concordances, index, listes de tréquences, tables comparatives*, Cedetec (Université Catholique de Louvain), Louvain-la-Neuve.
Gueroult, Martial: 1974, *Spinoza*, Vol. 2: *L'Âme*, Aubier, Paris.
Hadamard, J.: 1945, *An Essay on the Psychology of Invention in the Mathematical Field*, Princeton University Press, Princeton, New Jersey.
Hamelin, O.: 1982, 'La théorie de la certitude dans Spinoza', ed. by F. Turlot, *Bulletin de l'Association des amis de Spinoza*, No. 8.
Hoffman, Banesh: 1972, *Albert Einstein Creator and Rebel*, in collaboration with Helen Dukas, Viking, New York.
Hoffman, Banesh and Helen Dukas: 1979, *Albert Einstein, the Human Side: New Glimpses from His Archives*, Princeton University Press, Princeton, New Jersey.
Holton, Gerald: 1970, 'Mach, Einstein and the Search for Reality', in Robert S. Cohen and R. Seeger (eds.), *Ernst Mach, Physicist and Philosopher*, Reidel, Dordrecht, pp. 165—198.
Jakobson, R.: 1979, 'Einstein and the Science of Language', in Gerald Holton and Yehuda Elkana (eds.), *Albert Einstein, Historical and Cultural Perspectives: The Centennial Symposium in Jerusalem*, Princeton University Press, Princeton, New Jersey, pp. 139—150.
Kouznetsov, Boris: 1967, 'Spinoza et Einstein', *Revue de synthèse* **88**, sér. gén., Nos. 45—46, 3è série, janvier—juin, 31—52.
Lalande, A.: 1980, *Vocabulaire technique et critique de la philosophie*, 13è éd., Preses Universitaries de France, Paris.

Mach, Ernst: 1976, *Knowledge and Error*, transl. by P. Foulkes and T. J. McCormick, Reidel, Dordrecht.

Macherey, P.: 1979, *Hegel ou Spinoza*, Maspero, Paris.

Minkowski, Hermann: 1923, 'Space and Time', in H. A. Lorentz, A. Einstein, H. Minkowski, and H. Weyl (eds.), *The Principle of Relativity*, tr. by W. Perrett and G. B. Jefferey, Methuen, London.

Negri, A.: 1981, *L'Anomalia selvaggia, saggio su potere e potenzia in Baruch Spinoza*, Feltrinelli, Milan.

Paty, Michel: in press *a*, 'Albert Einstein', in *Dictionnaire des philosophes*, Presses Universitaires de France, Paris.

Paty, Michel: in press *b*, *La Materia trafugata*, Fultrinelli, Milan; and *La Matière dérobèe, l'appropriation de l'object de la physique contemporaine*, Archives Contemporaines, Paris.

Paty, Michel: 1979, 'Sur le réalisme d'Albert Einstein', *la Pensée*, No. 204 (avril), pp. 18—37.

Paty, Michel: 1981, 'Les contributions d'Einstein à l'éboration de la première théorie des quanta', *Bulletin de l'Union des physiciens*, No. 631 (février), 693—709.

Paty, Michel: 1982*a*, 'L'inséparabilité quantique en perspective', *Fundamenta Scientiae* **3**, 79—92.

Paty, Michel: 1982*b*, 'La notion de programme épistémologique et la physique contemporaine', *Fundamenta Scientiae* **3**, 321—336.

Poincaré, Henri: 1963 (originally published, 1913), *Dernières pensées*, Flammarion, Paris.

Poincaré, Henri: 1920, 'Les conceptions nouvelles de la matière', in H. Bergson *et al.* (eds.), *Le Matérialisme actuel*, Flammarion, Paris, pp. 48—67.

Rolland, Romain: 1931, 'L'Éclair de Spinoza', in *Empédocle d'Agrigente, suivi de l'Éclair de Spinoza*, Sablier, Paris.

Runes, Dagobert (ed.): 1951, *Spinoza Dictionary*, Philosophical Library, New York.

Schilpp, P. A.: 1951, *Albert Einstein, Philosopher-Scientist*, second edition, Library of Living Philosophers, Vol. 7, Tudor, New York.

Spinoza, Baruch: 1910, *Short Treatise on God, Man, and His Well-Being*, transl. by A. Wolf, A. and C. Black, London.

Spinoza, Baruch: 1925, *Spinoza Opera*, ed. by Carl Gebhardt, 4 vols., Carl Winter, Heidelberg. (Cited as 'Gebhardt' in the text)

Spinoza, Baruch: 1928, *Correspondence of Spinoza*, ed. by Abraham Wolf, London, George Allen and Unwin.

Spinoza, Baruch: 1951, 1955, *Chief Works of Spinoza*, transl. by R. H. M. Elwes, 2 vols., Dover, New York.

Spinoza, Baruch: 1963, *Earlier Philosophical Writings; The Cartesian Principles and Thoughts on Metaphysics*, transl. by Frank A. Hayes, Bobbs-Merrill, Indianapolis.

Tonnelat, Marie Antoinette: 1979, 'Einstein, mythe et réalité', *Scientia* **114**, 297—326.

Translated by
MICHEL PATY
and
ROBERT S. COHEN

PART VI

BIBLIOGRAPHY

ANNOTATED BIBLIOGRAPHY OF SPINOZA AND THE SCIENCES

Construed broadly, 'sciences' could be the rubric for almost the entire literature on Spinoza. Beginning at the border between physics and metaphysics, epistemology claims all the territory ever covered by the Greek '*episteme*', and both psychology and political science fall in line. But a bibliography resulting from such a broad construal would be little more (and probably much less) than an amalgamation of such standard sources as Oko and Wetlesen. Within the wide compass of the philosophy and history of science, I have focused on issues in physics, chemistry and biology; adding such social-scientific works as self-consciously attempt to be or to comment on the less social and more scientific; and adding finally what we have come to call 'applied sciences' such as medicine. Annotations are based on the cited articles and books themselves or the authors' abstracts; occasionally, these are supplemented by reviews and descriptive passages from others' works. (Lachterman, Wartofsky, and the anonymous chronicler(s) of the Association des Amis de Spinoza were particularly helpful in pointing to scientific aspects of articles that might otherwise have been overlooked.)

Agaësse, P.: 1969, 'Le Spinoza de M. Gueroult', *Archives de philosophie* **32**, 288—296.

Alexander, B.: 1928, 'Spinoza und die Psychoanalyse', *Chronicon Spinozanum* **5**.

Alexander, Samuel: 1921, *Spinoza and Time; Being the Fourth 'Arthur Davis Memorial Lecture' delivered before the Jewish Historical Society at University College on Sunday, May 1, 1921/Nisan 23, 5681*, George Allen and Unwin, Ltd., London. Partially reprinted in S. Paul Kashap (ed.), *Studies in Spinoza; Critical and Interpretive Essays*, University of California Press, Berkeley, 1972, pp. 68—85.

Allen, Harold J.: 1976, 'Spinoza's Naturalism and Our Contemporary Neo-Cartesians', in James B. Wilbur (ed.), *Spinoza's Metaphysics*, Van Gorcum, Assen, pp. 133—155.

Aron, W.: 1965, 'Baruch Spinoza et la médicine', *Revue d'histoire de la médecine hébraïque* **18**, No. 65, pp. 61—78 and No. 69, pp. 113—211.

Aron, W.: 1966—67, 'Freud et Spinoza', *Revue d'histoire de la médecine hébraïque* **19**, No. 3, pp. 101—116, and **20**, pp. 53—70, 123—130, 149—160.

Aster, Ernst von: 1922, *Raum und Zeit in der geschichte der philosophie und physik*, Rosl and Co., Munich.

Bachelard, Gaston: 1933, 'Physique et Métaphysique', in *Septimana Spinozana*, Nijhoff, The Hague.

Marjorie Grene and Debra Nails (eds.), Spinoza and the Sciences, 305—314.
© *1986 by D. Reidel Publishing Company.*

From the perspective of engineering science, ascribes to metaphysics the character of a *natura constructa* beyond Nature.

Baumann, Joh. Julius: 1868, *Die Lehren von Raum, Zeit und Mathematik in der neueren Philosophie*, Georg Reimer Verlag, Berlin.

Sections on Descartes, Spinoza, Newton, *et al.*

Bennett, Jonathan: 1980, 'Spinoza's Vacuum Argument', *Midwest Studies in Philosophy* **5**, 391—399.

Treats *Ethics* I, scholium to Proposition XV, and *Letter IV* to Oldenburg.

Bernard, Walter: 1964, 'Freud and Spinoza', *Psychiatry* **9**.
Bernard, Walter: 1972, 'Spinoza's Influence on the Rise of Scientific Psychology — A Neglected Chapter in the History of Psychology', *Journal of the History of the Behavioral Sciences* **8**, pp. 208—215.

Connects the 'most important' influence of Spinoza to the shaping of modern scientific psychology.

Bernard, Walter: 1977, 'Psychotherapeutic Principles in Spinoza's *Ethics*', in Siegfried Hessing (ed.), *Speculum Spinozanum 1677—1977*, Routledge and Kegan Paul, London, Henley, and Boston, pp. 63—80.

Gleans and briefly discusses eight principles from the *Ethics* thought to be of contemporary psychotherapeutic relevance.

Biasutti, F.: 1979, *La Dottrina della scienza in Spinoza*, Patron Editore, Bologna.

Discusses Spinoza's relation to the scientific revolution: Spinoza's doctrine of science and his historical situation; examines the correspondence and books known to have been in Spinoza's library.

Bickel, Lothar: 1931, 'Über Beziehungen zwischen Spinoza und der Psychoanalyse', *Zentralblatt für Psychotherapie und ihre Grenzgebiete* **4**. Reprinted in Lothar Bickel, *Probleme und Ziele des Denkens*, Humanitas Verlag, Zürich, 1939. Transl. by Walter Bernard ('On Relationships between Psychoanalysis and a Dynamic Psychology'), and reprinted in Siegfried Hessing (ed.), *Speculum Spinozanum 1677—1977*, Routledge and Kegan Paul, London, Henley, and Boston, pp. 81—89.

Discusses similarities among Spinoza's psychological ideas, psychoanalysis, and Constantin Brunner's psychological ideas.

Bidney, D.: 1940, *The Psychology and Ethics of Spinoza; A Study in the History and Logic of Ideas*, Yale University Press, New Haven. Second edition, Russell and Russell, New York, 1962.
Bluh, Otto: 1964, 'Newton and Spinoza', in *Proceedings of the International Congress on the History of Sciences, 10th, Ithaca, 1962*, 2 vols. Actes du dixième Congrès international d'histoire des sciences, Actualités scientifiques et industrielles 1307, Hermann, Paris, pp. 701—703.
Brunschvicq, Léon: 1923, *Spinoza et ses contemporains*, Presses Universitaires de France, Paris.

See especially Chapter 9.

Brunschvicq, Léon: 1933, 'Physique et métaphysique', in *Septimana Spinozana*, Nijhoff, The Hague, pp. 43—54.

A mathematician's approach: in accordance with geometrical and analytical physics, distinguishes between mechanistic and mathematical metaphysics.

Brunt, N. A.: 1955, *De wiskundige denkwijze in Spinoza's philosophie en in de moderne natuurkunde* [The Mathematical Mode of Thinking in Spinoza's Philosophy and in Modern Physics], Mededelingen Vanwege het Spinozahuis 12, E. J. Brill, Leiden.

Clay, J.: 1933, 'Physik und Metaphysik', in *Septimana Spinozana*, Nijhoff, The Hague.

Looks to Spinoza's metaphysics for logical possibilities confined by physics to the sphere of realities that can be perceived.

Coert, J. H.: 1938, *Spinoza's betrekking tot de geneeskunde en haar beoefenaren* [Spinoza's Relation to Medicine and its Practitioners], Mededelingen Vanwege het Spinozahuis 4, E. J. Brill, Leiden.

Corsano, A.: 1976, 'Spinoza e la scienza contemporanea', *Bollettino di Storia della Filosofia* 4, 51—58.

Cremaschi, S.: 1981, 'Concepts of Force in Spinoza's Psychology', in *Theoria cum Praxi; Zum Verhaltnis von Theorie und Praxis im 17. und 18. Jahrhundert,* Studia Leibnitiana Supplementa 20, Akten der III Internationalen Leibnizkongresses, Hannover, pp. 138—144.

Discusses *potentia, conatus, vis.*

Crommelin, Claude August: 1939, *Spinoza's natuurwetenschappelijk denken* [Spinoza's Thought on Natural Science], Mededelingen Vanwege het Spinozahuis 6, E. J. Brill, Leiden.

Curley, E. M.: 1973, 'Experience in Spinoza's Theory of Knowledge', in Marjorie Grene (ed.), *Spinoza; A Collection of Critical Essays*, Doubleday, Garden City, pp. 25—59.

Examines Spinoza's understanding of the structure of science, establishing the role of experimentation in relation to rational knowledge and intuitive knowledge.

Daudin, H.: 1948, 'Spinoza et la science expérimentale: sa discussion de l'expérience de Boyle', in *Revue d'Histoire des Sciences et de leurs Applications* 12, Ann. Tom. II, Paris.

De Vet, J. J. V. M.: 1983, 'Was Spinoza De Auteur Van "Stelkonstige Reeckening Van Den Regenboog" En Van "Reeckening Van Kanssen" ' [Was Spinoza Author of 'Algebraic Computation of the Rainbow' and of 'Calculation of Probabilities'], *Tijdschrift voor Filosofie* 45, 602—639.

Argues that Spinoza wrote neither treatise.

Deleuze, Gilles: 1968, *Spinoza et le problème de l'expression*, Éditions de Minuit, Paris, Arguments 37.

Deleuze, Gilles: 1983, *Spinoza, philosophie pratique*, Éditions de Minuit, Paris.

Discusses the geometrical method and its relation to optics.

Deregibus, Arturo: 1981, *Bruno e Spinoza. La realtà dell'infinito e il problema della sua unità*, 2 Vols., Giappichelli editore, Torino.

Lengthy discussion of Spinoza's work as marked by the mathematics of seventeenth century science.

Donagan, Alan: 1973, 'Spinoza's Proof of Immortality', in Marjorie Grene (ed.), *Spinoza; A Collection of Critical Essays*, Doubleday, Garden City, pp. 25—59.

Discusses the notions of time, duration, eternity.

Donagan, Alan: 1976, 'Spinoza and Descartes on Extension: A Comment', *Midwest Studies in Philosophy* **1**, 31—33.

Duchesneau, Francois: 1974, 'Du modèle cartésien au modèle spinoziste de l'être vivant', *Canadian Journal of Philosophy* **3**, 539—562.

Examines Cartesian mechanistic models in the hypothetical explanation of physiological functions, the theory of organism, and Spinoza's critique of the instrumental analogy underlying the automaton model.

Edgar, William J.: 1976, 'Continuity and the Individuation of Modes in Spinoza's Physics', in James B. Wilbur (ed.), *Spinoza's Metaphysics; Essays in Critical Appreciation*, Van Gorcum, pp. 85—105.

Escodi, J.: 1959, 'Semelhancas entre Spinoza e Freud', *Corpo e alma, Anais 3 Congresso Nacional de Filosofia*, Sao Paulo, pp. 403—410.

Espagnat, B. d': 1979, *À la recherche du réel*, Gauthier-Villars.

Espagnat, B. d': 1982, 'Spinoza et la physique contemporaine', Colloque Spinoza, Cerisy-La-Salle.

Foti, Véronique M.: 1982, 'Thought, Affect, Drive and Pathogenesis in Spinoza and Freud', *History of European Ideas* **3**, 221—236.

Explores the basis for linking Spinoza's psychology with psychoanalysis, finding that basis to be rather narrow.

Foucher de Careil, L. A.: 1862, *Leibniz, Descartes et Spinoza*, Librairie Philosophique de Ladrange, Paris.

Funkenstein, Amos: 1976, 'Natural Science and Social Theory: Hobbes, Spinoza and Vico', in G. Tagliacozzo and D. P. Verene (eds.), *Giambattista Vico's Science of Humanity*, Baltimore, pp. 187—212.

Giorgiantonio, Michele: 1954, 'Intorno ad un tentativo di ricostruzione della mecanica e della fisica di Spinoza', *Spinoza* **22**, 326—330.

Groen, J. J.: 1972, *Ethica en Ethologie: Spinoza's leer der affecten en de moderne psychobiologie* [Ethics and Ethology: Spinoza's Theory of the Affects and Modern Psychobiology], Mededelingen Vanwege het Spinozahuis 29, E. J. Brill, Leiden.

Gueroult, Martial: 1966, 'La lettre de Spinoza sur l'infini', *Revue de métaphysique et de morale* **71**, 385—411. Included as Appendix 9 in Martial Gueroult, Spinoza, Vol. 1: *Dieu*, Éditions Montaigne, Paris, 1968. Transl. by Kathleen McLaughlin ('Spinoza's Letter on the Infinite'), and reprinted in Marjorie Grene (ed.), *Spinoza, A Collection of Critical Essays*, Doubleday, Garden City, 1973.

On *Letter XII* to Louis Meyer.

Gueroult, Martial: 1974, *Spinoza*, Vol. 2: *L'âme*, Éditions Montaigne, Paris.

An analysis of Spinoza's physics and theory of cognition.

Hall, A. Rupert, and Marie Boas Hall: 1964, 'Philosophy and Natural Philosophy: Boyle and Spinoza', in *Mélanges Alexandre Koyré*, Histoire de la pensée 13, Vol. 2: *L'aventure de l'esprit*, pp. 241—256.

Hall, A. Rupert, and Marie Boas Hall: 1978, 'Le monde scientifique à l'époque de Spinoza', *Revue de Synthèse* **99**, 19—30.

Hallett, Harold Foster: 1930, *Aeternitas: A Spinozistic Study*, Oxford University Press, Oxford.

Hallett, Harold Foster: 1949, 'On a Reputed Equivoque in the Philosphy of Spinoza', *Review of Metaphysics* **3**, 189—212. Reprinted in S. Paul Kashap (ed.), *Studies in Spinoza; Critical and Interpretive Essays*, University of California Press, Berkeley, 1972, pp. 168—188.

Uses physiological and physical arguments to explore whether Spinoza employs the term 'idea' equivocally (as the mental correlate of the body's state, and as the objective essence of a thing extrinsic to that body).

Hallett, Harold Foster: 1957, *Benedict de Spinoza*, Athlone Press, London.

Hampshire, Stuart: 1961, *Spinoza*, Barnes and Noble, New York.

Includes observations on the biological aspect of Spinoza's metaphysics with special reference to modern developments in physical and biological science.

Hardin, C. L.: 1978, 'Spinoza on Immortality and Time', in Robert W. Shahan and J. I. Biro (eds.), *Spinoza: New Perspectives*, University of Oklahoma Press, Norman, pp. 129—138.

Argues, *contra* Donagan and Kneale, for Spinoza's expression of the relation between eternity and duration, attributing to Spinoza a theory of time wherein temporal passage is a sensory, and therefore confused, representation of the causal order.

Hassing, R. F.: 1980, 'The Use and Non-use of Physics in Spinoza's *Ethics*', *Southwestern Journal of Philosophy* **11**, 41—70.

Concentrates on *Ethics* II and selected letters, concluding that Spinoza's concept of *conatus* is not derived from physics.

Hedman, Carl: 1975, 'Toward a Spinozistic Modification of Skinner's Theory of Man', *Inquiry* **18**, 325—335.

Uses Spinoza's notion of *conatus* to modify the Skinnerian position without giving up the " 'scientific' approach to human behavior".

Hicks, G. Dawes: 1918, 'The "Modes" of Spinoza and the "Monads" of Leibniz', *Proceedings of the Aristotelian Society* **18**, 329—362.

(Hitchcock, Major General) United States Army: 1846, 'The Doctrines of Spinoza and Swedenborg Identified; in so far as They Claim a Scientific Ground', in *Four Letters*, Boston, pp. 187—192.

Published anonymously.

Hubbeling, H. G.: 1978, *Spinoza*, Verlag Karl Alber, Freiburg.

Maintains that Spinoza adopts Cartesian physics (see especially pp. 65—70).

Jacob, Pierre: 1974, 'La politique avec la physique à l'âge classique; Principe d'inertie et conatus: Descartes, Hobbes et Spinoza', *Dialectiques* **6**, 99—121.

Joachim, Harold Henry: 1940, *Spinoza's* Tractatus de Intellectus Emendatione, Clarendon Press, Oxford.

Jonas, Hans: 1965, 'Spinoza and the Theory of Organism', *Journal of the History of Philosophy* **3**, 43—58. Reprinted in Marjorie Grene (ed.), *Spinoza; A Collection of Critical Essays*, Doubleday, Garden City, 1973, pp. 259—278.

Describes the advance in the account of organic existence (beyond Cartsian dualism and mechanism) made possible by Spinoza's metaphysics.

Jonas, Hans: 1980, 'Parallelism and complementarity: On the Psycho-Physical Problem in the Succession of Niels Bohr', in Richard Kennington (ed.), *The Philosophy of Baruch Spinoza*, Studies in Philosophy and the History of Philosophy 7, Catholic University of America Press, Washington D.C., pp. 121—130.

Article in the present volume is a corrected reprint of this.

Kaplan, A.: 1978, 'Spinoza and Freud', (in Hebrew) in Z. Lavi *et al.* (eds.), *Spinoza Studies, Three Hundred Years in Memoriam*, The University of Haifa Press, Haifa, pp. 85—110.

Kayser, Rudolph: 1946, *Spinoza: Portrait of a Spiritual Hero*, transl. by Amy Allen and Maxim Newmark, Philosophical Library, New York. Reprinted by Greenwood, New York, 1968.

Introduction by Albert Einstein.

Kegley, Jacquelyn Ann K.: 1975, 'Spinoza's God and LaPlace's World Formula', in *Akten Des II Internationalen Leibnizkongresses*, Vol. 3, Steiner, Wiesbaden, pp. 25—35.

Describes Spinoza's God as 'an incorporated world formula' including determinism, objective unity of law yielding scientific explanation, and rejection of teleology; identifies LaPlace's formula and Spinoza's 'ideal Metaphysic' as containing the dominant scientific and philosophic ideas of the seventeenth and eighteenth centuries.

Klever, W. N. A.: 1982, 'Spinoza's Methodebegrip' [Spinoza's Concept of Method], *Algemeen Nederlands Tijdschrift voor Wijsbegeerte* **74**, 28—49.

Compares Spinoza's epistemology ("knowledge is acquired in an autonomous movement") to the contemporary methodology in science. Subsequent issues of the same volume carry an objection of Hein Bobeldijk and a reply from Klever.

Kneale, Martha: 1968—69, 'Eternity and Sempiternity', *Proceedings of the Aristotelian Society* **69**, 223—238. Reprinted in Marjorie Grene (ed.), *Spinoza; A Collection of Critical Essays*, Doubleday, Garden City, 1973, pp. 227—240.

Discusses the question whether an eternal object can or must be sempiternal as well, and suggests that Spinoza changed his mind on the issue.

Kneale, Martha: 1972, 'Leibniz and Spinoza on Activity', in H. G. Frankfort (ed.), *Leibniz; A Collection of Critical Essays*, Garden City, pp. 215—237.

Kouznetsov, Boris: 1967, 'Spinoza et Einstein', *Revue de synthèse* **88**, 45—46, 3è série, 31—52.

Lachterman, David R.: 1978, 'The Physics of Spinoza's *Ethics*', in Robert W. Shahan and J. I. Biro (eds.), *Spinoza: New Perspectives*, University of Oklahoma Press, Norman, pp. 71—112.

Offers an explication of Spinoza's physical conceptions (e.g., motion and rest, extension) in providing a foundation for his ethics, and looks beyond such conceptions for the models by which they are governed.

Lecrivain, André: 1977—78, 'Spinoza et la physique cartésienne', *Cahiers Spinoza*, Éditions Réplique, Paris, 1, pp. 235—265; 2, pp. 93—206.

Forms the basis for Lecrivain's article in the present volume which is a condensed redevelopment and extension of the former.

Lewis, Douglas: 1976, 'Spinoza on Extension', *Midwest Studies in Philosophy 1*, 26—31.

McKeon, Richard: 1928, *The Philosophy of Spinoza; The Unity of His Thought*, Longmans, Green and Co., New York.

Describes Spinoza's lively interest in experimental science and his epistemology. See especially pp. 130—157.

McKeon, Richard: 1965, 'Spinoza on the Rainbow and on Probability', in *Harry A. Wolfson Jubilee Volume*, 3 vols., American Academy for Jewish Research, Saul Lieberman, Jerusalem, pp. 533—559.

Matheron, A.: 1969, *Individu et communauté chez Spinoza*, Éditions de Minuit, Paris.

A thorough analysis of Spinoza's concept of a physical system.

Meerloo, Joost A. M.: 1965, 'Spinoza: A Look at His Psychological Concepts', *American Journal of Psychiatry* **121**, 890—894.

Meerloo, Joost A. M.: 1976, 'Intuition as a Cluster of Cognitive Functions', *Methodology and Science* **9**.

Mollenhauer, Bernhard: 1941, 'Spinoza and the Borderland of Science', *Personalist* **22**, 64—72.

Nails, Debra: 1979, '*Conatus* versus *Eros/Thanatos*; On the Principles of Spinoza and Freud', *Dialogue* **21**, 33—40.

Looks at descriptions of physical, biological, and psychological phenomena to establish that Spinoza's *conatus* does the work of both *eros* and *thanatos*.

Nesher, Dan: 1978, 'Methodological Changes in Spinoza's Concept of Science: The Transition from "On the Improvement of the Human Understanding" to *Ethics*', (in Hebrew) in Z. Lavi *et al.* (eds.), *Spinoza Studies, Three Hundred Years in Memoriam*, The University of Haifa Press, Haifa, pp. 53—64.

Nesher, Dan: 1979, 'On the "Common Notions" in Spinoza's Theory of Knowledge and Philosophy of Science', in M. Brinker, M. Dascal, and Dan Nesher (eds.), *Baruch de*

Spinoza, A Collection of Papers on His Thought, University Publishing Projects, Ltd., Tel Aviv, pp. 35—52.

Parkinson, G. H. R.: 1954, *Spinoza's Theory of Knowledge,* Clarendon Press, Oxford.

See especially pp. 12—14 and 157—162.

Pollock, Frederick: 1873, 'The Scientific Character of Spinoza's Philosophy', *The Fortnightly Review* **19** (N.S. 13), 567—585.

Pollock, Frederick: 1878, 'Benedict de Spinoza', *Popular Science Monthly,* supp. 11, 444—458.

Ramirez, E. Roy: 1981, 'Spinoza, en torno al movimiento', *Revista de filosofia* **19**, Universidad de Costa Rica, pp. 45—48.

Includes Spinoza's concept of conservation of motion in the historical and intellectual context that led to Newton's first law of motion.

Rensch, Bernhard: 1972, 'Spinoza's Identity Theory and Modern Biophilosophy', *Philosophical Forum* **3**, 193—207.

Uses selected scientific results of modern biology and physics to assess aspects of Spinoza's identity theory of mind and matter, and his determinism.

Rice, Lee C.: 1971, 'Spinoza on Individuation', *Monist* **55**. Reprinted in Maurice Mandelbaum and Eugene Freeman (eds.), *Spinoza: Essays in Interpretation,* Open Court Publ. Co., LaSalle, Illinois, 1975, pp. 195—214.

Examines the lemmas following Proposition XIII of Part II of the *Ethics* to sketch Spinoza's view of a physical system and to determine what Spinoza meant by '*individus*'.

Ritchie, Eliza: 1902, 'The Reality of the Finite in Spinoza's System', *Philosophical Review,* 16—29.

Rivaud, Albert: 1924—26, 'La physique de Spinoza', *Chronicon Spinozanum* **4**, 24—57.

Roth, Leon: 1929, *Spinoza,* E. Benn, London.

Runes, Dagobert (ed.): 1951, *Spinoza Dictionary,* Philosophical Library, New York.

Foreword by Albert Einstein.

Sachs, Mendel: 1976, 'Maimonides, Spinoza and the Field concept in Physics', *Journal of the History of Ideas* **37**, 125—131.

Semerari, G.: 1978, 'L'idea Della Scienza in Spinoza', *Annali della Facolta di Lettere e Filosofia* **21**, 205—218.

Stein, Ludwig: 1890, *Leibniz und Spinoza,* Georg Reimer Verlag, Berlin.

Includes a collection of letters by Leibniz about Spinoza and his work. See Die Spinozafreundliche Periode (1676—79), pp. 60—110.

Thijssen-Schoute, C. L.: 1954, *Lodewijk Meyer en diens verhouding tot Descartes en Spinoza* [Louis Meyer and his Relation to Descartes and Spinoza], Mededelingen Vanwege het Spinozahuis 11, E. J. Brill, Leiden.

Usmani, M. A.: 1964, 'Spinoza: A First Rate Scientific Rationalist', *Pakistan Philosophical Congress* II, 258—262.

van Deventer, Ch. M.: 1921, *Spinoza's leer De Natura Corporum beschouwd in betrekking tot Descartes' mechanika* [Spinoza's Doctrine *De Natura Corporum* Considered in Relation to Cartesian Mechanics], *Tijdschrift voor Wijsbegeerte* **15**, 265—293.

Maintains that Spinoza's theory of bodies has a purely Cartesian character.

van der Hoeven, P.: 1973a, *De Cartesiaanse Fysica in het Denken van Spinoza* [Cartesian Physics in the Thought of Spinoza], Mededelingen Vanwege het Spinoza-huis 30, E. J. Brill, Leiden.

van der Hoeven, P.: 1973b, 'Over Spinoza's Interpretatie van de Cartesiaanse Fysicaen Betekenis daarvan voor het System der Ethica' [On Spinoza's Interpretation of Cartesian Physics and Its Meaning for the Ethical System], *Tijdschrift voor Filosofie* **35**, 27—86.

The first part of the article is devoted to Spinoza's critique of Cartesian physics, particularly of Descartes's notions of an '*instant du temps*'; the second is an examination of the structural importance of Cartesian physics to Spinoza's *Ethics*.

van der Hoeven, P.: 1974, 'The Significance of Cartesian Physics for Spinoza's Theory of Knowledge', in J. G. van der Bend (ed.), *Spinoza on Knowing, Being and Freedom*, Assen, pp. 114—125.

van Os, C. H.: 1946, *Tijd, Maat en Getal* [Time, Measure and Number], Mededelingen Vanwege het Spinozahuis 7, E. J. Brill, Leiden.

On Spinoza's letter on the infinite (*Letter XII* to Louis Meyer).

VanZandt, Joe D.: 1973—75, 'On Substance', *Auslegung* **1—2**, 85—107.

Sets out conditions for philosophie and scientific understanding along Spinozistic lines.

von Dunin Borkowski, Stanislaus, S. J.: 1933, 'Die Physik Spinozas', in *Septimana Spinozana*, Nijhoff, The Hague, pp. 85—101.

Maintains that Spinoza denied in his physics the conception of extension as mass, and substituted for it the control of space by the field of power; discusses corpuscles of aetherial fluid.

Vygotskii, L. S.: 1970, 'Spinoza's Theory of the Emotions in Light of Contemporary Psychoneurology' (in Russian), *Voprosy filosofii*, No. 6, 119—130. Transl. by Edward E. Berg for *Soviet Studies in Philosophy* **10** (1972), 362—382.

Written 1932—34 and published posthumously. Identifies Spinoza as the first thinker to establish philosophically the very possibility of a true scientific explanatory psychology of human beings and to mark out a path for its subsequent development. Various contemporaries (e.g. Lange and Kilthey) are compared with respect to descriptive *versus* explanatory psychologies.

Wartofsky, Marx W.: 1973, 'Action and Passion: Spinoza's Construction of a Scientific Psychology', in Marjorie Grene (ed.), *Spinoza; A Collection of Critical Essays*, Doubleday, Garden City, pp. 329—353. Reprinted in Marx Wartofsky, *Models; Representation and the Scientific Understanding*, Boston Studies in the Philosophy of Science, Vol. 48, D. Reidel, Dordrecht, 1979, pp. 231—254.

Within the context of the relation between scientific theory and metaphysics, shows how Spinoza elaborates the physics of bodies to take into account the activities of human beings.

Wartofsky, Marx W.: 1977, 'Nature, Number and Individuals: Motive and Method in Spinoza's Philosophy', *Inquiry* **20**, 457—479. Reprinted in Marx Wartofsky, *Models; Representation and the Scientific Understanding*, Boston Studies in the Philosophy of Science, Vol. 48, D. Reidel, Dordrecht, 1979, pp. 255—276.

Concerned with individuation: the contradiction between (a) the view that substance (God and Nature) is simple, eternal, and infinite, and (b) the claim that substance contains infinite differentiation — determinate and finite modes, i.e., individuals.

Wetlesen, Jon: 1969, 'A Reconstruction of Basic Concepts in Spinoza's Social Psychology', *Inquiry* **12**, 105—132.

Wheeler, John A.: 1980, 'Beyond the Black Hole', in Harry Wolf (ed.), *Some Strangeness in the Proportion*, Addison-Wesley, Reading, Mass.

Wolf, A.: 1927, 'Spinoza's Conception of the Attributes of Substance', *Proceedings of the Aristotelian society*, N.S. 27, 177—192.

Wolff, E.: 1954, 'Spinoza and Cantor', *Revue de Synthèse* **75**, 161—163.

Wolfson, Harry Austryn: 1934, *The Philosophy of Spinoza*, Harvard University Press, Cambridge, Massachusetts. Reprinted 1983.

See especially Chapter 10, 'Duration, Time, and Eternity'.

Wood, L.: 1926, *Theories of Space and Time in Spinoza*, Ithaca, N.Y.

INDEX LOCORUM

In references to Spinoza's works, the abbreviation G, followed by a Roman numeral and page number (e.g. GII:9—11), refers to the Carl Gebhardt edition, *Spinoza Opera*, vols. I—IV (Carl Winter, Heidelberg, 1925). Authors other than Spinoza are included in this index only when the particular works cited are included in Gebhardt. For English translation citations of all passages, and for Gebhardt citations not included below, the reader should consult references in the text and notes.

I. Spinoza

Cogitata Metaphysica, 1663 (*Thoughts on Metaphysics*)
Pt. I, ch. 1:2 120
Pt. I, ch. 1:3—6 150n10(146)
Pt. I, ch. 1:6 103
Pt. I, ch. 1:7 21
Pt. I, ch. 1:11 23

Letters (square brackets indicate approximate dates)
IV (to Oldenburg [October 1661]) 306
VI (to Oldenburg [April 1662]) frontispiece, 7, 18, 56—57, 59n1(18), 103, 113, 115, 117, 121
IX (to De Vries [March 1663]) 6—7, 59n2(18), 103
XII (to Meyer, April 20, 1663) 19, 20—21, 34, 38, 73, 88n4(65), 96, 146, 150n6(137), 258, 262n11(255), 263n18(256), 263n20, 308—309, 313
XIII (to Oldenburg, July 17/27, 1663) 6, 38, 59n1(18), 590n10(38), 62, 88n4(65), 105, 115, 117—118, 122n11(117)
XXVI (to Oldenburg [May 1665]) 262n11(255)
XXVII (to Blyenbergh, June 3, 1665) 16
XXX (to Oldenburg [September 1665]) 122n7(109), 262n10—11(255)
XXXII (to Oldenburg, November 20, 1665) 40, 52, 57, 122n7(109), 262n10—11(255), 262n13
XXXVIII (to Van der Meer, October 1, 1666) 74, 96
XXXIX (to Jelles, March 3, 1667) 121n2(96)
XLI (to Jelles, September 5, 1669) 95
XLIII (to Ostens [February 1671]) 89n16(86), 186n35(179)
L (to Jelles, June 2, 1674) 96, 122n13(120), 150n9(146)
LIV (to Boxel [September 1674]) 285, 298n42
LVI (to Boxel [October 1674]) 285, 298n43
LVIII (to Schuller [October 1674]) 205
LX (to Tschirnhaus [January 1675]) 111
LXIV (to Schuller, July 29, 1675) 39, 41, 263n24

II. Other Authors

GENERAL INDEX

BOSTON STUDIES IN THE PHILOSOPHY OF SCIENCE

Editors:

ROBERT S. COHEN and MARX W. WARTOFSKY

(Boston University)

22. Milic Capek (ed.), *The Concepts of Space and Time. Their Structure and Their Development*. 1976.
23. Marjorie Grene, *The Understanding of Nature. Essays in the Philosophy of Biology.* 1974.
24. Don Ihde, *Technics and Praxis. A Philosophy of Technology.* 1978.
25. Jaakko Hintikka and Unto Remes. *The Method of Analysis. Its Geometrical Origin and Its General Significance.* 1974.
26. John Emery Murdoch and Edith Dudley Sylla, *The Cultural Context of Medieval Learning.* 1975.
27. Marjorie Grene and Everett Mendelsohn (eds.), *Topics in the Philosophy of Biology.* 1976.
28. Joseph Agassi, *Science in Flux.* 1975.
29. Jerzy J. Wiatr (ed.), *Polish Essays in the Methodology of the Social Sciences.* 1979.
30. Peter Janich, *Protophysics of Time.* 1985.
31. Robert S. Cohen and Marx W. Wartofsky (eds.), *Language, Logic, and Method.* 1983.
32. R. S. Cohen, C. A. Hooker, A. C. Michalos, and J. W. van Evra (eds.), *PSA 1974: Proceedings of the 1974 Biennial Meeting of the Philosophy of Science Association.* 1976.
33. Gerald Holton and William Blanpied (eds.), *Science and Its Public: The Changing Relationship.* 1976.
34. Mirko D. Grmek (ed.), *On Scientific Discovery.* 1980.
35. Stefan Amsterdamski, *Between Experience and Metaphysics. Philosophical Problems of the Evolution of Science.* 1975.
36. Mihailo Marković and Gajo Petrović, *Praxis. Yugoslav Essays in the Philosophy and Methodology of the Social Sciences.* 1979.
37. Hermann von Helmholtz, *Epistemological Writings. The Paul Hertz/Moritz Schlick Centenary Edition of 1921 with Notes and Commentary by the Editors.* (Newly translated by Malcolm F. Lowe. Edited, with an Introduction and Bibliography, by Robert S. Cohen and Yehuda Elkana.) 1977.
38. R. M. Martin, *Pragmatics, Truth, and Language.* 1979.
39. R. S. Cohen, P. K. Feyerabend, and M. W. Wartofsky (eds.), *Essays in Memory of Imre Lakatos.* 1976.
42. Humberto R. Maturana and Francisco J. Varela, *Autopoiesis and Cognition. The Realization of the Living.* 1980.
43. A. Kasher (ed.), *Language in Focus: Foundations, Methods and Systems. Essays Dedicated to Yehoshua Bar-Hillel.* 1976.
44. Trân Duc Thao, *Investigations into the Origin of Language and Consciousness.* (Translated by Daniel J. Herman and Robert L. Armstrong; edited by Carolyn R. Fawcett and Robert S. Cohen.) 1984.
46. Peter L. Kapitza, *Experiment, Theory, Practice.* 1980.
47. Maria L. Dalla Chiara (ed.), *Italian Studies in the Philosophy of Science.* 1980.
48. Marx W. Wartofsky, *Models: Representation and the Scientific Understanding.* 1979.
49. Trân Duc Thao, *Phenomenology and Dialectical Materialism.* 1985.
50. Yehuda Fried and Joseph Agassi, *Paranoia: A Study in Diagnosis.* 1976.
51. Kurt H. Wolff, *Surrender and Catch: Experience and Inquiry Today.* 1976.
52. Karel Kosík, *Dialectics of the Concrete.* 1976.
53. Nelson Goodman, *The Structure of Appearance.* (Third edition.) 1977.

54. Herbert A. Simon, *Models of Discovery and Other Topics in the Methods of Science.* 1977.
55. Morris Lazerowitz, *The Language of Philosophy. Freud and Wittgenstein.* 1977.
56. Thomas Nickles (ed.), *Scientific Discovery, Logic, and Rationality.* 1980.
57. Joseph Margolis, *Persons and Minds. The Prospects of Nonreductive Materialism.* 1977.
59. Gerard Radnitzky and Gunnar Andersson (eds.), *The Structure and Development of Science.* 1979.
60. Thomas Nickles (ed.), *Scientific Discovery: Case Studies.* 1980.
61. Maurice A. Finocchiaro, *Galileo and the Art of Reasoning.* 1980.
62. William A. Wallace, *Prelude to Galileo.* 1981.
63. Friedrich Rapp, *Analytical Philosophy of Technology.* 1981.
64. Robert S. Cohen and Marx W. Wartofsky (eds.), *Hegel and the Sciences.* 1984.
65. Joseph Agassi, *Science and Society.* 1981.
66. Ladislav Tondl, *Problems of Semantics.* 1981.
67. Joseph Agassi and Robert S. Cohen (eds.), *Scientific Philosophy Today.* 1982.
68. Władysław Krajewski (ed.), *Polish Essays in the Philosophy of the Natural Sciences.* 1982.
69. James H. Fetzer, *Scientific Knowledge.* 1981.
70. Stephen Grossberg, *Studies of Mind and Brain.* 1982.
71. Robert S. Cohen and Marx W. Wartofsky (eds.), *Epistemology, Methodology, and the Social Sciences.* 1983.
72. Karel Berka, *Measurement.* 1983.
73. G. L. Pandit, *The Structure and Growth of Scientific Knowledge.* 1983.
74. A. A. Zinov'ev, *Logical Physics.* 1983.
75. Gilles-Gaston Granger, *Formal Thought and the Sciences of Man.* 1983.
76. R. S. Cohen and L. Laudan (eds.), *Physics, Philosophy and Psychoanalysis.* 1983.
77. G. Böhme et al., *Finalization in Science,* ed. by W. Schäfer. 1983.
78. D. Shapere, *Reason and the Search for Knowledge.* 1983.
79. G. Andersson, *Rationality in Science and Politics.* 1984.
80. P. T. Durbin and F. Rapp, *Philosophy and Technology.* 1984.
81. M. Marković, *Dialectical Theory of Meaning.* 1984.
82. R. S. Cohen and M. W. Wartofsky, *Physical Sciences and History of Physics.* 1984.
83. E. Meyerson, *The Relativistic Deduction.* 1985.
84. R. S. Cohen and M. W. Wartofsky, *Methodology, Metaphysics and the History of Sciences.* 1984.
85. György Tamás, *The Logic of Categories.* 1985.
86. Sergio L. de C. Fernandes, *Foundations of Objective Knowledge.* 1985.
87. Robert S. Cohen and Thomas Schnelle (eds.), *Cognition and Fact.* 1985.
88. Gideon Freudenthal, *Atom and Individual in the Age of Newton.* 1985.
89. A. Donagan, A. N. Perovich, Jr., and M. V. Wedin (eds.), *Human Nature and Natural Knowledge.* 1985.
90. C. Mitcham and A. Huning (eds.), *Philosophy and Technology II.* 1986.
91. M. Grene and D. Nails (eds.), *Spinoza and the Sciences.* 1986.
92. S. P. Turner, *The Search for a Methodology of Social Science.* 1986.
93. I. C. Jarvie, *Thinking About Society: Theory and Practice.* 1986.
94. Edna Ullmann-Margalit (ed.), *The Kaleidoscope of Science.* 1986.
95. Edna Ullmann-Margalit (ed.), *The Prism of Science.* 1986.
96. G. Markus, *Language and Production.* 1986.